# Space Exploration For Dummies®

## A Timeline of Major Milestones during the Space Race

1957: *Sputnik 1,* the first artificial satellite, is launched by the Soviet Union

1957: *Sputnik 2,* launched by the Soviet Union, carries the first animal into orbit — a dog named Laika

1958: NASA (the National Aeronautics and Space Administration) is formed in the United States

1958: *Explorer 1,* the first American satellite, is launched

1959: First man-made object *(Luna 1)* orbits the Sun

1961: First human (Yuri Gagarin) orbits Earth

1961: First American (Alan Shepard) in space

1962: U.S. President John F. Kennedy gives a famous speech at Rice University committing federal support for the American space program

1962: First American (John Glenn) orbits Earth

1963: First woman (Valentina Tereshkova) in space

1965: First spacewalk performed by Soviet cosmonaut (Alexei Leonov); first American spacewalk performed by Ed White

1966: First spacecraft *(Luna 9)* performs a soft Moon landing

1966–1967: Five *Lunar Orbiter* spacecraft map the Moon, view the Earth from space, and search for lunar landing sites

1967: Death of three astronauts (Gus Grissom, Roger Chaffee, and Ed White) during a preflight test of the first Apollo mission

1968: First robotic spacecraft *(Zond 5)* returns to Earth after orbiting the Moon

1969: First human (Neil Armstrong) walks on the Moon

## A Handful of Space Exploration Web Sites

Here are some great resources for finding information on past, current, and future space missions:

- `www.space.com`: Space and astronomy news
- `science.nationalgeographic.com/science/space.html`: *National Geographic's* resource for all things space-related
- `history.nasa.gov`: NASA history
- `www.nasa.gov/missions/current`: Details on current NASA space missions
- `spaceflight.nasa.gov/shuttle`: All you ever wanted to know about the Space Shuttle
- `www.bnsc.gov.uk`: British National Space Centre
- `www.cnsa.gov.cn/n615709/cindex.html`: China National Space Administration
- `www.esa.int`: European Space Agency
- `www.isro.org`: Indian Space Research Organisation
- `www.asi.it/en`: Italian Space Agency
- `www.jaxa.jp/index_e.html`: Japan Aerospace Exploration Agency
- `www.roscosmos.ru/index.asp?Lang=ENG`: Russian Federal Space Agency

**For Dummies: Bestselling Book Series for Beginners**

# Space Exploration For Dummies®

Cheat Sheet

## Major Space Shuttle Missions

1977: First "free flight" of the Space Shuttle program by the atmospheric test orbiter *Enterprise*

1981: First Space Shuttle mission *(Columbia)*

1983: Sally Ride becomes the first American woman in space aboard *Challenger*

1986: *Challenger* explosion post-launch destroys the Shuttle and kills seven astronauts, including Christa McAuliffe (who would've been the first teacher in space)

1989: Launch of the Venus probe *Magellan* and the Jupiter probe *Galileo* from *Atlantis* (on two separate missions)

1990: Launch of the Hubble Space Telescope from *Discovery*

1995: First docking of the Space Shuttle *(Atlantis)* with Russia's Mir space station

1998: First International Space Station assembly mission *(Endeavour)*

2003: *Columbia* explosion during reentry destroys the Shuttle and kills six astronauts

2005: First Space Shuttle flight following the 2003 *Columbia* tragedy *(Discovery)*

2010: Last planned flight of the Space Shuttle program *(Endeavour)*

## Some Currently Active Spacecraft

Space agencies around the world have currently active spacecraft; the following is a sampling (see Parts IV and V for details on current and future missions):

- *Cassini:* In orbit around Saturn, studying the planet and its moons (NASA)
- *Chandrayaan-1:* In orbit around the Moon, studying its composition and geology (India)
- *Chang'e:* In orbit around the Moon, studying its surface composition (China)
- *Deep Impact:* Following a successful comet rendezvous, it's headed to a 2010 rendezvous with comet Hartley-2 (NASA)
- *Hayabusa:* En route back to Earth after collecting a sample from asteroid Itokawa; will reach Earth in 2010 (Japan)
- International Space Station: In orbit around Earth, studying extended human activity beyond Earth (NASA/Russia/others)
- Mars Exploration Rovers *Spirit* and *Opportunity:* On the surface of Mars, studying its geology and composition (NASA)
- *Mars Express:* In orbit around Mars, studying geology and surface composition (ESA)
- *Mars Odyssey:* In orbit around Mars, studying its geology and atmosphere (NASA)
- *Mars Reconnaissance Orbiter:* In orbit around Mars, studying its geology (NASA)
- *MESSENGER:* En route to Mercury; will reach orbit in 2011 and study its geology and composition (NASA)
- *New Horizons:* En route to Pluto; 2015 flyby to study geology (NASA)
- *SELENE:* In orbit around the Moon, studying its geology and gravity (Japan)
- *SOHO:* In orbit between the Sun and the Earth, studying the Sun (NASA)
- *Stardust:* Returned comet samples to Earth; will fly past Comet 9P/Tempel 1 in 2011 (NASA)
- *Venus Express:* In orbit around Venus, studying its atmosphere (ESA)
- *Voyager:* Heading out of the solar system after flybys of the outer planets, studying interstellar space (NASA)



For Dummies: Bestselling Book Series for Beginners

# Space Exploration For Dummies®

by Cynthia Phillips, PhD, and Shana Priwer

Wiley Publishing, Inc.

**Space Exploration For Dummies®**

Published by
**Wiley Publishing, Inc.**
111 River St.
Hoboken, NJ 07030-5774
www.wiley.com



Library of Congress Control Number: 2009926385

ISBN: 978-0-470-44573-0

Manufactured in the United States of America

10 9 8 7 6 5 4 3 2 1

WILEY

# About the Authors

**Cynthia Phillips** is a scientist at the SETI Institute in Mountain View, California. She works at the Carl Sagan Center for the Study of Life in the Universe, studying the geology of the solar system and the potential for life beyond Earth. Dr. Phillips served as an affiliate of the Imaging team for the *Galileo* spacecraft during its mission to Jupiter and has provided input for the design of a future mission to Jupiter's moon Europa. She received her BA in Astronomy, Astrophysics, and Physics at Harvard University (magna cum laude) and her PhD in Planetary Science, with a minor in Geosciences, from the University of Arizona. Dr. Phillips has taught astronomy, physics, biology, geology, math, and planetary science to students from junior high to college age and has given public lectures to general audiences on a variety of topics. She has extensive experience in explaining difficult health, scientific, and other concepts to diverse audiences.

**Shana Priwer** has an undergraduate degree in architecture from Columbia University and a graduate degree in architecture from Harvard University's Graduate School of Design. Her graduate thesis focused on using cutting-edge computer techniques to design buildings and produce 3-D renderings and animations of their interiors and exteriors. Priwer worked for an architecture firm for several years. She also has construction experience and has done carpentry and masonry on historic buildings. Priwer currently works for a software company that makes programs used by architects to produce two- and three-dimensional designs (the technology is also used by animation houses and others in the film industry). She enjoys writing on a variety of medical, technical, and other topics, for both children and adults.

Cynthia Phillips and Shana Priwer are the coauthors of 14 books on subjects including astronomy, Einstein, da Vinci, the physics of the built environment, and a variety of other topics.

# Dedication

To our children: Zoecyn, Elijah, Benjamin, and Sophia.

# Authors' Acknowledgments

The authors would like to thank the many people who made this book possible, from our agent, Barbara Doyen, to our acquisitions editor, Mike Lewis; our project editor, Georgette Beatty; our copy editor, Jen Tebbe; our technical reviewer, David Whalen; and all the other members of the staff at Wiley. Thanks also to Matt Harshbarger and Jon Prince of Precision Graphics for their work on the illustrations.

## Publisher's Acknowledgments

We're proud of this book; please send us your comments through our Dummies online registration form located at http://dummies.custhelp.com. For other comments, please contact our Customer Care Department within the U.S. at 877-762-2974, outside the U.S. at 317-572-3993, or fax 317-572-4002.

Some of the people who helped bring this book to market include the following:

***Acquisitions, Editorial, and Media Development***

**Senior Project Editor:** Georgette Beatty

**Acquisitions Editor:** Michael Lewis

**Copy Editor:** Jennifer Tebbe

**Assistant Editor:** Erin Calligan Mooney

**Editorial Program Coordinator:** Joe Niesen

**Technical Editor:** David J. Whalen

**Editorial Manager:** Michelle Hacker

**Editorial Assistant:** Jennette ElNaggar

**Art Coordinator:** Alicia B. South

**Cover Photo:** PhotoLink

**Cartoons:** Rich Tennant (www.the5thwave.com)

***Composition Services***

**Project Coordinator:** Katherine Crocker

**Layout and Graphics:** Samantha Allen, Reuben W. Davis, Melissa K. Jester, Mark Pinto

**Special Art:** Precision Graphics (precisiongraphics.com)

**Proofreaders:** John Greenough, Christopher M. Jones

**Indexer:** Potomac Indexing, LLC

---

**Publishing and Editorial for Consumer Dummies**

**Diane Graves Steele,** Vice President and Publisher, Consumer Dummies

**Kristin Ferguson-Wagstaffe,** Product Development Director, Consumer Dummies

**Ensley Eikenburg,** Associate Publisher, Travel

**Kelly Regan,** Editorial Director, Travel

**Publishing for Technology Dummies**

**Andy Cummings,** Vice President and Publisher, Dummies Technology/General User

**Composition Services**

**Gerry Fahey,** Vice President of Production Services

**Debbie Stailey,** Director of Composition Services

# Contents at a Glance

# Table of Contents

# Introduction

Since the dawn of time, humans have looked up at the skies and wondered what was up there. Amazingly, with all these centuries of sky-watching, it's only been in the last 50 years that humans have actually achieved the ability to visit those points of light in the sky. We've sent robotic spacecraft to take pictures, and we've also sent astronauts to experience seeing the Earth from the heavens. Humans have walked on the Moon, and future plans call for people to return to the Moon and continue on to Mars and beyond.

Space exploration is one of humanity's grandest adventures, filled with stories of amazing discoveries and narrow escapes. It's the result of incredible feats of technological innovation and has laid the groundwork for major advances here on Earth.

Perhaps the most-amazing thing about space exploration is that you don't have to be a rocket scientist to understand it! That's where *Space Exploration For Dummies* comes in handy. This book doesn't contain dry equations or long lists of space missions. Instead, it covers the greatest stories and discoveries of space exploration and highlights reasons why it's okay to dream about visiting the stars.

## About This Book

*Space Exploration For Dummies* isn't a textbook, a historical treatise, or an engineering manual. In this book, we take you on a lighthearted yet informative trip through the cosmos, beginning with humanity's first steps off this planet and continuing to our collective future in space. Along the way, you visit the planets of our solar system and see them as you've never seen them before. Expect to find out about the latest scientific discoveries from planetary probes that have traveled the far reaches of the solar system, from burning-hot Mercury to ice-cold Pluto. You also observe the latest designs for ships that will allow humans to continue their space journeys.

We follow a roughly chronological organization in this book, which means you can easily find out about the classic missions that started it all in the 1950s and 1960s, or you can jump right to the latest and greatest spacecraft. Although the main focus of *Space Exploration For Dummies* is NASA and the United States space program, we also include information on the space programs of other countries, including the former Soviet Union, China, and European nations.

One other note: Although human exploration has certainly been an exciting, and crucial, part of space exploration, robotic missions have also played a vital role in mankind's understanding of its place in the solar system. In this book, we cover both robotic and human spaceflight, primarily in separate chapters. Read just the astronaut parts, or just give in and read the whole thing — it's your choice.

## Conventions Used in This Book

Space exploration is a task that has been accomplished by both genders, but the phrase "manned spaceflight" is commonly used to differentiate astronaut flights from robotic missions. Although we use phrases such as "human spaceflight" or "robotic spacecraft" where possible, in some instances the term "manned" or "unmanned" is simply more historically appropriate. ***Remember:*** Both women and men have made important contributions to space exploration, so don't let the phraseology influence your ability to digest and enjoy the information presented here.

Following are a few other conventions we use to help you navigate this book:

- Keywords in bulleted lists appear in **bold.**
- *Italics* highlight new terms, emphasize certain words, and indicate when we're referring to spacecraft (as opposed to missions of the same name).
- Web sites appear in `monofont` to help them stand out.

When this book was printed, some Web addresses may have needed to break across two lines of text. If that happened, rest assured that we didn't put in any extra characters (such as hyphens) to indicate the break. If you want to visit one of these Web addresses, type in exactly what you see in this book (just pretend like the line break isn't there).

## What You're Not to Read

Granted, we'd love for you to read this book from cover to cover. But we're not above letting you know that some sections are by far more optional than others. Throughout the text, you'll see some paragraphs marked with a Technical Stuff icon — these are paragraphs that give some of the engineering or scientific details of the mission at hand (or just interesting asides that certainly won't harm you if you pass over them). Skip 'em if you want, or read them for an in-depth view.

Similarly, you'll find sidebars set off from the rest of the text in gray-shaded boxes. These elements provide some extra background information or other asides that we found particularly interesting but not essential to the story of space exploration as a whole. In some chapters, you'll encounter "Fast facts" sidebars that give standardized details on particular missions. Like the Technical Stuff paragraphs, you can skip these sidebars without missing out on any crucial bits of information.

## Foolish Assumptions

As we wrote this book, we made a couple assumptions about you, dear reader:

- You're fascinated with the science of space and want to know what it's like to be an astronaut.
- You revel in space missions and discoveries, and you look forward to hearing about what's to come in the world of space exploration.

Unlike the qualifications to be a real-live rocket scientist, you don't need any particular background in math, physics, space medicine, or any other field to read this book. (Want to brush up on some basic astronomy though? Glance over Chapter 2, which takes you on a whirlwind tour through the heavens.) All you need to read and enjoy this book is a comfortable chair, a sense of wonder, and perhaps a clear view of the sky.

## How This Book Is Organized

Space is huge, and there's lots of, well, space to be covered in a book all about space exploration. (We'll stop that running joke before it gets started.) But don't worry — we've arranged this book in loosely chronological order to cover why people should explore space, how we can get there, where and when humanity has gone into space, who has gone, and where we're going in the future. The following sections cover what you'll find in each part.

## Part I: Space Exploration: Where, Who, and How

This part gives you all the basics in one convenient location. Take a tour of astronomy or a quick course in rocket science. Check out all the details you've been waiting for on how to become an astronaut. But don't forget that space travel, although exciting, is also a dangerous pastime. Part I closes with a somber look at some of the tragedies and narrow escapes of the last 50 years of space exploration.

## Part II: The Great Space Race

It was the Cold War. Tensions between the United States and the Soviet Union had never been higher. Yet this time of great military rivalry spawned a technological competition that literally took the world to new heights, as you find out in Part II. Following the Soviet Union's early success with the *Sputnik 1* satellite, the U.S. played catch-up until its triumphant Apollo Moon landings. Robotic exploration also thrived during this golden age of space exploration and discovery.

## Part III: Second-Generation Missions

The Space Race was all well and good, and it accomplished its goals of putting "boots on the ground" (well, in this case, moon-boots in lunar dust!). But taking the tourist's view and spending a few hours or days on the Moon snapping pictures is one thing. Building a mature space program that can allow humans to really live and work in space is completely different. As you discover in this part, vehicles such as the Space Shuttle and destinations such as the Mir space station allowed society to start learning valuable lessons about space survival. Additionally, robotic missions grew more complicated as they explored both the inner and outer solar system, and Mars turned out to be a lot more difficult to explore than previously thought.

## Part IV: Current Space Exploration

If you think the history of space exploration is exciting, look at what's going on right now! From giant space telescopes that can view the far reaches of the cosmos to amazing discoveries about Earth's neighbor Mars, space exploration has yielded fascinating scientific results, as you find out in Part IV. The development of the International Space Station has allowed for a continual human presence in orbit for years now. The station itself has provided valuable experience in living, working, and doing science in orbit.

## Part V: The Future of Space Exploration

In this part, you discover that the future of the space program is optimistic given current NASA plans to send astronauts back to the Moon and then on to Mars. Upcoming missions include the retirement of the Space Shuttle and the production of a new capsulelike spacecraft called Orion that will take astronauts to the International Space Station and eventually beyond. Space tourism has opened up the skies to those with enough money to buy a seat. Space exploration is expensive, but it's well worth it in the grand scheme of things — the very future of humanity may depend on it.

## Part VI: The Part of Tens

The Part of Tens is a fun feature in all *For Dummies* books, and it gives you a quick way to find out some interesting facts and figures about space exploration. Here you can expect a brief summary of where to look for life beyond Earth and then consider ten ways in which space travel (unfortunately!) is nothing like your favorite movies or television shows. Finally, if you're still not convinced how cool, relevant, and necessary space exploration is, check out the list of ten types of everyday items that owe their very existence to NASA.

# Icons Used in This Book

Various icons are scattered throughout the text. Here's a quick rundown of what they look like and what they mean.

This icon indicates the most-important concepts and discoveries in space exploration.

The Technical Stuff icon is for concepts that are a tad more detail-oriented. Skip 'em if you want, or read on for some extra edification.

Whether you want to find out more about a particular nugget of space exploration history or keep up with an ongoing mission, keep your eyes peeled for this icon, which directs you to useful Web sites.

# Where to Go from Here

This book isn't a novel — you can start at the beginning and read it through to the end, or you can jump in right in the middle. Got a favorite planet? Head right to the chapter that describes missions to it. Itching to find out more about the International Space Station? Flip straight to Chapter 19. You can also just thumb through the book looking at all the pretty pictures. After all, taking those pictures is arguably half the point of space exploration.

If you're not sure where to start, or don't know enough about space exploration to even *have* a starting point in mind yet, that's great — that's exactly what this book is for. Just hop right in and get your feet wet. We recommend starting off with Chapter 2, though, if your basic astronomy is a bit rusty, or with Chapter 4 if you want to be an astronaut when you grow up (hey, we're all young at heart, right?). And make sure not to skip Chapter 25 if you want to find out all sorts of neat facts to annoy your friends with the next time you're watching a science fiction movie.

3 . . . 2 . . . 1 . . . Blast off!

# Part I

# Space Exploration: Where, Who, and How

## In this part . . .

Part I is set up to provide you with an overall view of what space exploration is all about. It begins with a brief introduction to astronomy, from our solar system to the universe as a whole. Exploring such a vast realm requires rockets. Lo and behold, we have a chapter on rocket science that covers the gamut from ancient fireworks to the giant rocket that launched astronauts to the Moon. Speaking of astronauts, they're the ones doing all that space travel, so discover what it's like to be an astronaut in this part! Of course, as exciting as space travel is, it can also be quite dangerous — that's why we close with coverage of some of the tragedies, and narrow escapes, of human space exploration.

# Chapter 1

# Space Exploration in a Nutshell

### In This Chapter

- Understanding the very basics of astronomy and rocket science
- Becoming an astronaut and learning from space accidents
- Competing in the Space Race
- Delving into current and future missions

In order to fully understand the stories and legends of space exploration, it's helpful to take a few steps back and see the big picture. Why have humans (and a few animals) gone into space? How has society's concept of the "frontier" changed over time, and why do we feel the need to keep pushing the boundaries of our known world? This chapter is your introduction to the study of space exploration; it examines what achievements have already been made in the field and gives you a few sneak peeks at what may be in the works.

## A Quick Spin around the Universe

From the time when humans first began making maps, they've had an urge to map the skies. Ancient Greek and Roman astronomers were among the first to formally name the constellations; Arabic, Chinese, and Native American cultures also had their own sky maps.

Theories, both religious and scientific, abound as to how the universe was formed. Astronomers generally agree, though, that an explosion of magnificent proportions was the starting point for our universe and that the Sun formed from a collapsing cloud of gas and dust. In the ensuing years, gravity pulled together the leftover gas, dust, ice, and other particles from the Sun's formation to create the planets in our solar system, which is roughly divided into inner (Mercury, Venus, Earth, and Mars) and outer (Jupiter, Saturn, Uranus, and Neptune) regions. The Sun and planets form one solar system, which, along with the many stars in the sky, all belong to the Milky Way Galaxy, one of billions of galaxies that comprise the known universe.

In Chapter 2, we provide information on how early humans viewed the sky and describe the solar system in detail; we also fill you in on extrasolar planets, stars, and galaxies.

## A Crash Course on Rocket Science

Rocket science, more formally known as *aerospace engineering* or *aeronautical engineering,* is the technology that allows spacecraft to make it into orbit. A spacecraft launched from Earth must be able to operate both inside the Earth's atmosphere and in the vacuum of space; typically, a launch vehicle such as a rocket is used to travel through the atmosphere to reach orbit. Unlike traditional airplanes, spacecraft have to deal with extreme temperature and pressure; they must not only endure the harshness of outer space but also conduct science and support human life out there. That's quite a tall order for technology.

Although ancient Greek and Chinese astronomers experimented with rocketry and fireworks, the real beginnings of rocket science can be traced to Leonardo da Vinci and his Renaissance-era sketches of flying machines. As technology, science, and engineering evolved in the following centuries, spaceflight advanced in leaps and bounds, culminating in the launch of the Soviet Union's *Sputnik 1* satellite in 1957, followed by the formation of the United States' National Aeronautics and Space Administration (NASA) in 1958.

All spacecraft are designed by a full team of specialists who are trained in different areas. Some engineers focus on propulsion technologies, for example, whereas others are experts in materials, ergonomics, aerodynamics, or other areas.

Flip to Chapter 3 for a summary of rockets in history, as well as details on how rockets lift off and travel in space, different sources of rocket and spacecraft power, and communication via radio telescopes.

## An Astronaut's Life and Times

Being an astronaut is no walk in the park. The requirements of the job necessitate not only very intensive training and knowledge but also the ability to function under living and working conditions that are often literally out of this world.

Earning the privilege to become a United States astronaut is a long and difficult road. Astronauts must possess advanced degrees, often in physics or another technical field, and must train constantly to stay in optimal physical condition. They endure grueling practice sessions in order to be prepared for survival both in space and upon their return to Earth. Astronauts train underwater so they can ready their bodies for performing activities in a low-gravity environment; they also work on all the different maneuvers that will be expected from them over the course of their journey.

When it comes to living in space, the International Space Station (ISS) may have the best view in town, but the accommodations and living conditions are a tad less than ideal.

- **Accommodations:** Until very recently, the ISS was a three-bedroom, one-bathroom house (and one of those bedrooms doubled as a laboratory). New living space and an additional bathroom were brought to the ISS, courtesy of Space Shuttle *Endeavour,* in November 2008.
- **Living conditions:** Astronauts aboard the ISS cope day in and day out with *microgravity,* a very low-gravity environment, which calls for some pretty specific changes in the way they do basic things. Activities that must be modified include eating (wouldn't want your food to float away from you!), sleeping, and using the restroom.

Check out Chapter 4 for the full scoop on the requirements to become an astronaut, the various titles of a space mission's crew, and the basics of living and working in space.

# Accidents in Space

With success comes the occasional mishap, and space exploration has had its fair share of accidents, as you discover in Chapter 5. These faux pas have ranged from the relatively minor (like the 1961 sinking of a Mercury capsule) to the near-miss-but-largely-salvageable (like the Apollo 13 mission to the Moon in 1970, which failed in most aspects but returned the crew back to Earth alive). However, they've also featured the utterly catastrophic, such as the explosion of Space Shuttle *Challenger* in 1986, which killed all astronauts aboard.

Space accidents come with a heavy price in terms of human life, tax-funded research dollars, and negative publicity for the space program. The saving grace of accidents in space though? Researchers learn from them and continually improve designs, technology, and safety with an eye toward future success.

# In the Beginning: The Great Space Race

The Space Race, which began with the launch of the Soviet satellite *Sputnik 1* in 1957, was the ultimate intercontinental throw-down. The world's major superpowers at the time, the Soviet Union and the United States, had similar goals in the realm of space exploration. Both countries were determined to prove their superiority in technology and spaceflight, resulting in fierce competition that drove the two nations to break boundaries and achieve goals previously thought to be impossible. The following sections highlight the major accomplishments of the Space Race.

## A duo of firsts: Sputnik 1 and animals in space

The first victory of the Space Race took just 98 minutes to secure. On October 4, 1957, the Soviet Union launched *Sputnik 1,* the first artificial satellite designed to orbit Earth, which returned a beeping signal that could be (and was!) tracked worldwide. The significance of this hour-and-a-half voyage went far beyond the orbit itself: The U.S. realized it had some immediate catch-up work to do and created NASA (the National Aeronautic and Space Administration) in order to get that work done.

Sending robotic satellites into space was one thing. Sending living, breathing human beings into space was something else entirely, and (despite earlier suborbital animal flights) it was a few years after the launch of *Sputnik 1* before Soviet and American engineers were confident enough in their designs to start using human test subjects. Consequently, a range of animals (including monkeys, dogs, and insects) became the world's first astronauts. Although not all of these animal astronauts survived, scientists were able to learn from early tragedies and eventually make human spaceflight possible. Flip to Chapter 6 for more information on the adventures of animals in space.

## The first people in space

The first people to travel into space all did so as part of two series of missions: the Vostok program in the Soviet Union and Project Mercury in the U.S. The Soviets claimed another early victory with the triumphant orbital flight of Yuri Gagarin in 1961; the Americans followed with flights by Alan Shepard in 1961 and John Glenn in 1962. Want more details? Be sure to read Chapter 7.

## Robots to the Moon

By the mid-1960s, both the U.S. and the Soviet Union were on track to send humans to the Moon, and robotic missions were needed to pave the way (see Chapter 8 for details).

- The Soviet Union made headway with its Luna missions in the 1960s and early 1970s, culminating with four missions that returned Moon soil samples to Earth. Two missions even included a robotic rover that drove over the lunar surface!
- The U.S. developed its own program beginning with the Ranger missions in the 1960s; the first few missions were failures, but *Rangers 7, 8,* and *9* returned many close-range photos of the Moon. The NASA Lunar Orbiter missions in the mid-1960s produced detailed lunar maps from orbit, and the Lunar Surveyor program in the late 1960s put the first American spacecraft on the Moon. Equipment was tested and techniques were refined, both of which led to the reality of human lunar landings.

## Human exploration of the Moon

Sending people to the Moon was, in many ways, the ultimate goal of the Space Race. Both the Soviets and the Americans desperately wanted to be the first to accomplish this Herculean goal — a goal only one nation would achieve.

Due to perseverance, ingenuity, and a little bit of Soviet bad luck, the Americans pulled ahead in the race to the Moon by 1968. In 1969, Americans Neil Armstrong and Edwin "Buzz" Aldrin became the first humans to explore the surface of the Moon when the Apollo 11 mission touched down on the Sea of Tranquility (see Chapter 9). Five other successful Apollo Moon landings followed, as we describe in Chapter 10, allowing astronauts to explore different parts of the lunar surface and bring back hundreds of pounds of precious Moon rocks. After a series of failures with its N1 rocket, the Soviet Union abandoned its goal of sending humans to the Moon in the early 1970s and refocused its sights on its space station program.

## Missions to other parts of the solar system

Exploring the Moon was of paramount importance to the Space Race, but it wasn't the only outlet for space exploration — on either side of the contest. Both the American and Soviet space programs sent space probes to Mars and Venus in the 1960s and 1970s, with highly variable rates of success. NASA also sent several missions to Jupiter and Saturn. Although more than a few of these missions failed, the successful ones sent back imagery that provided people worldwide with the first views of these planets and revealed new information about the planets' compositions and origins. Flip to Chapter 11 for full details on missions to other parts of the solar system.

## The end of the Space Race

A joint American-Soviet mission in 1975 marked the end of the Space Race (which had been slowly fizzling out since the success of Apollo 11, despite continued Cold War tensions throughout the 1970s). As you discover in Chapter 12, a series of Soviet Salyut space stations, and the NASA Skylab station, led to this joint mission that involved the docking of a Soyuz spacecraft and an Apollo capsule in orbit. This joint mission featured a handshake that symbolically ended the Space Race. Thus, although it began with a bang, the Space Race ended quietly and without fanfare as both countries took their space programs in new directions with the realization that continued success in space would require future collaboration.

# The Second Generation of Missions

American success with placing humans on the Moon in no way signaled the end of the space program. On the contrary: Now that the world knew what was possible, expectations began to rise for what other brave, new worlds could be explored. Several major series of second-generation missions, made possible by the successes of the Space Shuttle and Soyuz spacecraft, allowed astronauts to construct the world's first large-scale space habitations. New views of Mars flooded the news thanks to two spunky little rovers, and the mysteries of the inner and outer solar system began to unravel.

## The Space Shuttle and Mir

During the prime Space Race years, significant collaboration between the U.S. and Soviet space programs was impossible with these two superpowers competing directly against one another. After the Space Race concluded, though, pathways opened for such collaboration to take place.

Enter the Space Shuttle-Mir program, a joint venture whereby Russian cosmonauts traveled to the Russian space station Mir aboard an American Space Shuttle. American astronauts also traveled to Mir, learning from Russian experience about living and working in space for extended periods of time. In addition to knowledge transfer, the Space Shuttle-Mir program formalized a new spirit of international cooperation, replacing the previous years of bitter competition.

Check out Chapters 13 and 14 if you're interested in finding out more about Space Shuttle missions and the Mir space station.

## Journeys to Mars

Studying the Martian landscape and environment hasn't come easily. Although there've been a few great successes (including the Mars Pathfinder mission in 1996, which successfully placed the *Sojourner* rover on the surface of the Red Planet), there've been some rather spectacular failures. Several missions to Mars in the 1990s failed for a number of reasons, leading to dashed expectations and an increasing sense of futility regarding Mars research. Was Mars, a planet of incredible interest to scientists because of its striking similarities to Earth and its possibilities for life, going to remain shrouded in mystery? Within the next decade, these questions and more would be answered. Head to Chapter 15 for all the details.

## Exploration of the inner solar system

Although Mars remained an important goal for the American space program in the 1990s, NASA's attention turned to several targets in the inner solar system in the last decade of the 20th century and the first decade of the 21st century, as we explain in Chapter 16. The SOHO mission has returned valuable information about the Sun, and the Magellan and Venus Express missions have vastly increased scientists' knowledge of the surface and atmosphere of Venus. Other recent missions have returned to the Moon and sought to unravel the mysteries of asteroids and comets.

## Missions to the outer solar system

One of mankind's primary goals in exploring outer space is determining to what extent life, past or present, can be found. When looking for life, the NASA motto is "follow the water" — where there's water, there may be some form of life.

Scientists have several targets of primary interest in the search for life in the outer solar system, as we explain in Chapter 17. Jupiter and its moons were the focus of the Galileo mission in the 1990s, and significant discoveries about subsurface water came to light. From the late 1990s through the first decade of the 21st century, the Cassini mission, which carried the *Huygens* probe at one point, has made similar discoveries on Saturn's moons. The more that's known about life in the solar system, the closer scientists come to understanding more about the origins of humanity.

# Modern Space Exploration

We've been to the Moon (and back); we've explored the solar system; we've catalogued the cosmos — what else is there? Plenty, as it turns out. Space exploration in the 21st century continues the work of past missions, but it has also yielded groundbreaking new information about the universe. Powerful telescopes have captured breathtaking views of the cosmos, leading scientists to new understandings of humanity's place in the universe. The International Space Station is laying the groundwork for an enduring human presence in space, and Mars exploration has grown more successful, as you can see in the following sections.

## Space telescopes

Despite the success of the Apollo Moon missions, it's not currently practical for humans to make extended visits to other space destinations. Fortunately, space telescopes help provide access to previously inaccessible parts of the universe. The Hubble Space Telescope is one of the best-known because it has taken some of the most-amazing pictures of celestial objects ever seen, but other space telescopes (such as the Spitzer Space Telescope and the Compton Gamma Ray Observatory) have returned their fair share of amazing data as well. See Chapter 18 for more on telescopes in space.

## The International Space Station

Are space hotels but a glimpse into the far-distant future? Perhaps not — the International Space Station (ISS), a multicountry collaborative effort, has resulted in a series of modules that have supported a continuous human presence in orbit since late 2000.

Living and working in the ISS comes with a particular set of challenges, and the astronauts are making valuable scientific contributions with the experiments they conduct onboard. Although the station is decidedly unglamorous and can't accommodate untrained astronauts, it provides a basis upon which civilian forays into space might, in the future, be built. Flip to Chapter 19 for details on the ISS.

## The latest views of Mars

Despite some early failures in visiting Mars, missions in the 21st century have been able to successfully start exploring the Red Planet. The best publicized of these missions were the Mars Exploration Rovers, which brought two wheeled, fully mechanized rovers (*Spirit* and *Opportunity*) to the Martian surface. They began their task of documenting the planet's surface in unprecedented detail, and the phenomenal images they sent back home inspired even more interest in discovering the mysteries of Mars. Other ongoing Mars missions include orbiters that have produced stunning maps of the Red Planet's surface; check out Chapter 20 for the full scoop.

# Space Exploration in the Future

Even though NASA and other worldwide space agencies have designed and executed many successful missions, all of these accomplishments are but a fraction of the work involved in understanding the cosmos. New missions are constantly being planned — ones that will revisit interesting locations and explore new frontiers.

## The next space missions to watch out for

Some robotic missions to watch out for include a return visit to Jupiter's moon Europa, a mission on its way to Pluto, and a Japanese mission to return a sample of an asteroid to Earth. Missions are also planned to provide further details about Mars and to study the Moon with an eye toward future astronaut missions.

On the human spaceflight side, Project Constellation will replace the Space Shuttle with a new Crew Exploration Vehicle dubbed Orion. This capsule-based system will allow astronauts to visit the International Space Station and eventually return to the Moon or head to Mars. Chapter 21 has details on many upcoming robotic and manned missions.

## Increased access to space

If that yearly ski trip is starting to seem old and tired, you may soon have new options on the horizon. For those with the means (and we're not talking chump change), purchasing a ticket into space is becoming a reality for the first time in the history of space exploration. Several wealthy individuals have done just that by paying millions of dollars, training and preparing with real astronauts, and taking a ride into space. Although space travel is beyond the means of most folks, prices may eventually come down enough to make spaceflight, or even a stay at a commercial space "hotel," a possibility.

Of course, increased access to space doesn't just refer to the advent of space tourism. As more and more countries over the years have been bitten by the space exploration bug, a third nation has emerged as a true space power. China launched its first *taikonaut* (the Chinese term for astronaut) into orbit in 2003, followed by a two-person mission in 2005 and a mission with a spacewalk in 2008. China has plans to develop its own space station and eventually send astronauts to the Moon, helping spark some renewed interest in lunar exploration back in the U.S. Could a new Space Race be in the works? Probably not, but a little healthy competition can be good for scientific innovation! Chapter 22 has all the details on space tourism, commercial spaceflight, and China's new space program.

## Reasons for continuing to explore space

Why continue to explore space? Why this fascination with the heavens when there are enough problems and diversions here on Earth? Space exploration can be seen as a continuation of humanity's need to explore, to push the limits of the frontier. With future population pressures and the speedy consumption of natural resources, space exploration and colonization may one day become a necessity. In the far-distant future, when the Sun becomes unstable, space exploration may be vital to ensure the survival of the human race. Chapter 23 goes into more detail on the many reasons why space exploration is still a worthy pursuit.

# Chapter 2

# A Tour of the Universe

### In This Chapter

- Observing the sky
- Examining the solar system's design
- Understanding extrasolar planets, stars, and galaxies

What is a planet? How did early humans view the night sky? How did people determine mankind's place in the solar system, and the universe, in an era before space exploration began? And what's beyond the solar system? This chapter helps answer those basic questions and much more. We also describe the basic organization of the solar system and touch on the status of Pluto, once the ninth planet and now one of a group of bodies known as dwarf planets.

## Planets as Wanderers: How Early Humans Viewed the Sky

In the earliest days of humanity's fascination with the heavens, eagle-eyed observers noticed that although the patterns of the stars never changed, some points of light seemed to travel through them on a fixed path. These objects were called planets, from the Greek root *planasthai,* meaning "wanderers." Such initial observations helped ancient astronomers track the seasons and, eventually, allowed for the revolutionary understanding of how the solar system is arranged. The Earth's place in that solar system was, as you can imagine, one of the keenest points of interest.

In the following sections, we delve into the patterns of stars that early humans observed (now known as constellations) and a few important early discoveries about the layout of the solar system.

## Constellations: Pictures in the sky

*Astronomers,* scientists who study the universe, organize the sky according to fixed *constellations,* or groups of stars and other heavenly bodies that are arranged in a particular way. The International Astronomical Union (IAU) recognizes 88 "official" constellations, which form a map of the entire sky when combined together (Figure 2-1 shows some constellations).

The basic European northern constellations originate in Greek mythology and were said to have been created by Zeus and his divine cohorts (according to myriad ancient Greek writers, that is). Homer, a poet who lived in the seventh century BCE, wrote of arrangements of stars resembling certain shapes, such as the Ram. Later, other ancient Greek authors wrote of star arrangements based on the gods and their antics. Not wanting to be left out, Roman authors such as Ptolemy (of the second century CE) catalogued the stars into at least 48 different constellations, some of which involved Roman mythology. Although this list was limited to those constellations that could be seen in the Northern Hemisphere, it formed the basis for today's categorization.

In the following sections, we list a few of the most-famous constellations and describe the constellations of the zodiac.

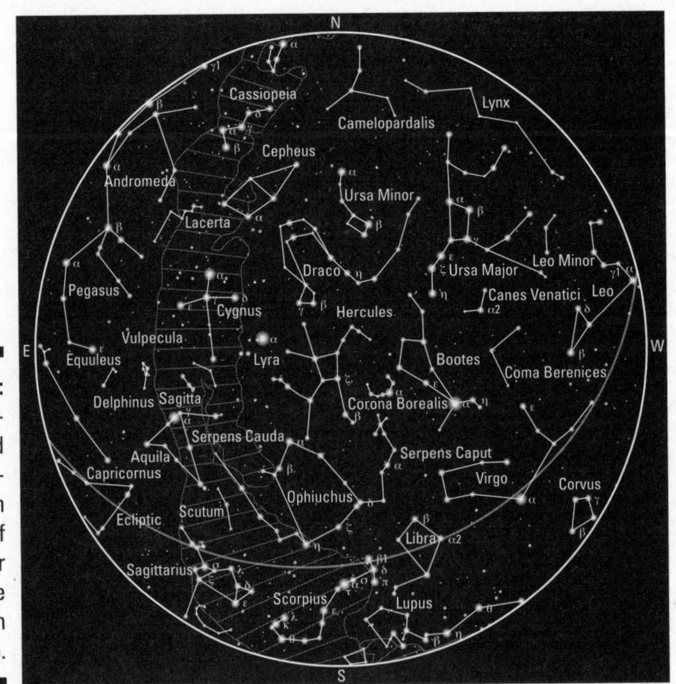


**Figure 2-1:** These IAU-recognized constellations form a map of the summer sky in the Northern Hemisphere.

## Why don't constellations move?

The stars in constellations are always fixed with respect to each other. The whole sky, complete with constellations, appears to rise and fall each night with the rotation of the Earth, but the shapes of the constellations themselves remain constant. The orientation of the stars, as seen from the Earth, doesn't change, because the stars are so far away from the Earth. They form a fixed backdrop against which the objects in our own solar system — the planets — move each night.

### *Zeroing in on a few famous constellations*

Some of the better-known Northern Hemisphere constellations include

- **Cassiopeia:** This constellation is visible year-round. Cassiopeia is named after the queen who nearly sacrificed her daughter Andromeda for her own vanity.
- **Orion:** Named after a mythological hunter born with the aid of Zeus and killed by a scorpion sent from Zeus's jealous wife Hera, this constellation is identifiable by the three-starred Orion's Belt. It's visible high in the winter sky.
- **Pegasus:** This summer constellation consists of a number of stars that form the shape of a horse. Pegasus is named for the winged creature that sprung from the head of Medusa.
- **Perseus:** Named for the Greek mythological hero who slew Medusa, this constellation is visible in the winter sky. It's related to fellow constellations Cassiopeia, Andromeda, Cepheus, and Cetus.
- **Ursa Major:** This constellation is visible year-round. Called the Great Bear in Navajo mythology, Ursa Major also includes the Big Dipper, an unofficial star grouping that has a very recognizable shape.

To find out more about the constellations, pick up the latest issue of *Sky & Telescope* magazine for its monthly star map, or visit the magazine's Web site, `www.skypub.com`, for an interactive sky chart. The book *Astronomy For Dummies* is also a great resource for lots more info about the stars and how to observe them.

### *Traveling through the zodiac*

When ancient astronomers first observed the planets, they also noticed that the wanderers seemed to travel through a set of 12 constellations. These 12 constellations made up the ancient Babylonian system of navigation in which each station was given an animal sign. This system was later called the *zodiac,* which meant "circle of animals" in Greek. Over time, these cultures created a calendar where each constellation corresponded approximately to one month of the year.

Different countries around the world have different zodiacs. For example, the western one is unlike the traditional Chinese zodiac, which is also unlike the Hindu or Celtic versions. The 12 signs of the western zodiac are as follows:

- Aries
- Taurus
- Gemini
- Cancer
- Leo
- Virgo
- Libra
- Scorpio
- Sagittarius
- Capricorn
- Aquarius
- Pisces

These 12 signs of the zodiac make up what's called the *Ecliptic,* which is the plane of the solar system as seen on the sky. The solar system is laid out like concentric circles on a flat plane, with the Sun at the center and the planets moving in ever-increasing circles around it. When someone on Earth looks to the sky and observes the solar system's other planets, those planets are much closer than the fixed stars that make up the constellations. The planets therefore appear to move with respect to the constellations and drift through the Ecliptic plane, whose background is the 12 signs of the zodiac.

## Is Earth the center of the universe? Early discoveries about the solar system

Early views of the solar system's design assumed that the Earth was at the center of the cosmos. After all, the Sun, Moon, planets, and stars all appeared to rise and set each day and seemed to circle the Earth. Why shouldn't the Earth be front and center on the universe's stage?

At the time, such a layout made theological sense because it reinforced humanity's view of the Earth as "special." This *geocentric,* or Earth-centered, view of the universe was supported by both mainstream science and (thanks to various powerful churches of the 16th century) religion. However, several scientists in the 1500s and 1600s disputed this view with scientific discoveries; we describe two of the most-important findings in the following sections.

### Copernicus: Constructing a heliocentric view

One of the founding fathers of modern astronomy, Nicolaus Copernicus, made waves for his rejection of the Earth as the literal center of the universe. Based on an elegant theoretical model of the geometry of the solar system and the motions of the planets, he eventually came up with the idea that the

Sun was at the center of the solar system, not the Earth. Copernicus first presented his *heliocentric theory* around 1510 in a preliminary thesis that was never published.

He later expounded upon this thesis in a work called *De revolutionibus orbium coelestium (On the Revolutions of the Celestial Spheres),* which wasn't published until 1543, the year of his death. Most people think that Copernicus delayed publishing his ideas because he was afraid to go against the powerful religious forces of the day — then again, he could also have just feared looking foolish.

### Galileo: Building on Copernicus's work

Astronomers after Copernicus continued his work, but none did so more faithfully than Galileo Galilei, born in 1564. An Italian astronomer, mathematician, physicist, and all-around scientist, Galileo was one of the first scientists to provide a mathematical description for the laws of nature. He also contributed significantly to society's understanding of the physics of motion.

Galileo designed enhanced telescopes (think of 'em as the Hubble Space Telescope of the Renaissance period) that were the most powerful of the day and that allowed him to perform the most-detailed astronomical observations of the time, including a study of Jupiter's moons. By using his telescopes, Galileo was able to identify lunar topography, observe the phases of Venus, and make detailed notes about stars and their positions over time. His studies also convinced him that the geocentric theory of planetary orbit was wrong, and he expressed vocal support for the heliocentric theory espoused by Copernicus. However, the Catholic Church persecuted him for teaching these beliefs and committed him to house arrest toward the end of his life.

## The Organization of the Solar System

If you want to understand how the solar system is organized, we suggest you begin at the beginning — the formation of the Sun and planets. The modern scientific view of the origin of the solar system (and universe) isn't quite up to biblical standards, but it's still spectacular.

Astronomers believe it all began 4.6 billion years ago, when a giant molecular cloud of gas and dust began to collapse. What triggered the collapse of this cloud is still in question, but it may have been a shock wave caused by the explosion of a nearby star in the molecular cloud that started another corner of the cloud collapsing.

What *is* known is that after parts of the molecular cloud began sticking together, the whole process started to pick up speed due to the influence of gravity. Gravity is a fairly weak force over long distances, but as the cloud collapsed and became denser and denser, the gravitational attraction of the particles in the cloud, now called the *presolar nebula,* made this collapse increase in speed. As the nebula caved in, it began to spin to conserve *angular momentum* (we'll spare you the nitty-gritty details of this property of physics!). The presolar nebula subsided into a flat, spinning disk of gas and dust called a *protoplanetary disk,* with our solar system's protosun at the center.

It probably took about 100,000 years, very short on cosmic timescales, for the gas and dust at the center of the protoplanetary disk to balance the force of gravitational collapse with the pressure of the hot gas. After 50 million years or so, the protostar's center reached extreme enough conditions of temperature and pressure to allow nuclear fusion to occur, and the protostar became a full-fledged star, fusing hydrogen molecules together to make helium.

As the cloud collapsed, material in the protoplanetary disk began to clump together (you can thank gravity again for that one). There wasn't enough material to make another star, but there was plenty to make a solar system full of bodies called *planetesimals.* These smaller bodies kept crashing into each other and sticking together, forming larger and larger bodies. By the time the Sun was near the conditions required for it to burn hydrogen, its energy output was blowing much of the leftover material out of the solar system.

Over time, the objects in the inner solar system (near the Sun) formed the four terrestrial planets you know and love — Mercury, Venus, Earth, and Mars. Conditions in the inner solar system were still very hot, due to the bright young Sun, and it was too hot for materials such as ices and other easily evaporated compounds to stick around. For this reason, the planets of the inner solar system are made mostly of rock and metal.

Out somewhere beyond the orbit of Mars is what astronomers call the *snow line,* the distance from the early Sun at which ice could finally condense out of the remnants of the protoplanetary disk (now called the *solar nebula*). At this great distance from the Sun, ices formed rapidly and were swept up into what became the giant planet Jupiter, which quickly grew large enough to start capturing hydrogen gas from the solar nebula. Jupiter, composed primarily of hydrogen gas, is the largest planet in the solar system. Saturn, the next planet out from the Sun, is similar to Jupiter in composition but much smaller.

Beyond Saturn lie Uranus and Neptune, which are sometimes called *ice giants* to distinguish them from the *gas giants,* Jupiter and Saturn. Uranus and Neptune contain much more water, ammonia, and methane ices than Jupiter and Saturn, because they were formed in regions farther from the Sun and thus are even colder.

Past Neptune are the farthest reaches of the solar system, which contain primordial remnants from the solar nebula. Pluto, formerly known as the ninth planet, lurks in this dim region, accompanied by a large number of cold, icy bodies known as Kuiper Belt Objects. (We clue you in to the Pluto saga later in this chapter.) Beyond the Kuiper Belt lies an even more remote region of the solar system: the Oort Cloud, a swarm of icy bodies that envelops the flat solar system like a blanket.

Both the Kuiper Belt and the Oort Cloud are the sources of the occasional comet that wanders into the inner solar system and beautifully lights up the night sky. As these icy objects approach the Sun, some of their ice gets heated up and turns into gas, which trails behind the small ice-ball like the train of a bridal gown in a spectacular finish. Not bad for such small, icy bodies from the unfashionable reaches of the solar system!

This formation process resulted in the modern, orderly solar system in which planets travel around the Sun on nearly circular orbits and basically lie in a flat plane. The solar system is made up of 8 planets, 3 (and probably more) dwarf planets, a large number of small bodies (which include asteroids and comets), and more than 130 *satellites* (objects that orbit the planets). Check out the solar system's organization in Figure 2-2 and discover more about the different areas of the system in the following sections.

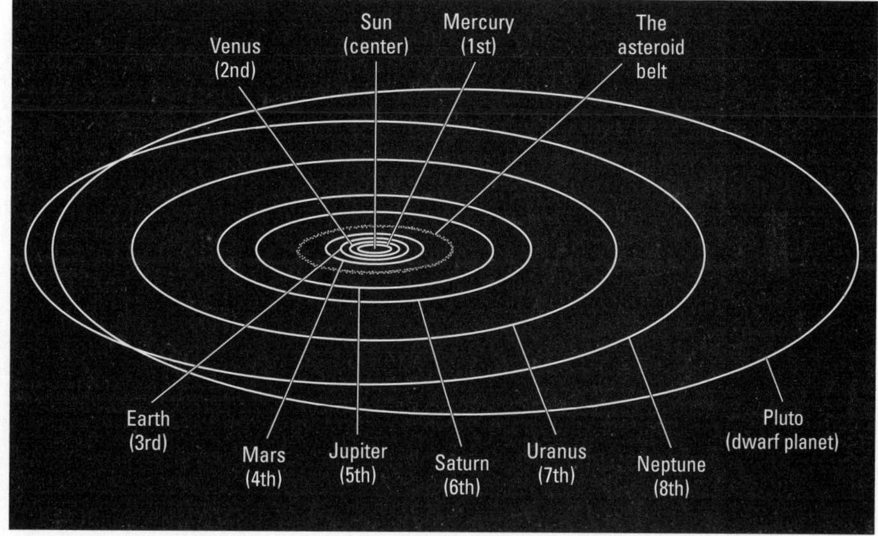


**Figure 2-2:** The organization of the solar system — not drawn to scale!

## Rock on: The inner solar system

The solar system features four inner, rocky planets — Mercury, Venus, Earth, and Mars —that are sometimes called the *terrestrial planets* because of their basic similarity to Earth. But even though these planets are of similar sizes, they vary in appearance. Following are some key details on the planets of the inner solar system:

- **Mercury:** The closest planet to the Sun, Mercury is named after the Roman travel and messenger god. Its hot surface is scarred with ancient lava flows and covered with impact craters. It rotates slowly in what's called a *resonance* — Mercury rotates three times on its axis (days) in two trips around the Sun (years). It has a very high density, which suggests that only heavy metallic elements could've condensed into the material that eventually formed Mercury so near to the Sun. Mercury is the smallest of the eight planets; it's even smaller than the moons Titan and Ganymede.
- **Venus:** Sometimes referred to as Earth's sister planet, Venus is named after the Roman goddess of beauty due to its bright appearance. However, its thick clouds hide a surface that's anything but gentle and welcoming: The planet's thick atmosphere means that surface temperatures are hot enough to melt lead! You won't find water on the surface of Venus, just mountains, lava flows, and a few impact craters.
- **Earth:** Home to plants, animals, people, and an insane number of reality TV shows, Earth is the largest terrestrial planet. From space, its appearance is dominated by clouds and oceans, with scattered continents. Earth is located at such a distance from the Sun that water, a key requirement for life as we know it, is actually stable in all three phases — solid ice, liquid water, and gaseous water vapor. Earth's active surface is covered with many different geological features due to the workings of water, air, volcanoes, and plate tectonics. Earth has one large satellite, the Moon, which scientists think formed when a Mars-sized object smashed into the Earth early in the solar system's history, sending material out into space that eventually formed the Moon.
- **Mars:** Named after the Roman god of war for its red appearance, Mars is larger than Mercury but smaller than Venus. A cold, old world today with a thin atmosphere that makes water stable only in the polar ice caps, Mars may once have been much warmer and wetter. Spacecraft images show that Mars's surface is covered by features that appear to have been carved by liquid water. Mars is home to the largest volcano in the solar system and a huge canyon that would stretch from California to the East Coast of the United States if placed on Earth. Mars also has two tiny satellites, Phobos and Deimos, which are likely *captured asteroids* (former residents of the asteroid belt that strayed a bit too close to the gravity of the Red Planet).

## Asteroids: Filling in the gaps

Between the orbits of Mars and Jupiter lies the main asteroid belt. *Asteroids* are mostly small, primitive bodies left over from the formation of the solar system. These small bodies possess regular orbits around the Sun but are much too small to be planets. A few hundred thousand asteroids have been discovered so far, but when added all together, their mass is still less than that of the Moon.

Many asteroids are irregularly shaped, but some are sizable enough to be largely round. Most meteorites that hit the Earth come from the asteroid belt, and scientists can compare the chemical composition of meteorites in a laboratory to observations of asteroids taken with a telescope to try and determine where different meteorites came from.

## Great balls of gas (and ice): The giants of the outer solar system

Beyond the asteroid belt lie the outer planets: Jupiter, Saturn, Uranus, and Neptune. Sometimes referred to as the *giant planets,* these large balls of gas and ice are much larger than the inner planets and contain most of the non-solar mass of the solar system. Here's the scoop on the various outer planets:

- **Jupiter:** Named after the Roman king of the gods, Jupiter is certainly the king of the planets — it contains more mass than all the rest of the planets in the solar system combined! Made up mostly of hydrogen with about 10 percent helium, Jupiter probably has a small rocky core near its center. The planet is known for the Great Red Spot, a huge storm (larger than the Earth), that has been seen from Earth for more than 300 years. Like any good king, Jupiter has its subjects in the form of at least 63 moons. Four of them, the *Galilean satellites,* are large and regular; the rest are small.
- **Saturn:** The solar system's second-largest planet, Saturn, is named after the Roman god of agriculture. It's made mostly of hydrogen and helium and is so low in density that it would actually float in water if you could find enough. Saturn's rings are its most-breathtaking feature. They spread to twice the planet's diameter but are only about 1 kilometer thick and consist mostly of small particles of water ice and ice-coated rock. Saturn has at least 34 satellites, the most interesting of which are Titan (with its thick atmosphere) and Enceladus (which has geysers).
- **Uranus:** The seventh planet from the Sun, Uranus is named after the Greek sky god and is in fact as blue as the sky. It's larger (but less massive) than Neptune and made up of rock and ice. Its hydrogen-and-helium atmosphere is much less immense than the atmospheres of Jupiter and Saturn. Uranus is the only planet in the solar system to rotate on its side: The rotation axis of most planets is perpendicular to the plane of the solar system, with North pointing up, but Uranus's rotation axis is parallel to the plane of the solar system. The planet has 21 known moons and a small ring system.

- **Neptune:** Named for the Roman god of the ocean, Neptune has a deep blue color to match and a composition that's similar to Uranus's. The planet has a dark ring system and one large moon, Triton, as well as at least 12 other moons that are small and irregular. Triton orbits around Neptune backwards from most other moons in the solar system and is thought to probably be a captured object from the farthest reaches of the solar system. Triton even has small volcanic plumes!

## Talk about drama: The great Pluto debate

Any schoolchild of a certain age will tell you that the solar system is made up of nine planets. Heck, she may even be able to name them all in order. However, the definition of a planet isn't quite as obvious as it seems, especially at the small end of the spectrum. The structure of our solar system features four small inner plants, four large outer planets, and an assortment of leftover debris beyond the orbit of Neptune. These leftovers range in size from the planet formerly known as Pluto down to much smaller objects.

For 75 years, Pluto, as the largest, brightest, and closest of these far-outer-solar-system objects, was recognized as a planet. In the first decade of the 21st century, however, studies of the outer solar system revealed that one or more large objects may exist beyond the orbit of Pluto that could be Pluto-sized or even larger. Astronomers rushed to redefine what constitutes a planet, and in 2006, they came up with a decision that has forced people around the world to revisit some basic facts about the solar system.

In the long run, this reclassification will provide a more-consistent picture of the planets. For now, we describe Pluto's history and the category of dwarf planets in the following sections.

### Flashing back through the history of Pluto

Pluto was on astronomers' radar for about 90 years before it was officially discovered. Early astronomers noticed something in the area of Neptune and Uranus that appeared to be interfering with their orbits. They couldn't define the source of that interference but suspected it might be a planet (considering Neptune had previously been found due to interference with the orbit of Uranus).

In 1906, astronomer Percival Lowell, from a rich manufacturing family, took up the challenge to search for the ninth planet. He built a large observatory, called Lowell Observatory and located near Flagstaff, Arizona, and devoted himself to searching the heavens. (On a side note, Lowell was known for his observations of Mars, where he thought he saw canals built by intelligent life.) Lowell died in 1916, but the observatory staff continued searching the heavens, and in 1930, Clyde Tombaugh definitively confirmed the presence of what was first called Planet X. This object was eventually renamed Pluto and became the solar system's ninth planet.

In keeping with the mythological names given to other planets in the solar system, Pluto is named after the ancient Roman god of the underworld (known as Hades in Greek mythology). Another reason for the choice may have been that Pluto contains the initials of Percival Lowell, initials that are also reflected in Pluto's symbol, which includes the letters P and L.

Ironically, although Pluto was found after models of planetary orbits predicted that something was influencing the orbit of Neptune, Pluto itself was too small in size to have had the predicted effect. Astronomers later determined that Neptune's mass had been overestimated — with the proper mass of Neptune, no extra planet is necessary. Pluto's discovery was thus a lucky mistake.

### Examining what it means to be a planet — and why Pluto doesn't make the cut

Although the definition of a planet seems intuitively obvious, astronomers have had a surprisingly difficult time coming up with a planetary definition that includes only what are known to be planets and doesn't include extra objects such as asteroids and comets.

The definition of a planet has changed substantially over time. Ancient Greek astronomers defined planets as *wandering stars,* or celestial bodies that moved consistently during the year. The Greeks were aware of what would later be named Saturn, Jupiter, Mars, Venus, and Mercury, but they also considered the Sun and Moon to be planets. Renaissance astronomy, particularly the heliocentric theory that we present earlier in this chapter, removed the Sun and Moon from the list of planets. In later years, astronomers added Neptune, Uranus, and Pluto to the solar system's planetary lineup.

The primary reason why Pluto currently fails to qualify as a planet has to do with its orbit. Pluto's orbit is elliptical, meaning it's shaped like an egg rather than a circle. According to the International Astronomical Union (IAU)'s 2006 definition, planets must have orbits that don't overlap with those of other planets. In other words, a planet must orbit the Sun and also be big enough to dominate the immediate area surrounding its orbit. Pluto's orbit crosses Neptune's orbit, so according to the IAU logic, Pluto could no longer be considered a planet.

Another scientific requirement for a planet is that it possesses sufficient internal gravitation to be round. A planet with enough mass reaches a *hydrostatic equilibrium,* meaning everything on its surface is at about the same distance from the center. Pluto is relatively round, but perhaps not enough so to meet this new, more-stringent definition. In fact, this particular point met with great contention from astronomers who wanted to maintain Pluto's planetary status.

### Defining dwarf planets

So if Pluto is no longer a planet, what is it? It's a dwarf planet! In its 2006 planetary definition, the IAU also created a new dwarf-planet category. *Dwarf planets* are objects that orbit the Sun, are large enough to be round but are smaller than Mercury, and can orbit the Sun in a zone with lots of other objects. Three dwarf planets are currently known to exist:

- Pluto, which orbits in the Kuiper Belt
- Ceres, which orbits in the asteroid belt
- Eris, which is located in the Kuiper Belt

Here's some info that may help you win your hometown trivia night some day (and if you do, we'd be happy to go in on the winnings with you): Eris, discovered in 2006, was the object that caused all the hubbub in planetary definitions. Turns out Eris is actually larger than Pluto! For a while, Eris was dubbed the tenth planet, but astronomers soon started worrying about how many more planets might be lurking in the far reaches of the solar system. The new dwarf-planet category handily creates a place to put Pluto, Eris, and any future objects discovered in the region.

# Extrasolar Planets and Stars and Galaxies, Oh My!

When you look up at the night sky in a dark place, you can see many points of light. They may all look the same to the naked eye, but astronomers have determined that these objects can be very different from each other. The planets in our solar system circle the Sun, which is a star like many of the other stars seen in the night sky. Guess what? Many of those stars may have their own planets circling them, too. Every star you see in the sky is part of the Milky Way Galaxy, which is made up of 300 billion stars and is about 100,000 light years across. The Milky Way Galaxy is but one of billions of galaxies that comprise the universe. Find out more about extrasolar planets, stars, and galaxies in the following sections.

## Extrasolar planets

The *solar system* is the Sun and other objects (planets, dwarf planets, asteroids, comets, and other small bodies) that are connected to it by gravity. *Extrasolar planets,* also called exoplanets, are planets found outside the solar system. Currently, astronomers recognize more than 300 extrasolar planets orbiting other stars in the Milky Way Galaxy. Many of these planets are thought to be enormous gaseous balls similar in nature to Jupiter, but some are as small as only a few times the mass of the Earth.

Knowledge and understanding of extrasolar planets is critical to space exploration because it allows astronauts to discover more about the extent of the universe. Since the time of Copernicus and Galileo (whom we touch on earlier in this chapter), humanity has had a sneaking suspicion that "we" (meaning the solar system) can't possibly comprise the full extent of the universe. The entire field of extrasolar-planet research, in fact, relies on the assumption that this solar system isn't the only one out there. Discovering extrasolar planets both confirms that assumption and provides insight into where they are and what they're made of.

As recently as ten years ago, our solar system was the only example of a planetary system that astronomers knew about with absolute certainty. Recently, an explosion of observational data has begun to definitively detect planets surrounding other stars. The first extrasolar planets found were mostly large planets located close to their stars, because larger planets are easiest to detect from Earth. More-recent studies have been finding smaller and smaller planets. Future dedicated spacecraft, like the upcoming Kepler mission (see Chapter 21), hope to finally detect Earth-like planets around other stars.

## Homes of the stars

The Sun is a familiar sight to anyone who has ever stepped outside. It's readily visible in the daytime sky on Earth, but it's actually a star like the other stars in the night sky. It seems to be much bigger and brighter than those other stars, and that perception results from the fact that it's so much closer to the Earth.

So what exactly are stars? At their most fundamental, stars are balls of gas that maintain their cohesiveness via gravity and generate energy through nuclear fusion. Stars are very hot in the center where they burn hydrogen into helium and give off electromagnetic radiation, some of which can be seen as visible light. Stars vary in color due to this difference in heat energy; some appear whitish, whereas others appear bluish or yellowish. Stars look to be in fixed positions, but they're actually in constant motion, just like the planets.

Stars are formed in gas clouds called *nebulae*. As turbulence builds up inside a nebula, gas at the center of the turbulence heats up and the cloud begins to fold in on itself. The cloud eventually collapses, forming a star. The extra dust around the star may become part of the star, or it may go on to become an asteroid, comet, or planet. Growing a star isn't a quick process; the Sun took about 50 million years to reach the level of an adult star, and scientists think it'll live a long adult life — about 10 billion years.

## Galaxies galore

If you step outside on a dark night, particularly in a remote location far from cities and streetlights, you may see a bright band of stars spanning the sky. A familiar sight to prehistoric cultures, these days that band has an official name — the Milky Way Galaxy — because of its white, diffuse appearance. This bright band of light and stars is home to Earth. It's flat, like a disk, and has a spiral structure when viewed from above. The Sun is located in one of the spiral arms of the Milky Way Galaxy, which is just one of many galaxies in the universe.

*Galaxies* can be loosely defined as stars, dust, black holes, gas, and other forms of matter that are united by gravity. A significant number of galaxies exist in the universe; early Hubble Space Telescope studies place the estimate at around 125 billion. (See Chapters 14 and 18 for more about Hubble.) Galaxies can range in size from small (perhaps 10 million stars) to large (with a trillion or more stars), and they come in several different shapes:

- Spiral galaxies have whiplike, curved arms.
- Elliptical galaxies are oblong.
- Irregular galaxies can have a range of other shapes.

The Milky Way Galaxy contains the Sun and planets, plus about 300 billion additional stars. Included among these extra stars are *star clusters* (either bound or related groups of stars) and *nebulae* (collections of gas and dust that can be the birthing places for stars). Known nebulae and star clusters in the galaxy are listed in the *Messier database,* a listing first created by an astronomer named Charles Messier; he was searching for comets but ended up making a list of everything he observed that *wasn't* a comet. All the objects he identified received a Messier number, and these numbers are still used today for nebulae, clusters, galaxies, and other star groupings.

Modern astronomers have determined that many of the fuzzy patches of light originally called nebulae are in fact distant galaxies. One of the most famous, formerly the Andromeda Nebula, is now known to be the Andromeda Galaxy, one of the closest galaxies to the Milky Way. It's 2.5 million light years away, meaning it takes light 2.5 million years to travel from Andromeda to Earth. Check out an image of the Andromeda Galaxy in the color section; note that the spiral arms, made up of millions of stars, give it a similar appearance to our own Milky Way Galaxy (where our solar system is located in a spiral arm).

# Chapter 3

# Yes, It's Rocket (And Spacecraft) Science After All

### *In This Chapter*

- Detailing the history of rockets
- Understanding the forces acting on rockets
- Fueling rocket journeys
- Powering spacecraft
- Using radio signals to communicate

For a successful space mission, you need three key elements:

- Something to get your spacecraft into orbit
- Some way to move the spacecraft to its destination after it reaches space
- Some way to power the spacecraft during its journey

Rockets, those elegant and powerful cylinders, are what guide people and robotic spacecraft into space. They also traditionally serve as propulsion systems for spacecraft post-liftoff so the satellite, Space Shuttle, you name it, can reach its destination. Rockets represent some of mankind's most-advanced technological creations. From their relatively humble beginnings in the realm of ancient firecrackers to the larger-than-life Saturn V, rockets may have changed in scale and application but they still rely on the same basic principles of physics.

Of course, the spacecraft that rockets carry into orbit around Earth or beyond have their own power requirements. And given the distances that spacecraft must travel, calling home is a lot more difficult than a long-distance phone call. Communications between space and Earth require an advanced network of radio antennae and, fortunately, such a system already exists today.

In this chapter, we briefly describe the history of rockets and look at the basics of how rockets and spacecraft work. By the time you're done reading it, you'll be a certified rocket scientist (well, almost).

## Back in Time: A Quick Tour of Rockets in History

From ancient Chinese fireworks to modern ballistic missiles, the development of rockets has had both military and scientific implications. Here's a quick rocket-history timeline (we describe the major landmarks in detail in the following sections):

1st century CE: The ancient Greeks create a steam-powered, rocketlike engine.

12th century CE: The Chinese invent fireworks and military rockets.

1608: Galileo creates his first telescope (see Chapter 2 for details).

1687: Isaac Newton formulates laws of motion, which describe how rockets can move forward by ejecting mass backward at high speed.

1820: Rockets are used militarily throughout Europe.

1926: Robert Goddard, an American, launches the first liquid-fueled rocket.

1942: Germans launch the first V-2 rocket, otherwise known as the first suborbital ballistic missile, which was used in World War II.

1949: Americans launch the first Viking rocket, which was based on designs of captured V-2 rockets.

1953: NASA engineers complete the Redstone rocket.

1957: Soviets launch *Sputnik 1,* the world's first artificial satellite.

1958: Americans launch *Explorer 1,* the first U.S. satellite sent around Earth.

1961: Soviets launch the Vostok program and send the first human to space.

1961: NASA's Project Mercury sends the first American into space.

1966: Soviets achieve the first soft lunar landing with *Luna 9.*

1968–1972: Americans use Saturn V rockets to send Apollo spacecraft to the Moon.

1973: Americans launch Skylab, the first U.S. space station.

1981–2010: NASA Space Shuttles employ reusable solid rocket boosters.

## The earliest creations and uses

The modern rocket was a long time coming. The first known device similar to a rocket dates back to the first century CE. It was then that Hero, a Hellenistic Greek from Alexandria, invented a steam-powered engine called the *aeolipile,* which was similar to a rocket engine. It's unclear, however, whether this device was ever fully implemented.

As time progressed, military needs in particular drove early rocket development. For instance, the Chinese used rockets for warfare during the 12th century CE. Rockets also played a role during wars of the 20th century CE, particularly during World War I, when they were mounted to airplanes in order to deliver explosives to ground- or air-based targets.

The main predecessor to the modern space rocket was developed by an American named Robert Goddard. He designed and launched the first liquid-fueled rocket in 1926 because he was interested in space travel. Although this first rocket (named Nell) made it a mere 41 feet (12 meters) into the air, Goddard continued making advances in both speed and distance until his death in 1945.

Experts in other countries, including Wernher von Braun and Dr. Hermann Oberth in Germany; Fridik Tsander and Sergei Korolev in Russia; and many others, conducted experiments with liquid-fueled rockets in the 1940s. Such experiments led to the development of the German V-2 rocket, which was capable of reaching the edge of space and served as a formidable weapon during World War II.

## Advances in the 1950s, 1960s, and 1970s

One of the most-significant early military applications of rocketry was the Redstone rocket, whose first launch occurred in 1953 at Cape Canaveral Air Force Station, Florida. Working out of Huntsville, Alabama, German scientist Wernher von Bruan and his team created this liquid-fueled missile that used oxygen and alcohol to generate a 75,000-pound thrust. (*Thrust* is the force necessary for an object to take flight, defy gravity, and move through space; we describe a rocket's thrust more fully later in this chapter.) The Redstone measured approximately 70 feet (21 meters) long and had a speed of about 3,800 miles (6,100 kilometers) per hour during *burnout,* the period at which *propellant* (the combination of a fuel and an oxidizer that powers a rocket) has been completely exhausted. Intended as a defense mechanism during the Cold War years, the U.S. Army started using the Redstone rocket in 1958.

As technology and science continued to advance, new and better rockets made their way into use, and peaceful applications were developed for military missiles. The first American satellite sent around the Earth, *Explorer*

*1,* was launched in 1958. Designed by the Jet Propulsion Laboratory in California, *Explorer 1* was launched by the Jupiter-C rocket vehicle, a modified version of the earlier Redstone rocket and an assembly that consisted of ballistic missile technology. The Jupiter-C rocket was assembled in four stages, with the largest and heaviest (Stage 1) having a thrust of 83,000 pounds and a burn time of 155 seconds. (*Burn time* refers to how long the rocket burns its propellant.)

Later years saw rockets, specifically the Saturn series of rockets, used for human space exploration. Primarily developed in the 1960s by scientists who had emigrated from Germany to the U.S., the Saturn rockets were first proposed for launching military satellites. They went on to become the main launch vehicles for Project Apollo, NASA's foray into lunar exploration.

The largest rocket built during this time was the Saturn V. Nicknamed the Moon Rocket, the Saturn V was a liquid-fueled rocket used between 1967 and 1973 by both Project Apollo and Skylab, the first U.S. space station, and it remains the world's most powerful rocket ever by *payload* (the cargo that the rocket carries into space). The Saturn V was built in three *stages,* separate portions of a rocket that burn to completion and are then discarded. Standing an impressive 363 feet (110 meters) tall, the Saturn V helped make human spaceflight possible.

During the 1950s and 1960s, rockets changed greatly in size. Figure 3-1 shows the progression of rocket sizes throughout the years.

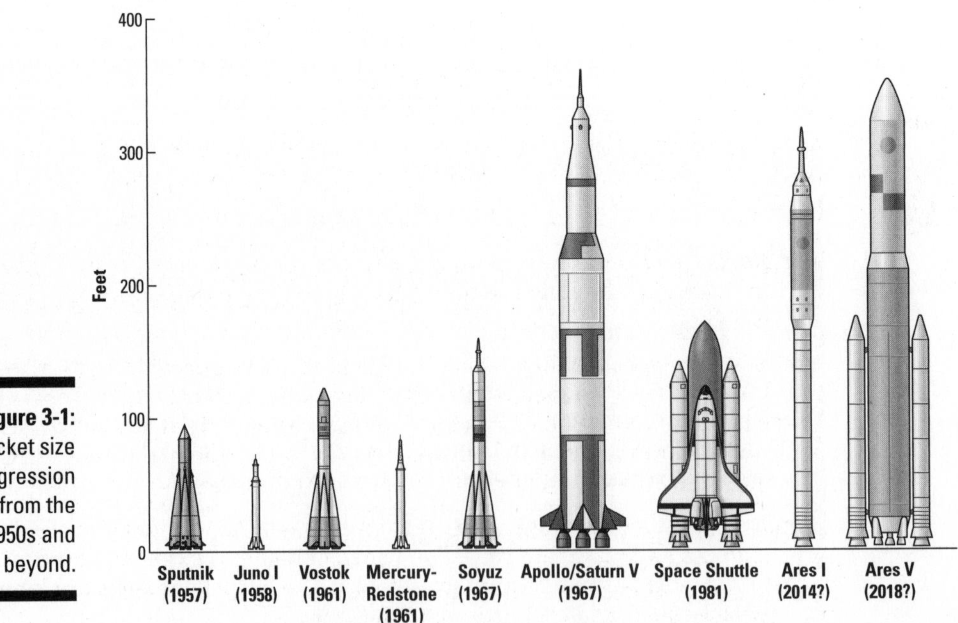


**Figure 3-1:** Rocket size progression from the 1950s and beyond.

## Starting to recycle in the 1980s

Rockets were traditionally disposable items designed to be discarded after a single use. However, the Space Shuttle, first launched in 1981, was designed with some reusable parts. Its smaller solid rocket boosters are designed to fall into the ocean after their fuel is exhausted; these boosters are retrieved, cleaned, refilled, and used again on future missions. Each booster can be used for several missions before it has to be retired. Employing reusable boosters theoretically allows a spacecraft to be launched more often and at a lower cost than using rockets that must be replaced completely after each launch.

The Space Shuttle orbiter itself is, of course, reusable as well. Those in the current NASA fleet have a lifetime of many dozens of launches. In fact, the oldest Space Shuttle still in the fleet, *Discovery,* has been traveling into orbit since 1984 — well beyond its expected lifetime.

# Let's Get Physics-al: How Rockets Lift Off and Travel through Space

All things being equal, spacecraft would prefer to stay right where they are on the ground, thank you very much. They need substantial initial energy to overcome gravity and lift off of the Earth's surface. That's where rocket engines come into play. In the following sections, we explain the forces that propel (and hinder) a rocket's liftoff from Earth and its movement through space.

Knowing how a rocket is built can help you understand how it lifts off and travels through space. Figure 3-2 shows the following essential parts of a rocket:

- **Rocket engines:** These powerful engines, usually located at the bottom of the rocket, provide the thrust that allows the rocket to escape Earth's gravity.
- **Booster rockets:** When a main rocket engine can't provide enough thrust for the rocket to reach orbit, booster rockets (also called *strap-ons*) can be added to provide some extra oomph, particularly at liftoff. Booster rockets are visible as the smaller rockets to the side of the main rocket in Figure 3-1, in particular the Space Shuttle and Ares V.
- **Propellant tanks:** Rockets burn through an amazing amount of fuel and oxidizer (which create *propellant* when combined) extremely quickly. Consequently, they require giant tanks. The propellant tanks are labeled as "fuel tank" and "oxidizer tank" in Figure 3-2.

- **Tail fins:** These small wings at the base of the rocket help stabilize it on its upward journey through the atmosphere.
- **Nose cone:** Shaped like a rounded cone, this part of a rocket sits at the very top and helps make the rocket's shape more aerodynamic during launch. The nose cone is located at the top of the escape rocket and tower in Figure 3-2.
- **Payload:** This is the satellite, spacecraft, or other component (in other words, the cargo) that the rocket carries into space. A rocket is usually much larger than the payload it carries. In Figure 3-2, the payload is the small Mercury capsule near the top.

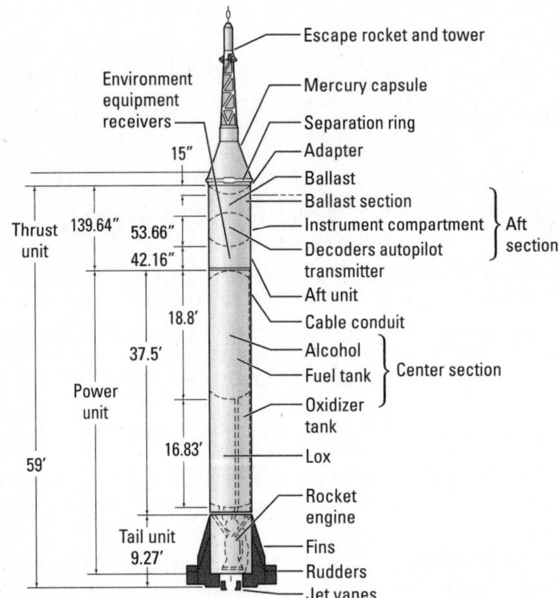


**Figure 3-2:** The most-important parts of a rocket.

## Understanding the main thrust of the issue

The main physics concept underlying a rocket engine is *thrust,* or the force necessary for an object to take flight, defy gravity, and move through space. A thrust force must be generated (usually by a propulsion system) in order for motion to occur. In the case of a rocket engine, gases are accelerated to the point where they can be expelled from the bottom of the engine. In English, that means massive amounts of propellant are burned and then discharged at a very high speed. This process generates the high level of thrust required for a rocket to achieve vertical motion.

Rocket propellants are designed to burn fast — without exploding, mind you, because an actual explosion would be counterproductive at best and disastrous at worst. All rocket propellants are based on combining a fuel and an oxidizer, which burn in the vacuum of space. Missiles and fall-away rocket boosters may use solid propellants, which can be efficient for certain purposes (such as long storage) but don't allow much control over the engine itself. After the rocket is ignited, it essentially can't be turned off until all the propellant is used up. Liquid propellants are generally used in cases where stopping and starting the engines is necessary; they combine a fuel source, such as liquid hydrogen, with an oxidizing agent, such as liquid oxygen. (For more on the different types of propellant, see the related section later in this chapter.)

## Examining the forces of nature that affect liftoff

Five forces come into play when a rocket engine is attempting liftoff, and most of them are trying their best to keep that rocket on the ground:

- **Weight:** The first force that must be overcome is the weight of the rocket plus all of its propellant. Weight always points down toward the center of the Earth because it's related to the force of gravity.
- **Gravity:** Gravity is the natural attraction of all objects toward the center of the Earth (or whatever planet you happen to be standing on), and it's what keeps your feet firmly planted on the ground until you choose otherwise.
- **Thrust:** Thrust, which counteracts weight, is what's produced when the rocket engine starts. It's related to how much propellant is flowing through the engine, as well as the specific chemical properties of the fuel-and-oxidizer combination. The thrust force is lined up with the long axis of the rocket. (See the preceding section for the full scoop on thrust.)
- **Drag:** Drag is the force you've experienced if you've ever gone swimming; it's a motion-dependent force exerted against any objects that move through fluid or air. Drag acts against the object's motion, meaning it serves to impede the motion of an airborne rocket. Larger objects, as in those with more surface area, experience increased drag.
- **Lift:** The lift force — usually not a factor in rocket operations — helps stabilize the upward movement of an object, such as an airplane, bird, or anything else that flies. Lift operates perpendicularly to drag, and it works to help control a rocket's motion.

If you want to know more about the forces that act on a rocket, check out this great Web site: `www.grc.nasa.gov/WWW/K-12/rocket/rktfor.html`.

## Escaping Earth and navigating the cosmos

A rocket flight, whether it's into the Earth's orbit or to a farther destination, such as the Moon, has a number of different phases:

- First, the rocket must leave Earth's surface. The familiar countdown of "3 . . . 2 . . . 1 . . . Blast off!" heralds the rocket carrying the spacecraft lifting off from the launch pad. Slow at first, the rocket quickly picks up speed and is soon gone from sight.
- For a vehicle traveling at an angle or without sufficient speed to escape Earth's gravity, the rocket travels on what's known as a ballistic trajectory and curves in a graceful arc until it eventually returns to the Earth's surface. (Usually two or three stages are needed for the rocket to be able to leave Earth's gravity and make it into space.) For a vehicle destined to orbit the Earth, the rocket must reach escape velocity, which is about 25,000 miles (40,000 kilometers) per hour for the Earth.
- After the rocket engines stop firing, the rocket is basically in free-fall around the Earth. For travel to the Moon, another upper-stage firing puts the spacecraft onto the right orbit so it can reach its destination. Such a firing of the rockets is called a *burn.* During the successful Apollo Moon landing missions, for example, each Apollo spacecraft performed an *orbit insertion* by firing its rockets to go into orbit around the Moon. The Lunar Module then separated and used small retrorockets to counter the Moon's lower gravity, allowing the spacecraft to land slowly and safely. Rockets were used at the end of the mission to lift off from the lunar surface and dock with the orbiting Command Module, which then fired its rockets to leave lunar orbit and return to Earth on a trajectory that would allow it to be captured by the Earth's gravity, reduce its velocity by atmospheric drag, and splash down safely in the ocean. Figure 3-3 shows all of these stages.

  For a robotic spacecraft destined to travel to Mars or another planet, a large burn is usually conducted shortly after leaving the Earth's atmosphere in order to send the spacecraft onto the correct trajectory. Over the many months or years it can take to travel to another planet, the rocket makes small course adjustments with its thrusters and sometimes uses flybys of other planets to get a gravitational boost to help it reach its final destination.

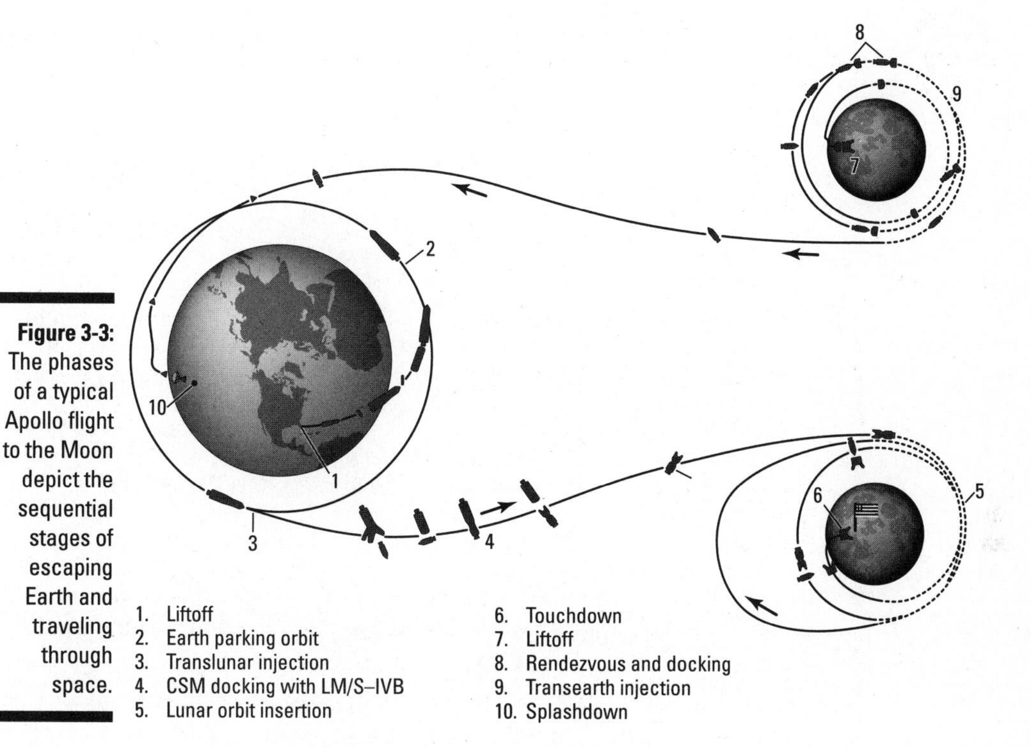


**Figure 3-3:** The phases of a typical Apollo flight to the Moon depict the sequential stages of escaping Earth and traveling through space.

# Start It Up: Different Sources of Rocket Power

To provide the thrust rockets need to lift off from Earth (we fill you in on thrust earlier in this chapter), rockets need propellant — and lots of it! Different types of propellant are used in different types of rockets, and some multistage rockets may use a different propellant in each stage. Although current rockets use traditional propellants, rockets of the future may propel themselves using interesting and exotic methods.

## Fueling the fire

Conventional rockets use a combination of a fuel and an oxidizer, which together are referred to as *propellant.* When large amounts of these two materials are mixed, the resulting chemical reaction, if properly contained, can provide enough thrust to send a spacecraft into orbit!

In the next couple sections, you find out how much propellant it takes to send a rocket into space, as well as the several types of propellant available for modern rockets.

### *The amount of propellant needed*

Traveling in space requires an inordinate amount of propellant. The Space Shuttle, for example, uses more than 1 million pounds (450,000 kilograms) of solid rocket propellant in each of its solid rocket boosters, and its external tank contains upward of half a million gallons (1.9 million liters) of liquid propellant (composed of liquid hydrogen and oxygen.)

If you think that's a lot, remember that this amount is only enough to get low-Earth-orbiting vehicles such as the Space Shuttle into orbit. For a spaceship to exit the Earth's gravitational pull, like spacecraft used in the Apollo Moon missions, much more propellant is necessary. Apollo's Saturn V (which we describe in the earlier section "Advances in the 1950s, 1960s, and 1970s") had three stages:

- The first stage had five engines that consumed 15 tons (13.6 metric tons) of propellant per second during the rocket's two-and-a-half-minute burn, requiring a total of about 4.5 million pounds (2 million kilograms) of liquid oxygen and kerosene fuel.
- The second stage, which burned liquid hydrogen and liquid oxygen, consumed 900,000 pounds (400,000 kilograms) of propellant in less than seven minutes.
- The third, smaller stage used 230,000 pounds (100,000 kilograms) of propellant.

With human spaceflight, you not only have to bring enough propellant to get to where you're going but you also need to save some so you can get back home. The huge Saturn V rocket enabled the Apollo missions to escape Earth's gravitational pull and travel the 239,000 miles (385,000 kilometers) to the Moon over the course of a little less than four days and still return home safely. A future human mission to Mars, however, would require even more propellant both to enter and exit the Martian orbit, simply because Mars is both larger and farther from Earth than the Moon.

Robotic spacecraft guided into space by rockets aren't constrained by having to come back home when their mission is finished. Such spacecraft still need to bring enough propellant with them to enter orbit around their target destination, however. The *Cassini* spacecraft, which was launched in 1997, traveled 2.2 billion miles (3.5 billion kilometers) and spent seven years on its circuitous journey from the inner solar system (the four planets closest to the Sun, as explained in Chapter 2) to the planet Saturn. When *Cassini* was launched, its total mass was 12,593 pounds (5,712 kilograms); of that mass, more than half (6,905 pounds, or 3,132 kilograms) was propellant. The spacecraft used a little less than half of this propellant in a single, 96-minute burn when it arrived at Saturn in order to slow itself down enough to successfully enter orbit.

### Types of propellant

Types of rocket propellant include the following:

- **Liquid chemicals:** Liquid rocket propellants (think kerosene, hydrazine, or liquid hydrogen as the fuel; and liquid oxygen, liquid fluorine, or nitrogen tetroxide as the oxidizer) have greater energy density per volume than solid propellants, so they can carry more payload into space. They're more difficult to handle and store than solid ones though, so a more-sophisticated system is required. Liquid rocket propellants can be adjusted or turned off once started. The liquid oxygen/kerosene combination is the most commonly used rocket propellant for lower stages; liquid oxygen and liquid hydrogen are used in upper stages and on the Space Shuttle.
- **Solid chemicals:** Solid rocket propellants (such as ammonium perchlorate as the oxidizer and hydroxyl terminated polybutadiene or polybutadiene acrylonitrile as the fuel) are simple, relatively cheap, and can be stored for long periods of time. However, the amount of thrust can't be changed once ignited, and rockets using solid propellants can't be turned off once started (although design variations do allow some flexibility). Solid-fueled rockets are typically employed for military applications (such as Intercontinental Ballistic Missiles); they're also used as the boosters on the Space Shuttle.
- **Hybrid chemicals:** Hybrid-fueled rockets use a solid rocket fuel combined with a liquid or gas oxidizer so as to get the benefits of both rocket types. Theoretically, such a rocket has the stability and cost advantages of a solid-fueled rocket with the thrust control of a liquid-fueled rocket. *SpaceShipOne,* the first privately funded spacecraft to reach space, was powered by a hybrid rocket (see Chapter 22 for more on *SpaceShipOne* and other commercial space missions). However, hybrid rockets are still under development, primarily within the private space industry.

## Forecasting the future of rocket propulsion

After a spacecraft has ridden into space on a rocket, it then needs a propulsion system to take it where it needs to go, whether that's into Earth orbit or onward to a distant planet. Instead of bringing along a rocket with a huge tank of propellant, future missions may use more-sophisticated propulsion systems. The systems described in the next two sections are in varying stages of development and may be completely impractical for certain missions. With that said, they can also represent enormous savings in certain areas.

### Ion propulsion uses xenon for a kick

One promising possibility for the future of rocket propulsion is ion propulsion. A traditional rocket creates thrust from the exit of exhaust at high velocity, which pushes the spacecraft in the opposite direction. Rather than rocket propellant, this new technology uses a gas called xenon. The xenon gas is given an electrical charge that causes it to *ionize,* a process in which electrons are removed from an atom and become charged. The ions are then accelerated through a series of grids and ejected from the end of the spacecraft, pushing it in the opposite direction just like a conventional rocket. Spacecraft such as *PamAmSat 5* in 1995 and NASA's Deep Space 1 mission in 1999 have successfully tested the ion propulsion system.

One disadvantage of the ion propulsion system is that it generally has a very low thrust, meaning it can build up great speeds but only over a relatively long time. This characteristic is problematic when trying to enter orbit around a planet because the rocket has to stop accelerating halfway there in order to begin the gradual process of slowing down enough to enter orbit instead of relying on the long burn of a conventional rocket. Another disadvantage is that ion propulsion systems also require an external energy source, such as solar panels.

### Solar sails for the modern-era Columbus

Another futuristic propulsion option is a solar sail in which a large array of thin, gossamer membranes is kept taut by a series of frames. These sails capture radiation pressure from photons given off by the Sun and use this tiny force over long periods of time to build up motion away from the Sun. Similar to sailboats on the ocean moving with energy from the wind, a spacecraft guided through space via solar sails can theoretically adjust the sail orientations with respect to the solar wind in order to change direction. Like ion propulsion, however, solar sails require a long time to accelerate or decelerate. They also require a very large area to carry a relatively small spacecraft mass.

What are the requirements for solar sail materials? They must be

- Lightweight
- Very reflective
- Able to withstand high temperatures
- Able to resist tears from small meteor impacts

A spacecraft with a solar sail, *Cosmos 1,* was launched in 2005 to test the technology. It was built by two private space exploration advocacy groups, Cosmos Studios and the Planetary Society. Unfortunately, the Russian Volna rocket that was to have carried *Cosmos 1* into orbit failed on launch; the spacecraft never reached its intended orbit and was lost. The Planetary Society is currently raising money to build a new solar sail spacecraft and try again.

# Keeping the Lights On in Orbit: Spacecraft Power

A spacecraft needs power sources to run life-support equipment and onboard instruments. However, bringing fuel into space to generate power is a costly endeavor in terms of both finances and storage. Fuel tanks aren't small objects by any stretch of the imagination, and by definition they have a finite storage capacity. Fortunately, other ways exist for powering spacecraft. We describe a couple of these options in the following sections.

## Harnessing the Sun's power

It wouldn't make sense for a robotic mission to the inner solar system to try and bring along enough liquid or solid fuel for the entire mission. Why? Because the satellite has the solar system's cheapest power source — the Sun — right at its disposal. Most spacecraft, except those destined for Jupiter or beyond, bring along large arrays of solar panels to capture the Sun's energy for use as an onboard power source.

In the following sections, we describe how solar panels work and list some famous spacecraft that have used these panels.

### How solar panels work

Solar panels consist of lightweight flat panels covered with *solar photovoltaic cells,* cells whose purpose is to capture the Sun's energy and convert it into usable electricity. Solar cells are usually made from a semiconducting material such as silicon. They absorb the photons delivered in sunlight and produce electricity by separating the electrons from the rest of the atom. The cells are arrayed into rows, which are then connected to each other and set into the frame of the solar panel. Solar cells used in space may also be made of *gallium arsenide,* a compound of gallium and arsenic, because it may be a more-efficient semiconductor than silicon.

Most robotic missions to the inner solar system use solar panels. The power gained from the Sun can be used for heating and cooling, as well as for electric spacecraft propulsion. The solar panels on a spacecraft are similar to those used here on Earth, but the cells are usually packed more tightly together and cover as much of the panel as possible in order to maximize the amount of electricity produced. They're often fanned out (to take advantage of the Sun's position) into winglike shapes and typically positioned so as not to cast shadows onto each other. Space solar panels are designed to rotate so they can always face the Sun, and many use a type of tracking system to automatically orient themselves correctly. In the weightless environment of space, large, delicate solar panel arrays can be unfurled that would collapse under their own weight on the surface of Earth or another planet.

Solar panels have one distinct downside: Their usefulness on spacecraft is largely limited to missions in the inner solar system. In the outer solar system (Jupiter and beyond, as described in Chapter 2), the Sun is much fainter, and there isn't enough sunlight to make solar panels an effective choice of power supply. (See the later section "Processing plutonium for Pluto" for details on the power source that spacecraft headed to the outer solar system use.)

### Notable spacecraft powered by the Sun

Many famous spacecraft have used solar panels, including

- **The *Vanguard 1* satellite:** Launched in 1958, *Vanguard 1* (which orbited Earth) was the first spacecraft to use solar panels. It's still in orbit today, but it's no longer sending information back to Earth.
- **The Mars Exploration Rovers:** *Spirit* and *Opportunity,* the Mars Exploration Rovers, first made the Mars skyline famous in 2004 with their solar-array extensions. These extensions continue to provide plenty of power to keep the robots moving, as long as they don't get too dusty! (Check out these solar panels in the color section.)
- **Most satellites that orbit the Earth:** From the weather satellites that help give you tomorrow's forecast to the telecommunications satellites that help you call your sibling in another country, most Earth-orbiting satellites use solar panels to generate cheap, renewable energy.
- **The International Space Station (ISS):** The ISS relies on an array of solar panels that currently spans 4,000 square feet (375 square meters) and is 190 feet (58 meters) long. The arrays are motorized to track the position of the Sun so as to maximize the amount of solar radiation they can capture. When fully completed, the station's solar array will cover nearly an acre.

New space projects, such as the Orion spacecraft (part of NASA's Project Constellation, which you can read about in Chapter 21), will make heavy use of solar power. Designed to serve as an eventual replacement for the Space Shuttle, Orion will one day transport both cargo and people to the ISS and other destinations beyond the Earth, such as the Moon. It'll use solar wings to provide power for its service module. These wings not only represent an enormous fuel savings but they also add another performance requirement to the spacecraft.

## Processing plutonium for Pluto

Because the Sun's energy isn't as strong in the outer solar system, and because toting along a conventional power station is impractical due to the weight of the fuel, journeys to Jupiter and beyond rely on a radioactive power supply: the radioisotope thermoelectric generator (RTG). Essentially, an *RTG* is a type of electrical power generator that uses radioactive decay as its energy source.

As radioactive materials decay, they release heat. This heat is then translated into usable electricity via *thermocouples* (temperature sensors that calculate the difference in temperature between conductors). In the process of measuring this differential, the circuit between conductors is completed, generating voltage. When used in an RTG, thermocouples are situated inside the radioactive fuel container and generate electricity as the fuel decays and creates heat.

Why use radioactive materials rather than safe, reliable solar power? RTGs make sense because they require a relatively small mass of plutonium to act as a fuel source, and they can provide a stable source of power over relatively long time periods. The three RTGs on the *Cassini* spacecraft, for example, were originally meant to provide power over an 11-year period, during which the power from the RTGs would drop from 276 watts to just 216 watts. If all continues to go well and the *Cassini* spacecraft is awarded an extended mission, the RTGs would still be at about 199 watts at the end of 16 years.

RTGs are used to provide power for robotic space exploration, particularly for missions and probes that are so far away from the Sun that the solar panels NASA normally uses just don't provide enough power. Space missions that have used RTGs include robotic spacecraft such as *Galileo* and *Cassini.*

A current shortage of RTGs may affect the design of upcoming missions to the outer solar system. Security issues at the U.S. Department of Energy labs that process plutonium necessitated the shutdown of the entire plutonium processing program, putting the agency behind schedule. NASA's *New Horizons* spacecraft, currently en route to Pluto, didn't receive all the plutonium fuel it had originally been allocated due to the shortage. Power constraints may also affect the upcoming Outer Planet Flagship Mission that's in the planning stages for a trip to study Jupiter's moons.

# Phone Home: Communicating via Radio Telescopes

How do scientists on Earth talk to spacecraft far from Earth? They use a system of three giant radio telescopes called the Deep Space Network, or DSN (shown in Figure 3-4). The telescopes that make up the DSN are intentionally spaced around the world, with antennae in California, Spain, and Australia. They're all operated by (mostly local) contractors who are supervised by NASA.

**Figure 3-4:** The Deep Space Network is made up of three huge radio telescope dishes, like this one, spread around the globe.

*Courtesy of NASA*

What are the fundamental requirements for a radio telescope?

- A transmitter to emit radio signals
- A parabola-shaped antenna to catch signals
- A receiver to process signals

The DSN allows at least one antenna to be able to track a spacecraft just about anywhere in the solar system at any time, and communications can be passed from one antenna to the next as the Earth rotates. The telescopes use radio signals to transmit commands to spacecraft, and they can also listen to the various responses from spacecraft. The DSN can listen to only one target at a time though, so sometimes different missions compete for use of the DSN at the same time, kind of like when you vie for control of the telephone line with the most talkative member of your family. Available DSN time also occasionally limits the amount of data that a spacecraft can send back.

Radio signals travel at the speed of light, so depending on how far away you are, there can be a significant time lag when calling home to Earth. For example, it takes about 16 minutes for a transmission to travel from Earth to Mars, or vice versa. This time delay makes real-time operations basically impossible, which is why most robotic missions are autonomous — programmers and scientists design routines, and the spacecraft run them on their own. Even the Mars Exploration Rovers must be programmed ahead of time with the day's events. If they had to call Earth every time they had to make a decision, they'd have a 32-minute round-trip time delay that would significantly reduce the amount of time they had to do scientific research. Whenever astronauts eventually travel to Mars, they'll be subject to the same 32-minute time delay in sending and receiving messages to and from Earth, so they'll also need to act autonomously most of the time.

The technological demands on the DSN are significant.

- Not only must it be carefully scheduled to accommodate the ever-growing number of NASA space missions but it must also continue to support old missions as well.
- Detecting a signal from a spacecraft hundreds of millions of miles away is incredibly difficult. Due to onboard power constraints, spacecraft usually send out very low-power signals (about the same wattage as a refrigerator light bulb). The signal is focused into a very tight beam and is sent directly to Earth, but as it travels over millions or hundreds of millions of miles, it spreads out and becomes weaker. By the time the antennae of the DSN are able to detect the signal, it may be as weak as a billionth of a billionth of a watt.
- The DSN antennae are designed to have a very narrow beam, so they must be pointed directly at a spacecraft's position to send or receive messages. The signal is captured by the antenna or antennae, amplified, and eventually reconstructed into the latest images of Saturn's satellites or the composition of a Martian rock.

# Chapter 4

# The Life and Times of an Astronaut

### In This Chapter

- Undergoing astronaut training
- Getting the scoop on astronaut job descriptions
- Working and living in space

Many children dream of becoming astronauts, and for most of them, this challenging career path remains but a dream. That doesn't make the thought of traveling in space any less exciting though. Many aspiring astronauts apply, but few are chosen for training — and trust us, the elite few who *are* selected have their work cut out for them! Astronaut training involves extensive academic, physical, and survival-related work, as well as time in simulators and experience in a mock-weightless environment. After training, commanders, pilots, and various specialists live and work in space. You find out all the exciting details in this chapter.

Today's astronauts follow in the footsteps of greatness. From the first American to take flight to the catastrophic deaths of the *Challenger* and *Columbia* crews, all astronauts have made great sacrifices (both personal and professional) to follow their dreams. The job for tomorrow's astronauts promises to be no less thrilling, difficult, or history-making.

## Preparing to Become an Astronaut

One of the most-popular questions asked of astronaut training programs worldwide is this: What makes a good astronaut? Generally, a solid candidate is one who's proficient in her field and in good physical condition. She's also able to work as part of a team and can communicate easily with different types of people. Being able to hold up well under adverse conditions and possessing strong leadership abilities are also valued qualities.

## Educational and physical requirements

Astronauts must hold at least a bachelor's degree in physical science, mathematics, biological science, or engineering from an accredited college or university. They must also have at least three years of work experience after college (the good news is advanced degrees count toward this required work experience). Picking a specialty in a space-related area of science can't hurt either; NASA encourages potential astronaut candidates to focus on any area in which they can excel.

If you think you want to become a United States astronaut, the time to start acting on that desire is while you're in school. Taking as many high school chemistry, physics, and other science classes as you can will help you with the preparations for more-advanced science classes later on.

Both civilians and military personnel can become astronauts. Pilot training is beneficial for many of the positions on a Space Shuttle crew and is an absolute necessity for securing a piloting job. (We describe a pilot's job in more detail later in this chapter.)

In addition to the various educational and skill-related requirements, astronaut candidates also have to be physically able to do their jobs. They must meet the following distinct physical requirements, as well as other medical requirements:

- **Eyesight:** All astronauts have eyesight that's correctable to 20/20 with the aid of glasses or contact lenses. As of 2007, laser eye surgeries have been accepted for vision correction as well.
- **Blood pressure:** The maximum blood pressure an astronaut can have is 140/90.
- **Height:** Don't apply if you're shorter than 62 inches (1.57 meters) or taller than 75 inches (1.93 meters).

You may be interested to know that age isn't a restricting factor. The average age of astronaut candidates is currently 34.

United States citizenship is most definitely a requirement to be a NASA astronaut, though payload specialists (which we cover later in this chapter) may be citizens of other countries, depending on their qualifications.

Check out `astronauts.nasa.gov` for more info about the requirements to become a U.S. astronaut.

## Training programs

Ultimately, applying to NASA's Astronaut Candidate Program is the step that leads toward the start of a two-year training period that can result in becoming a U.S. astronaut. The selection process is extremely competitive: Typically, more than 4,000 people apply every two years, and only about 20 are accepted into the training program, with just 10 to 15 becoming qualified astronauts.

The training received by both astronaut candidates and those people accepted into the astronaut program falls into two main categories: academic and physical. After completing the two-year training period, a lucky few of the candidates are selected to join the astronaut program. These folks continue with the same basic academic and physical training they engaged in as candidates — it just becomes more specialized. After being chosen for a specific mission, astronauts typically receive ten months of training (or more!) before taking flight.

### Academic training for astronauts

Although training to become an astronaut may seem like the most fun job ever, it's still hard work. The academic side of the training involves taking classes in rocket and shuttle science, in addition to mathematics, meteorology, oceanography, physics, and other sciences.

Additional academic training for astronauts can consist of

- Foreign language courses (particularly Russian)
- Required reading
- Space Shuttle training (everything from the propulsion systems to the power and communications systems)
- Tests
- Trainer plane acclimation
- Workbooks

### Physical training for astronauts

The physical training for astronauts can be quite arduous and covers a range of skills that must be acquired in order to be certified for space travel. Astronauts are trained to expect the worst, so their physical training covers a lot of ground, as you can see from this list:

- Aviation physiology
- Ejector seat training
- Flight training
- "Ground School" plane navigation
- Land navigation
- Land survival

- Lunar and Martian gravity simulation
- Parachuting
- Reduced-gravity simulation
- Water survival
- Wearing an advanced crew escape suit (ACES)

Replicating the weightless environment of space travel here on Earth is rather difficult. Astronauts can train in a specially equipped airplane, affectionately nicknamed the Vomit Comet, that flies through a series of dives and steep climbs on a *parabolic trajectory* (a smooth curve that turns over at the top). During the fast dives, astronauts inside the plane experience short periods of weightlessness — and we mean short! These weightless periods last for less than 30 seconds, which doesn't give the astronauts time to do much training.

Fortunately there's also the underwater option. NASA's Neutral Buoyancy Lab (shown in Figure 4-1 and part of the astronaut training facility at Johnson Space Center in Houston, Texas) is the world's largest swimming pool. Technicians place mockups of astronaut mission targets, such as the International Space Station, underwater. Wearing specially designed spacesuits that function much like the real thing, astronauts train for spacewalks that may include space station construction, satellite deployment, or repair missions. Training underwater is the closest that astronauts can come on Earth to the actual space environment, and during a *spacewalk* (which is an *extravehicular activity* [EVA] performed in space) where every minute counts, astronauts appreciate the endless underwater drills they endured during their training.

**Figure 4-1:** Astronauts train underwater at NASA's Neutral Buoyancy Lab.

*Courtesy of NASA*

## International programs

The U.S. is, of course, not the only country to produce astronauts. The Chinese have *taikonauts,* and the Russians have *cosmonauts.* Yet no matter the country of origin or the name they're called, all astronauts are trained to high standards to ensure both their safety and the success of their missions. The training specifics tend to differ from country to country though. Russian cosmonauts, for example, train using Russian spacecraft (such as the Soyuz) rather than the American spacecraft that U.S. astronauts study. Each country's astronauts also train for familiarity with their own spacesuits and vehicles.

The Russian cosmonaut program has some additional core differences from the U.S. astronaut program. Russian cosmonauts receive extra training in the English language and space-specific terminology. Also, although any American civilian with the proper scientific background can apply to be an astronaut, most Russian cosmonauts are chosen from either the Russian Air Force, IMBP (the State Research Center of The Russian Federation), or RSC Energia (the S.P. Korolev Rocket and Space Corporation).

In China, the government has focused on training experienced Air Force pilots for its budding taikonaut program. The candidates selected for China's first batch of taikonauts are all required to have at least 1,000 hours of flying time, as well as a master's degree (at minimum) in a technical field such as physics, engineering, or biology. Because no cooperative agreement currently exists between the Chinese space program and the Russian or American programs, no language training is required for taikonauts. (You can read more about China's space program in Chapter 22.)

# Introducing the Crew of a Space Mission

United States astronauts can fill one of several positions aboard a spacecraft: pilot, commander, payload specialist, mission specialist, or education specialist. Each position usually requires specific training, because being an astronaut is just like any other specialized job in that there's more than one type of astronaut. Someone who was an expert in flying a four-seater airplane, for example, wouldn't one day decide to fly a commercial jetliner with 500 seats. He'd need specific training for takeoff, landing, maneuvering, and the host of other skills required to safely fly a plane. Astronauts have similar basic training, but their particular skill sets qualify them for certain positions.

## Commanders and pilots are at the wheel

The *commander* is the person responsible for the overall safety and success of the mission, as well as for the space vehicles themselves. The *pilot* works with the commander to operate the spacecraft, deploy satellites, and perform other major functional tasks while in space.

A commander must possess an advanced degree in engineering, physics, mathematics, or a relevant scientific field and have experience working in that field. Physically, pilots need to be able to fit into their seats comfortably. Consequently they have to be between 62 inches (1.57 meters) and 75 inches (1.93 meters) tall. Most pilot astronauts have military flying experience or, if not, plenty of flight hours logged. Regardless of their civilian or military status, they must possess a pilot's license because they're the ones in the driver's seat.

## Mission and education specialists bring additional expertise

The *mission specialist* is responsible for ensuring the success of the specific mission to which he's assigned. Mission specialists can lead spacewalks, run experiments in space, operate all the equipment for a mission, and otherwise assist with the functioning of a mission. They must have at least a bachelor's degree in a math- or science-related field, as well as professional experience in that field. They also need to meet the same height requirements as other types of astronauts.

Some missions have a *mission specialist-educator* (MSE) in addition to a mission specialist. This person performs a similar function as the mission specialist, but he also has other educational duties. MSEs have a degree in teaching and must have worked as teachers for several years. (For more info on the new Educator Astronaut Program, which all MSEs must complete, flip to Chapter 5.)

## Payload specialists meet specific technical needs

*Payload specialists* are the technical experts for the particular payload on a mission (the *payload* is the cargo that a spacecraft carries into space; in this context, "payload" usually refers to equipment and scientific experiments). They may be specially trained in satellites, for example, and are frequently responsible for the scientific components of a mission. Most payload specialists are selected for a particular mission, and that's often their only mission in space. Mission specialists and pilots, on the other hand, often qualify for multiple flights.

Payload specialists usually aren't NASA astronauts; they may be physicians, researchers, or technicians who are experts in using a particular piece of equipment, for example. Payload specialists don't need to meet the U.S. citizen requirement. Many payload specialists are astronauts from other countries, particularly Russia. All payload specialists have to receive NASA approval, meet the same general fitness and physical requirements as other astronauts (we describe these requirements earlier in this chapter), and go through training for space travel.

Commercial payload specialists were common in the early days of the Space Shuttle program, but they have become less common. Commercial companies have reduced the use of the Shuttle to launch commercial payloads due to increases in cost, delays, and perceived risk following the *Challenger* explosion in 1986 (see Chapter 5). After this loss, astronaut crews were composed primarily of career astronauts, and experiments were operated by mission specialists rather than payload specialists.

# Living and Working in Space

Just like the rest of us back here on the ground, astronauts need to take care of the basics of living. However, activities such as eating, drinking, bathing, and exercising become a lot more difficult when they're performed in the microgravity of space! Check out the adaptations that astronauts have to make to their daily routines in the following sections.

## Eating and drinking

Everyone needs to eat, including astronauts. The foods they eat are packed with nutrients. They're also relatively dense so they can't float away (that means no salt or sugar — bummer), crumble (which can lead to clogged instruments — never a good thing), or take up too much space. Space food has come a long way from what the early Mercury astronauts had to endure in the 1960s; most of their meals consisted of cubed food, and semiliquids sucked from toothpaste-like tubes. Later innovations led to better packaging, eliminating the unpopular squeeze tubes; hot water for rehydrating food; and foods that can be eaten from bowls by using utensils.

## A sample menu of astronaut food

The food astronauts eat is designed to provide energy, meet nutritional needs, and keep the astronauts happy. Meeting all of these goals required some new engineering of previous space foods, but the current menu offers more variety than in years past. It also allows the astronauts to eat more of the foods they're familiar with while they're away from home.

Common astronaut beverages:

- Coffee (sweetened or unsweetened)
- Tea (sweetened or unsweetened, with or without lemon)
- Lemonade
- Fruit juices and punch
- Apple cider

Typical astronaut breakfast:

- Cereal
- Roll
- Fruit
- Breakfast drink
- Coffee or tea

Typical astronaut lunch:

- Peanut butter and jelly burrito
- Chicken with salsa
- Macaroni and cheese
- Buttered rice
- Fruit juice

Typical astronaut dinner:

- Shrimp cocktail
- Steak or chicken
- Fruit cocktail
- Fruit drink
- Tea with sugar or lemon

The galleys in today's Space Shuttles have equipment for dispensing water and heating food but not for storing cold foods, meaning all food must be either fresh, dehydrated, irradiated, or thermostabilized (such as those shelf-stable foil packages) in order to travel well. Soups and beverages are drunk through straws or served in squeeze bottles. More solid meals can be eaten with utensils, which are generally held down to the astronauts' meal trays with the aid of magnets and straps. Without the straps, an astronaut's lunch would float away before she got to her second bite! (You can see modern astronaut food in Figure 4-2.)

Meat, breads, cheeses, fruits, vegetables, nuts, desserts, and drinks are all common staples aboard a spacecraft. Astronauts get to pick from a range of meals, and the amount of food each astronaut is allocated is calculated according to his or her body weight and type. Due to the international nature of space travel, foods that appeal to people from different countries are now offered. And what about that gift-shop staple, dehydrated "Astronaut Ice Cream"? This product was actually developed for NASA and flew on Apollo 7, but it proved unpopular with the crew and was discontinued. It's much more popular on the ground than in space! You can find out the details of astronaut meals in the nearby sidebar, "A sample menu of astronaut food."

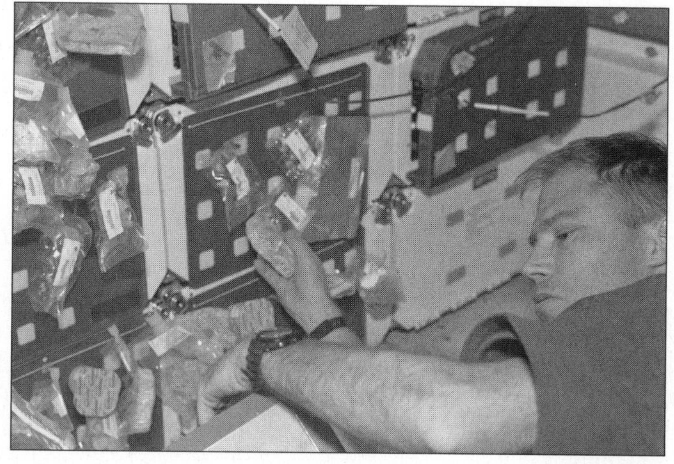

**Figure 4-2:** Astronaut food has come a long way since the 1960s.

*Courtesy of NASA*

## Sleeping soundly

When it's time for sleep, astronauts have the option of crawling into strapped-down sleeping bags, climbing into the commander or pilot seats, or making use of several bunk beds. Because astronauts are weightless in gravity-free space, they must be attached to some part of the spacecraft before settling down for the night. Additionally, many astronauts wear sleep masks — you probably would too if the Sun rose every 90 minutes right outside your window!

Astronauts are usually scheduled for eight hours of sleep at the end of each day. The crews of the early capsule-based Gemini and Apollo missions attempted staggered sleep schedules, but the crowded quarters made this setup difficult. Nowadays, all astronauts on the Space Shuttle typically sleep and wake at the same time. (Although on some missions, the crew may be divided into two teams with staggered sleep and work periods to get more done.) NASA's Mission Control Center monitors the Space Shuttle at all times and can immediately alert the crew to any problems that arise while they're asleep.

## Staying clean

Hot shower while aboard a Space Shuttle? Not happening. Astronauts on a spacecraft take sponge baths while in orbit (unless of course they're lucky enough to be aboard the International Space Station — it has a shower!). Believe it or not, some of that water actually comes from the astronauts' breath. Most living creatures exhale and sweat, both of which contribute to onboard humidity. Special equipment helps convert that extra moisture in the air back into water. Nevertheless, astronauts do wash their hair (with rinseless shampoo), brush their teeth, shave, and otherwise keep themselves clean using the tools in their personal hygiene kits.

Space shuttles have toilets, but they don't use water. Instead, these space commodes employ a fan and vacuum system to suck waste and urine into the toilet. And yes, astronauts must strap themselves down before using the facilities because a strong seal between the astronaut and the toilet keeps the waste from floating up into the cabin.

Unlike most other things on a spacecraft, waste isn't recycled; it's stored and disposed. During the early days of the space program, liquid waste was vented into space. In fact, one Apollo-era astronaut reportedly quipped that the most beautiful sight he saw in space was a urine dump at sunset. Because water is heavy and expensive to bring into space, NASA is currently working on a new system to completely recycle all water collected from astronauts, including urine, into clean drinking water.

## Exercising in orbit

Exercise is as important in space as it is on Earth — perhaps even more so because living in microgravity reduces both bone and muscle strength. Following are the many different ways the human body changes while in space:

- Muscles atrophy.
- Bone mass decreases.
- Calcium levels increase.
- The face gets puffier.
- Heart rate, red blood cell counts, and blood plasma levels all decrease.
- The spine straightens and expands.
- The immune system becomes compromised, leading to an increased risk of infection.

Astronauts can also experience these less-than-desirable conditions:

- Claustrophobia
- Clogged noses and sinuses ("space sniffles")
- Sleep deprivation and disruption of circadian rhythm

- Space Adaptation Syndrome (sometimes called “space motion sickness” or just “space sickness;” it results from the body’s adaptation to microgravity and, in most cases, lasts for the first few days of an astronaut’s journey)
- Temporary disorientation and vertigo

The International Space Station has a treadmill and an exercise bicycle, but the astronauts there still have to strap themselves to the equipment before using it. Astronauts may exercise for as much as three to four hours per day!

## Dressing well

During takeoff and landing, astronauts wear Launch and Entry Suits that can maintain air pressure for the astronauts in case the spacecraft has a leak. The suits are also insulated, which can help keep astronauts warm if they need to swim in cold ocean water during an emergency escape. At other times in space, astronauts can wear more-comfortable clothing, much like what they can wear here on Earth. Typical clothing choices include one-piece coveralls, shorts, and either a T-shirt or a rugby shirt in crew colors.

Essential components of a spacesuit include

- Stable pressurization
- Insulation and temperature regulation systems
- An oxygen supply
- Hydration equipment
- A communications carrier system
- Micrometeoroid protection
- Flame-retardant material
- Flexible joints
- Tether points to the spacecraft
- Ultraviolet radiation shielding
- Gas and liquid expulsion systems
- A waste collection system
- Arm, helmet, glove, and torso assemblies

## Taking spacewalks and doing other work

When work needs to be done outside of the spacecraft or space station, a spacewalk may be in order (head to the color section of this book to see an image of a spacewalk). *Spacewalks,* or *extravehicular activities* (EVAs) performed in space, involve a battery-powered Extravehicular Mobility Unit (EMU). Spacewalks can be *tethered,* meaning the astronaut is connected to the spacecraft by an oxygen-carrying "umbilical cord," or *untethered,* meaning the astronaut must have some means of propulsion that can return her safely to the spacecraft after the task at hand is completed.

Russian cosmonaut Alexei Leonov executed the first spacewalk in the history of space travel in 1965, when he spent ten minutes outside the airlock of the *Voskhod 2* spacecraft (see Chapter 7 for details).

Onboard a spacecraft, astronauts have a full schedule of activities that can involve running scientific experiments, troubleshooting or fixing malfunctioning systems, or performing scheduled maintenance activities to keep their spacecraft in top working condition.

Astronauts train for months or years before a mission, and because time in space is incredibly valuable (and pricey!), every minute counts. After early missions during which astronauts felt they were completely overscheduled and rebelled against instructions from Mission Control, NASA has realized that astronauts, like everyone else, need some downtime built into their schedules to read, talk to friends and relatives back on Earth, or just stare out the window. With an out-of-this-world view, no astronaut has ever complained about that option!

# Chapter 5

# Tragedies in Space

### *In This Chapter*

- Facing the flood during Mercury-Redstone 4
- Beginning badly: Apollo 1
- Looking at losses in the Soviet Soyuz program
- Solving Apollo 13's problem
- Remembering the destruction of the Space Shuttle *Challenger*
- Recalling the disastrous reentry of the Space Shuttle *Columbia*

Space travel is dangerous, and the triumphs of space exploration have been marred by periodic failures.

- The 1961 flight that launched the second American into space, Mercury-Redstone 4, almost ended in tragedy when the capsule began to sink following its landing in the ocean. Astronaut Virgil "Gus" Grissom narrowly escaped.
- The astounding success of the early days of American space exploration was brought to an end by the Apollo 1 fire in 1967. Three years later, a tragedy during the Apollo 13 mission was narrowly averted.
- The Soviet Union didn't escape disaster either: One cosmonaut died during the Soyuz 1 mission in 1967, and three died during the Soyuz 11 mission in 1971.
- The early years of the Space Shuttle program were marred by the *Challenger* explosion in 1986, which killed all seven astronauts onboard. The *Columbia* explosion in 2003, which also killed the entire crew, prompted a wholesale reevaluation of NASA's Space Shuttle program.

We provide greater detail about these tragedies in this chapter. Although space exploration certainly has its risks, as evidenced by the brave men and women who risked or gave their lives in these incidents, each failure also helped make future flights — and future generations of space explorers — safer because safety enhancements were made from the many lessons learned.

# Mercury-Redstone 4 (1961): A Flooded Capsule

In the early days of the American space program, NASA was just figuring out how to send astronauts to the edge of space on *suborbital* flights. These flights didn't orbit the Earth. Instead, they sampled the conditions of space for a few minutes before returning to the ground. The flight tests in this series were designed to further NASA's research into how humans would react and handle situations during human spaceflights; rocket and flight research, although obviously important, were secondary to this goal of furthering the budding idea of putting humans in space.

NASA's Mercury-Redstone 4 (M-R4) mission in 1961 was the second manned U.S. suborbital flight (and the fourth mission in this series of tests). Astronaut Virgil "Gus" Grissom piloted M-R4 in a Mercury capsule, named *Liberty Bell 7,* atop a Redstone rocket. This test flight differed from its predecessor, Mercury-Redstone 3 (the flight that made Alan Shepard the first American in space; see Chapter 7 for details), in two major ways:

- *Liberty Bell 7* had an additional viewing window, which was put in at the request of the Mercury astronauts. The side windows in the previous Mercury spacecraft didn't allow for an extensive viewing area, and the new window increased the astronauts' field of view to more than 30 degrees in both directions.
- *Liberty Bell 7* also had an explosively actuated side hatch. What does "explosively actuated" mean, you ask? Well, the Mercury-Redstone 3 spacecraft had a mechanical hatch that was quite heavy (69 pounds, or 31 kilograms, to be exact) and had to be operated manually by the astronaut inside. The new hatch design in *Liberty Bell 7* weighed just 23 pounds and operated "explosively." Literally. When the astronaut was ready to emerge and hit a trigger switch, charges would detonate and break the titanium bolts holding the hatch onto the spacecraft. The hatch would then fly open without the astronaut having to lift it off.

Other changes from the Mercury-Redstone 3 capsule included new rate stabilization controls, a change in how the rockets were fired, a redesign of the instrument panel, and improvements to the pilot seat.

### Fast facts about Mercury-Redstone 4

Following are the basic facts about the Mercury-Redstone 4 mission:

- **Launch date and site:** July 21, 1961; Cape Canaveral Air Force Station, Florida
- **Spacecraft name and mass at launch:** *Liberty Bell 7;* 2,840 lb (1,286 kg)
- **Launch vehicle:** Redstone rocket
- **Size of crew:** 1 astronaut
- **Number of Earth orbits:** 0 (suborbital flight only)
- **Mission duration:** 15 minutes, 37 seconds

## Grissom's narrow escape

The actual M-R4 mission went off without a hitch, starting with its launch on July 21, 1961. (Technically the mission was supposed to launch two days earlier, but that plan was postponed due to bad weather.) Liftoff was successful, and the spacecraft was aloft for 15 minutes and 37 seconds.

The problem occurred during landing. After splashing down in the ocean, the Mercury capsule was still watertight, and Grissom logged the cockpit data before requesting a pickup from the waiting recovery helicopters. According to Grissom, it was then that the newly designed hatch activated prematurely, causing the hatch door to fly open and water to start pouring into the capsule. Within minutes, the capsule flooded with water. Grissom narrowly averted death by climbing out of the *Liberty Bell 7* and struggling in the ocean for a few minutes while his flight suit filled with water from an open neck seal. Fortunately, Grissom was quickly picked up by a recovery helicopter.

Though it tried, the recovery team wasn't able to prevent the *Liberty Bell 7* from sinking. The spacecraft wasn't seen again until 1999 when a commercial crew of divers located it 300 nautical miles (550 kilometers) off the coast of Florida, under 15,000 feet (4.5 kilometers) of water. Although the recovery didn't help solve the question of why the hatch blew too early, it did provide a valuable historical record: The capsule is currently on display in a Kansas museum.

## The hatch release changes that followed Liberty Bell 7

Following the loss of the *Liberty Bell 7,* a number of changes were made to future Mercury capsules. Initial reports cast suspicion that Grissom somehow opened his hatch prematurely, but later analysis suggested that the external release lanyard could've come loose, considering it was held on by a single screw aboard the spacecraft. On future Mercury flights, the release was attached more securely.

Another change proved more tragic. Because *Liberty Bell 7* had shown that an explosive-release hatch had the potential to release prematurely, the designers of the Apollo spacecraft didn't initially include a quick-release door in that spacecraft. Unfortunately, that decision led to the demise of the three Apollo 1 astronauts, as we explain in the next section.

# Apollo 1 (1967): A Lethal Fire

Originally dubbed the AS-204 mission, the 1967 Apollo 1 mission would've been the first launch of NASA's new Apollo spacecraft, paving the way for an eventual Moon landing. A Saturn 1B rocket would've launched the Command Module and the following three astronauts into orbit:

- Lieutenant Colonel Virgil "Gus" Grissom
- Lieutenant Colonel Edward White
- Lieutenant Commander Roger Chaffee

On January 27, 1967 — one month before the scheduled launch date — all three men were onboard the spacecraft for a test run on the launchpad when events took a horrible turn.

## Death on the ground

The simulation test was what's known as a *plugs-out test,* or a test designed to see whether a spacecraft can be operated completely on internal power after being disconnected from external cables and umbilical tubes. After the spacecraft's cabin was sealed, it filled with pressurized oxygen, and communications problems caused the launch simulation to be delayed several times. At some point during this lull, a spark led to fire erupting in the cockpit.

The overpressurized Command Module turned into a furnace as the pure-oxygen atmosphere fanned the flames. In a matter of seconds, the capsule ruptured from the pressure. It took outside engineers five minutes to get the spacecraft's hatch open, during which time the astronauts were trapped inside the cabin. All three men died in the fire, which ravaged the interior of the capsule (see Figure 5-1).

**Figure 5-1:** The Apollo 1 capsule, as seen postfire.

*Courtesy of NASA*

## A tangle of flaws that contributed

Although the actual source of the fire wasn't isolated in the reports and investigations that followed the fatal test run, several potential causes may have led to the destruction of the Command Module:

- The atmosphere in the module consisted of 100-percent oxygen at higher than normal pressure, which basically allowed almost anything to burn — and burn quickly at high temperatures. NASA officials chose to use this atmosphere rather than a more Earth-like mixture of oxygen and nitrogen to prevent decompression sickness as pressure in flight was reduced. (Pure-oxygen atmospheres had been used safely on earlier Mercury and Gemini spacecraft, giving them no reason to think such an atmosphere wouldn't work for the new Apollo spacecraft.)
- Flammable materials were present in several areas of the cockpit. With the high pressure of oxygen in the cabin, even materials like the aluminum instrument panels and the cabin's Velcro lining were highly flammable.
- The spacesuits worn by the first Apollo astronauts contained nylon, which may have created static electricity when rubbed against the seats. The suits also melted in the fire.

Nylon, a synthetic fiber made from coal, air, and water, will certainly melt in a high-temperature, oxygen-fanned inferno. It's used frequently in clothing, carpets, and rope, and it found early military applications in parachutes and tires. The melting point of nylon is about 509 degrees

Fahrenheit, or 265 degrees Celsius; a typical nonchemical fire, such as that found in a candle flame, burns between 1,472 and 1,832 degrees Fahrenheit (or 800 and 1,400 degrees Celsius). Clearly nylon wouldn't have been able to withstand the fire the Apollo 1 astronauts faced.

- The hatch in the Command Module opened inward rather than outward. The Apollo 1 astronauts had requested the design be changed so that the hatch would open outward, but NASA kept the inward design (thinking that resulted, ironically, from Grissom's Mercury-Redstone 4 disaster in which the explosive hatch opened prematurely; we cover that mission earlier in this chapter). However, during the fire, the pressure inside the cabin was too great for the astronauts to be able to open the inward-swinging door.
- A myriad of other electrical, plumbing, and wiring flaws may also have played a role, according to the findings of a NASA review panel.

The official cause of death for the astronauts was listed as smoke inhalation. The men had also suffered burns over much of their bodies, as evidenced by Grissom's nearly completely melted spacesuit. After this tragedy, NASA renamed the AS-204 mission Apollo 1 in honor of the deceased astronauts.

## Spacecraft design changes made in the aftermath

In the aftermath of the fire, several design changes were made to Project Apollo before it was allowed to continue:

- The spacecraft's hatch was redesigned to open outward; the explosive actuation was added back in; and a requirement that the hatch must be able to open in less than ten seconds was introduced.
- All flammable materials were exchanged for ones that could extinguish flames rather than fuel them, and the nylon spacesuits were replaced with suits made from a nonflammable fabric.
- Takeoff cabin atmosphere was redesigned to be more like that on Earth at sea level (60 percent oxygen and 40 percent nitrogen), with more-gradual pressure changes.
- All wires and tubes were coated with a fire-retardant insulation, and all known wiring problems (more than 1,000 in all) were fixed.

**Fast facts about Apollo 1**

Following are the basic facts about the Apollo 1 mission:

- **Intended launch date and site:** February 21, 1967; Cape Canaveral Air Force Station, Florida
- **Spacecraft name and mass:** *CM-012;* 45,000 lb (20,412 kg)
- **Launch vehicle:** Saturn 1B rocket
- **Size of crew:** 3 astronauts
- **Intended number of Earth orbits:** 200
- **Intended mission duration:** 14 days

# Soyuz 1 (1967) and Soyuz 11 (1971): Failures on Landing and in Orbit

The United States wasn't alone in experiencing early tragedies in its space program. The Soviet Union's space program experienced them too, but that program's administrators were generally much more secretive than NASA officials. Early tragedies, such as the loss of cosmonaut Valentin Bondarenko due to a 1961 fire in a simulator on the ground, were concealed from the masses. Tragedies that took place in the sky, however, quickly became public knowledge. These disasters included the loss of one cosmonaut upon landing on *Soyuz 1* in 1967 and the loss of three others in orbit on *Soyuz 11* in 1971.

## Soyuz 1

One of the first major tragedies of the Soviet Union's (now Russia's) space program occurred aboard the *Soyuz 1.* In the 1960s, this new spacecraft was designed to be the flagship vehicle of the Soviet Union's space program. Not only that but it was also intended to pave the way for human flights to the Moon. *Soyuz 1* carried one cosmonaut, Colonel Vladimir Komarov, and it was scheduled to be part of a manual docking experiment that was to be followed the next day by another Soyuz crew with three additional cosmonauts. Two of these men would've performed a *spacewalk,* otherwise known as an *extravehicular activity* (EVA) performed in space, to the *Soyuz 1.*

The launch proceeded according to plan on April 23, 1967, but almost immediately afterward, problems began. A solar panel failed to deploy correctly, causing a power shortage for the *Soyuz 1* and wrapping itself around the spacecraft. After several orbits, it became clear that the *Soyuz 1*'s stabilization system wasn't functioning. The second Soyuz crew initially began preparing to join Komarov in space and help fix the *Soyuz 1,* but weather conditions at the launch site, Baikonur, and news of more-severe problems aboard the *Soyuz 1* halted those plans.

Due to *Soyuz 1*'s problems, the spacecraft was recalled for an early return to Earth. Although Komarov's control over the spacecraft was limited, he did manage to fire the retrorockets and deorbit more or less on schedule. Komarov deployed his drogue chute to slow the spacecraft's speed to below the speed of sound, but after that, the main parachute didn't deploy due to a mechanical sensor failure. The backup parachute did deploy but got tangled up with the drogue chute, which hadn't detached from the spacecraft like it was supposed to. *Soyuz 1* came crashing to the Earth at around 90 miles (145 kilometers) per hour and exploded on impact. Komarov was found dead inside.

Parachute problems can be extremely dangerous, because lives may depend on their proper functioning. Spacecraft absolutely count on their parachutes being in good working order because often when rockets descend, they deploy parachutes to slow down as they approach Earth. Different types of parachutes can be used to slow a spacecraft in stages. For example, *drogue* parachutes are longer and thinner than the type of parachute used for humans; they're designed for deployment from very fast-moving objects and are a typical part of a spacecraft's equipment list.

Of unfortunate significance is the fact that the Soyuz 1 mission represented the first fatality during a spaceflight. The *Soyuz 1* spacecraft was rushed to launch to meet a political deadline and therefore hadn't been properly tested or checked. The *Soyuz 1* crash set the Soviet space program back by about 18 months while engineering and design flaws (including those regarding the parachutes) were addressed, though politics and competition in the Space Race kept the Soviets pushing forward.

## Fast facts about Soyuz 1

Following are the basic facts about the Soyuz mission:

- **Launch date and site:** April 23, 1967; Baikonur Cosmodrome, USSR
- **Spacecraft name and mass at launch:** *Soyuz 1;* 14,220 lb (6,450 kg)
- **Launch vehicle:** Soyuz rocket
- **Size of crew:** 1 cosmonaut
- **Number of Earth orbits:** 18
- **Mission duration:** 1 day, 2 hours, 47 minutes, 52 seconds

## Soyuz 11

Soyuz 11 is celebrated as the first time a crew successfully docked with and visited a space station, namely Salyut 1, the first space station (see Chapter 12 for more info on Salyut 1). The crew consisted of three cosmonauts: Commander Georgi Dobrovolski, Flight Engineer Vladislav Volkov, and Test Engineer Viktor Patsayev. They docked with Salyut 1 on June 7, 1971, a day after launch, and remained on the space station for 22 days.

The cosmonauts encountered a surprise when they first entered Salyut 1: The air was smoky and the ventilation system was in need of serious repairs. The crew replaced parts and then retreated back to their Soyuz capsule until the space station air was more breathable. After making history as the longest-lasting crew to live in space at that time, the men headed back home and landed on Earth on June 30.

However, a grisly site awaited the engineers who opened the *Soyuz 11* capsule — all three cosmonauts were dead. The ensuing investigation showed that the men had asphyxiated due to an opened breathing ventilation valve, one that came loose during reentry when the explosive bolts connecting the descent and service modules fired at the wrong times. Air pressure inside the cabin plummeted to zero until the capsule broke through the Earth's atmosphere. One of the cosmonauts' sensors showed that he perished within about 40 seconds of the air pressure in the cabin dropping.

Even had the men known what was wrong, they would've been unable to reconnect the loose valve because it was positioned behind their seats. Ground crews attempted resuscitation on landing, but they were too late to do any good.

### Fast facts about Soyuz 11

Following are the basic facts about the Soyuz 11 mission:

- **Launch date and site:** June 6, 1971; Baikonur Cosmodrome, USSR
- **Spacecraft name and mass at launch:** *Soyuz 11;* 14,970 lb (6,790 kg)
- **Launch vehicle:** Soyuz rocket
- **Size of crew:** 3 cosmonauts
- **Number of Earth orbits:** 383
- **Mission duration:** 23 days, 18 hours, 21 minutes, 43 seconds

As a result of this tragedy — the only time people have died while actually in space, beyond the atmosphere — the Soyuz capsule was redesigned to carry just two cosmonauts. With the space saved by eliminating the third cosmonaut, those onboard could wear pressure suits for the launch and landing periods. The three crew members of *Soyuz 11* didn't wear spacesuits, because there wasn't enough room for them.

# Apollo 13 (1970): A Successful Failure

One of the most-sensational accidents in space, which was brought to new prominence thanks to the self-titled 1995 movie, the Apollo 13 mission to the Moon would've been the third such crewed mission. The crew consisted of Commander James Lovell, Lunar Module Pilot Fred Haise, and Command Module Pilot Jack Swigert. The astronauts intended to explore part of the Moon known as the Fra Mauro Formation, but they never got the chance. The launch proceeded relatively uneventfully, but on April 14, 1970 — just two days following the launch — disaster struck.

## "Houston, we've had a problem"

Apollo 13 contained two tanks of liquid oxygen that were used for the astronauts to breathe, as well as for running the spacecraft's power-generating fuel cells. A thermostat lay inside each tank, and it (plus a heater) kept the oxygen at the correct temperature. Crew members stirred these tanks regularly in order to keep the oxygen properly mixed. (The contents of the tanks had a tendency to settle out into layers, so periodic stirrings helped ensure accurate tank readings. And no, the astronauts didn't go out there with a big spoon — the stirring process was automated.)

When ground controllers requested a stirring of the tanks, the astronauts turned on the tank fans. However, after one such stirring about two days into the mission, the Number 2 oxygen tank unexpectedly sparked and exploded. In the process, the other oxygen tank and the Service Module itself were damaged; in fact, one entire side panel of the module was ejected in the explosion. The astronauts reported the explosion back to Mission Control, at Johnson Space Center in Texas, with the famously understated line, "Houston, we've had a problem."

Following the explosion, the Service Module became unusable due to the loss of power and oxygen. The Command Module was still in good shape and had a ten-hour battery supply remaining, but the crew needed to save those batteries for the return trip to Earth. The astronauts shut down the Command Module and moved to the Lunar Module for the duration of their trip. Because the Service Module was no longer fully functional, the crew had to abandon its exploration of the Moon and use the Moon's gravity to enter a trajectory that could send the spacecraft back toward Earth. Flight controllers back at Mission Control assisted the crew with determining these course corrections, but not without the watchful eyes and ears of a fearful American public thanks to the ongoing television coverage of the event.

Apollo 13's return to Earth wasn't without its difficulties, the most prominent of which was the fact that the Lunar Module was designed to support only two people for two days. (With a little ingenuity that allowed the men to retrofit a carbon dioxide "scrubbing" apparatus from the Command Module, the Lunar Module ended up supporting all three crew members for four days; see Figure 5-2). Little clean drinking water, failures in the remaining oxygen tank gauges, and a host of other potential problems plagued the mission, but the entire crew made it home safe and sound on April 17.

**Figure 5-2:** The Apollo 13 Lunar Module with its retrofitted carbon dioxide filter (the square device in the center).

*Courtesy of NASA*

## The causes of the explosion

What problems led to the explosion of Apollo 13's oxygen tank?

- The shelf that carried the oxygen tanks had been intended for another Apollo mission and had been dropped at one point, causing no exterior damage but perhaps some internal damage to the tank that eventually exploded.
- The thermostat inside the oxygen tank wasn't upgraded to handle a 65-volt supply after a design change raised the requirement from the original 28 volts. All other components aboard the Apollo spacecraft were retrofitted to accept the new voltage level.
- The temperature readout thermometer only displayed temperatures up to 100 degrees Fahrenheit (38 degrees Celsius); higher temperatures were mistakenly read at 100 degrees Fahrenheit.

A fundamental problem with the tank piping was actually uncovered during prelaunch testing. Due to this problem, a heater was required to fully empty the oxygen tank. Because the thermostat was capable of handling only 28 volts, even though the actual power supply used was 65 volts, the excess electricity welded the thermostat closed, and the actual temperature inside the tank ended up being much higher than the 100 degrees Fahrenheit that the thermostat could read. Oxygen gas evaporated much faster than anticipated, meaning those high temperatures had a chance to burn the coating off of the electrical wires inside the tank.

- The stirring procedure set electricity charging through exposed wires, which led to sparking and the subsequent explosion.

## The consequences of Apollo 13

As a result of the Apollo 13 mission, NASA had to make certain changes to avoid a repeat mishap. Technical problems with the tanks, thermostats, and heating systems had to be addressed and resolved (although these fixes didn't happen until several years later).

One change that was made right away involved increasing the distance between the oxygen tanks and installing both a third oxygen tank and an emergency battery in a different location on the Service Module. NASA practices were also adjusted to reflect the need for testing all systems and equipment under a wider range of conditions. One factor that didn't change, though, was the high level of training that astronauts and Earth-bound engineers received; it was universally recognized that their creativity and cleverness played an enormous role in the astronauts' survival.

### Fast facts about Apollo 13

Following are the basic facts about the Apollo 13 mission:

- **Launch date and site:** April 11, 1970; Kennedy Space Center, Florida
- **Command and Service Module name and mass at launch:** *CM-109* (call sign "Odyssey"); 63,800 lb (28,945 kg)
- **Lunar Module name:** *LM-7* (call sign "Aquarius"); 33,590 lb (15,235 kg)
- **Launch vehicle:** Saturn V rocket
- **Size of crew:** 3 astronauts
- **Number of lunar orbits:** 0 (lunar landing canceled)
- **Mission duration:** 5 days, 22 hours, 54 minutes, 41 seconds

# *The Space Shuttle Challenger (1986): A Launch Ending in Disaster*

The Space Shuttle had a healthy track record when the January 28, 1986, flight of the *Challenger* was scheduled for launch. *Challenger* was the second vehicle of the Space Shuttle fleet to be launched into orbit, and it had already completed nine successful missions. This tenth voyage was extra-special because Christa McAuliffe, a New Hampshire–based teacher and the first astronaut to take flight as part of NASA's Teacher in Space Program, was aboard. (The *Teacher in Space Program* was designed to interest more of the general public in the space program.)

## *The disaster's cause: An O-ring seal*

After several delays due to weather, liftoff of the Space Shuttle *Challenger* on mission STS-51-L proceeded as planned. What happened next was pure tragedy. An O-ring seal, located in the right solid rocket booster (SRB) failed, creating a breach in the SRB's joints. Hot gases from the SRB escaped through the seal, burning the SRB, the external tank, and the *Challenger* orbiter itself. The SRB ultimately separated from the rocket.

Exactly 73 seconds after liftoff, the orbiter broke into fragments, killing all seven crew members onboard when they hit the ocean minutes later (see Figure 5-3). Because of the popularity of the Teacher in Space Program, the launch was televised live around the world, so millions of people watched helplessly as the *Challenger* crew perished.

**Figure 5-3:** The wing, fuselage, and engines completely separated from one another as *Challenger* fragmented and returned to Earth in a fiery blaze.

*Courtesy of NASA*

## Fast facts about Space Shuttle *Challenger*'s final mission

Following are the basic facts about the fatal STS-51-L mission of Space Shuttle *Challenger:*

- **Launch date and site:** January 28, 1986; Kennedy Space Center, Florida
- **Spacecraft name and mass at launch:** Space Shuttle *Challenger;* 268,829 lb (121,939 kg)
- **Size of crew:** 7 astronauts
- **Number of Earth orbits:** 0
- **Mission duration:** 73 seconds

The final crew of the Space Shuttle *Challenger* consisted of

- Pilot Michael Smith
- Commander Dick Scobee
- Mission Specialist Lt. Col. Ellison Onizuka
- Mission Specialist Ronald McNair
- Payload Specialist Gregory Jarvis
- Mission Specialist Judith Resnick
- Educator in Space Christa McAuliffe

## The repercussions: A shutdown of the Space Shuttle program

After *Challenger*'s demise, the Space Shuttle program was stopped for 32 months. A governmental commission called the Rogers Commission was appointed by President Ronald Reagan to figure out what had happened. The conclusions took the commission beyond the O-ring failure and brought to light flaws in how NASA addressed known problems, as well as the agency's unwillingness to heed engineering and other warnings. These problems were rectified before the Space Shuttle program resumed in 1988.

The Teacher in Space Program, however, couldn't be resurrected. The program, along with other attempts to send private, nonastronaut citizens into space, was abandoned after the *Challenger* disaster. Only in recent years has the idea of sending educators into space received serious, renewed interest, such as the private Teachers in Space program started by the Space Frontier Foundation in 2006. (Check out `spacefrontier.org` for the full scoop on the Space Frontier Foundation.)

## Sending teachers to space

NASA's Teacher in Space Program was launched in 1984 as a means to create greater general interest in math, space, and science among America's youth. The program was quite popular: It had 11,000 applicants in its first year. Christa McAuliffe was the chosen one, slated to be the first teacher to launch into orbit aboard the Space Shuttle *Challenger.* Barbara Morgan, a teacher from Idaho, was McAuliffe's backup and went on to become a full-fledged NASA astronaut.

All was not completely lost when the program closed down as a result of the *Challenger* disaster. NASA instituted a replacement program in the 1990s called the Educator Astronaut Program. Instead of training teachers as temporary astronauts, the program recruited them to become permanent astronauts. After their space journeys, participants gain employment through NASA (usually as mission specialist-educators [MSEs]) instead of returning to their teaching careers. The first of these MSEs are scheduled for Space Shuttle missions in early 2009.

The first group of MSEs to come from the Educator Astronaut Program consisted of

- Joe Acaba
- Ricky Arnold
- Dottie Metcalf-Lindenburger

On the more-formal front, NASA now has a new program called the *Educator Astronaut Program* that gives astronaut candidates with teaching backgrounds a permanent place in the astronaut corps as mission specialist-educators (see the nearby sidebar for more info).

# The Space Shuttle Columbia (2003): Reentry Disintegration

The *Challenger* mishap in 1986 occurred shortly after liftoff. Although many people think that if something's going to go wrong in a spaceflight, it's probably going to happen at the beginning of the flight, they're wrong. Ultimate disaster can strike at any time — even at the very end of a mission. The disintegration of Space Shuttle *Columbia* in 2003 is one of the most-prominent examples of such a mishap.

## Airborne foam creates a hole in the left wing

Although the liftoff of Space Shuttle *Columbia* on mission STS-107 appeared to go well on January 16, 2003, engineers studying the details of the launch noticed a problem. During the launch, a piece of the Space Shuttle's foam insulation broke loose from the external propellant tank and hit the left wing about 82 seconds after liftoff, creating a hole of about 6 to 10 inches (15 to 25 centimeters) in diameter in the spacecraft's left wing. It may not seem like flying foam could cause much of a problem, but remember that the foam was flying very quickly: It weighed a little more than a pound (0.45 kilograms) but was traveling at around 800 feet (or 244 meters) per second.

*Columbia* was able to continue its mission, but the wing damage resulted in a compromise of the Space Shuttle's thermal protection system. NASA engineers had seen problems with foam coming loose during launch before, but historically such launches had been allowed to continue, so there was a certain amount of confidence that reentry would proceed as planned. Even if the astronauts onboard had realized the problem, there was little they could've done in the way of repairs.

As *Columbia* reentered the Earth's atmosphere at the end of its mission, the compromised wing was destroyed by the heat caused by atmospheric drag. Normally, Space Shuttle wings get very hot from frictional heat as the spacecraft rapidly descends through the Earth's atmosphere. In this case, however, the high-temperature gases penetrated into the interior of the damaged wing, leaving it unprotected by the broken heat shielding.

The Space Shuttle started losing its thermal tiles, and ground observers witnessed a number of bright flashes. The entire spacecraft soon broke apart as it hurtled toward the Earth. People on the ground reported hearing a loud boom as the sky filled with smoke and flying debris.

Following are the names of the final crew members of Space Shuttle *Columbia:*

- Pilot William McCool
- Payload Specialist Ilan Ramon (from Israel)
- Payload Commander Michael Anderson
- Mission Specialist Kalpana Chawla
- Mission Specialist Laurel Clark
- Mission Specialist David Brown

## The Space Shuttle program stops again

As with the *Challenger* accident in 1986, *Columbia*'s demise caused the Space Shuttle program to cease operations for more than two years. During this time, organizational and decision-making procedures at NASA were overhauled, and technical design changes were made that would help limit foam detachment in later missions. Special cameras were also added to later Space Shuttle missions for purposes of surveying the orbiters' thermal tiles for damage.

This delay in Space Shuttle operations led to a halt of progress on the International Space Station, because the Space Shuttle was its primary means of receiving supplies and new components. Basic equipment and new crew members were supplied by the Russian Federal Space Agency during that interim.

## Fast facts about Space Shuttle *Columbia*'s final mission

Following are the basic facts about the fatal STS-107 mission of Space Shuttle *Columbia:*

- **Launch date and site:** January 16, 2003; Kennedy Space Center, Florida
- **Spacecraft name and mass at launch:** Space Shuttle *Columbia;* 263,700 lb (119,615 kg)
- **Size of crew:** 7 astronauts
- **Number of Earth orbits:** 255
- **Mission duration:** 15 days, 22 hours, 20 minutes, 32 seconds

# Part II
# The Great Space Race

## In this part . . .

Part II reveals the story of space exploration's driving force in the second half of the 20th century: the Space Race. The political struggle between the United States and the Soviet Union manifested itself in the skies as the two sides competed on an ever-escalating playing field.

First the Soviet Union launched *Sputnik 1* (the first satellite), followed shortly by the first human in orbit. The U.S. played catch-up, sending its Mercury and Gemini crews into Earth orbit. But landing on the Moon was the ultimate prize. After both nations sent robotic spacecraft to orbit and land on the Moon, the U.S. gained a decisive victory with its Apollo Moon landings. Robotic exploration continued with missions that took the first close-up views of Mars, Venus, and the planets of the outer solar system, and astronauts remained in Earth orbit, where they visited the first space stations. Then in 1975, the Space Race came to a symbolic end with a handshake in space on the joint American and Soviet Apollo-Soyuz mission.

## Chapter 6

# A Small Beeping Sphere and the Dawn of the Space Age

### *In This Chapter*

- Fueling space research through international tension
- Sending Soviet and American satellites to new heights
- Launching animals into space

With the launch of two satellites — *Sputnik 1* and *Explorer 1* — in the 1950s, space became a place to go rather than a place to just look at. The dawn of space exploration led to a psychological transformation in the way humans viewed space. It was no longer merely a place to observe; for the first time in the history of humanity, it was a viable destination.

In this chapter, we run through the first artificial satellites, including a detailed description of the Soviet *Sputnik 1* satellite and its effect on science and technological development in the U.S. — in particular, the launch of *Explorer 1.* We also cover early experiments with sending animals (think dogs and monkeys) into orbit.

## Cold War Tensions and the Space Race

Following the end of World War II, two nations rivaled each other for global dominance: the Soviet Union and the United States. One of the major areas in which the two countries sought to outdo each other was in the area of space exploration.

Why space? Well, a combination of technological and political factors allowed space exploration to be the subject of competition between the two superpowers:

- Technological readiness was the first factor. Advances in rocketry and engineering, along with the upcoming field of computers, meant that for the first time in human history, humanity had the skills to leave the planet.

- Politically, space exploration turned out to be a golden opportunity. As a relatively new field with little human achievement to date, there was an enormous opportunity for someone (and his nation) to take the credit for many "firsts": first human to orbit the Earth, first person to walk on the Moon, and so on. The U.S. and the Soviet Union both wanted to claim these honors and earn the respect of the entire world, so both countries made heavy investments in space science during the years of the Space Race (1957–1975).

The superpower rivalry between the Soviet Union and the U.S. was described as the *Cold War,* a term that encompasses all the tensions, hostilities, and outright competition between the world's greatest nations following World War II. The Soviet Union and the U.S. were both Allied powers during the war itself, but they had differing worldviews on what should happen beyond the war. That sticky little thing called Communism didn't help unite the countries either.

Although the Cold War had its downsides (accusations of treason, spying, a costly American involvement in the Korean and Vietnam conflicts, and a massive propaganda machine), it undeniably produced some serious innovation in technology and industry. The buildup of both the Soviet and American defense systems was another result of these tensions, as was a race to produce and stockpile nuclear arms. The militaristic aspect of the Cold War was frighteningly intense, with a series of treaties, wars, and other events that solidified the opposition between the two countries. War was never officially declared, hence the term Cold War, but these countries and their allies soon had divisive political and technological agendas to meet.

Several decades passed before the end of the Cold War; between Soviet leader Mikhail Gorbachev's introduction of *glasnost* (transparent and open government) and *perestroika* (economic restructuring) and President Ronald Reagan's influence, the Cold War petered out in the late 1980s and early 1990s. The collapse of the Soviet Union in 1991 brought further realizations of capitalism and a free market to the former Soviet states, which opened the door for real collaboration between Russia and the U.S.

## Sputnik 1: The First Satellite in Orbit (1957)

The opening act of the Space Race was the competition to put a satellite into orbit around the Earth. Both superpowers were invested in being the first nation to send a satellite into Earth's orbit, but the first achievement of the Space Race went to the Soviet Union.

Why a satellite? Good question! The idea sprang from the *International Geophysical Year,* the time frame from July 1, 1957, through December 31, 1958, that scientists designated as a period for coordinated observations of the Sun and other bodies in space. Activities included studies of aurora, cosmic rays, geomagnetism, and solar activity. Although studying these effects from the ground was possible, it occurred to scientists as they set the dates for the International Geophysical Year in 1952 that measurements could be made much more effectively with an instrument that was actually *in* space.

Previously, scientists made measurements of the upper atmosphere using balloons, which could only reach about 25 miles (40 kilometers) above the Earth's surface. Sounding rockets could also take suborbital measurements, closer to the edge of space at 63 miles (100 kilometers) or even beyond, but these small rockets remained aloft for just a few minutes before falling back to Earth. In contrast, scientists thought an orbiting satellite would be able to take measurements for hours or days rather than mere minutes. Suddenly, scientific concerns matched up directly with political reality, and both the Soviet Union and the United States began working to build such a satellite.

In the following sections, we describe the creation of *Sputnik 1* and its successful launch in more detail.

## Developing the little metal ball that changed history

Soviet scientists recognized that sending a satellite into orbit would give them much-needed information in the pursuit of human spaceflight, so in 1954, they began the design and development work that led to the *Sputnik 1* satellite. (In fact, their efforts were speeded by news that the U.S. was embarking on a similar pursuit.)

*Sputnik 1,* the world's first satellite, was a relatively simple device that looked very different from modern satellites (check it out in Figure 6-1). It was a metallic sphere that measured 23 inches (58.5 centimeters) across and was covered with a bright, shiny heat shield made from aluminum, magnesium, and titanium. Four long antennae extended nearly 10 feet (3 meters) from one side of the sphere. The satellite's design was primarily chosen for its simplicity, strength, and ease of construction; the spherical shape was echoed in many future Soviet and Russian spacecraft, such as Soyuz.

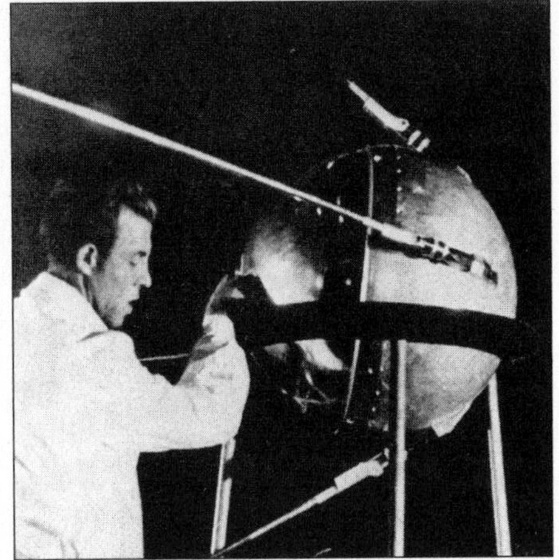

**Figure 6-1:** With its simple ball shape, *Sputnik 1* looked very different from modern satellites.

Courtesy of NASA

Inside the satellite were a battery-operated power supply, very basic environmental controls, and a radio transmitter for sending data back to Earth. The radio signal supplied useful information to Earth-bound scientists who used its strength to study the Earth's *ionosphere* (the very top levels of the atmosphere) as well as atmospheric density. Scientists also monitored the temperature and pressure inside *Sputnik 1,* primarily to see whether the spacecraft had been hit by any meteorites.

After many tests, and a few failures, Soviet space engineers decided to launch *Sputnik 1* from a modified R-7 Sputnik rocket. This rocket was actually designed as an Intercontinental Ballistic Missile (ICBM) consisting of parallel boosters and a sustainer that gave its base a somewhat flowery appearance. The R-7 Sputnik is a massive rocket (98 feet [30 meters] tall and weighing 590,000 pounds [267,000 kilograms]), and although it proved to be a failure as an ICBM, its large payload mass made it perfect for launching spacecraft.

## Fast facts about Sputnik 1

Following are the basic facts about the Sputnik 1 mission:

- **Launch date and site:** October 4, 1957; Tyuratam range (now part of Baikonur Cosmodrome), USSR
- **Spacecraft name and mass at launch:** *Sputnik 1;* 184 lb (84 kg)
- **Launch vehicle:** R-7 Sputnik rocket
- **Number of Earth orbits:** 1,440
- **Length of time in Earth orbit:** 3 months

The R-7 Sputnik rocket that launched *Sputnik 1* went on to enjoy a long and healthy life. It was used into the 21st century for launching spacecraft, and it also launched many required components to the International Space Station (which we describe in detail in Chapter 19).

## Celebrating success

The launch of *Sputnik 1* proceeded as planned on October 4, 1957, thus cementing the Soviet Union's place in history as the first country to launch an artificial satellite into orbit. The Soviet Union notified observers around the world to watch for the satellite, which could be seen from the ground through binoculars or telescopes as it passed through the night sky. Additionally, the radio signal from *Sputnik 1* could be heard with a common shortwave radio. This beeping signal quickly became a symbol of the Soviet Union's technological prowess.

Although *Sputnik 1* didn't return images or other visual data to Earth, the satellite did provide scientists with information about the temperature both inside the satellite and on its exterior surface. This data was downlinked after the satellite had orbited the Earth once, allowing scientists to gain a better understanding of temperatures in space.

The batteries on *Sputnik 1* allowed transmission to continue for 22 days. The satellite's orbit gradually deteriorated after that, and it burned up in Earth's atmosphere in early January 1958, after spending about three months in orbit.

Other robotic spacecraft followed as part of the Soviet Union's Sputnik program, which lasted until it was replaced by the Cosmos spacecraft in 1961. However, the most-important achievement of *Sputnik 1,* that polished, metallic sphere in the sky, was, simply, its success. The door was now open for the making of history, and the world wouldn't have long to wait.

# Explorer 1: The U.S. Response (1958)

As part of its planning for the International Geophysical Year (explained earlier in this chapter), the United States was starting to lay the foundations for a satellite program of its own in 1954. Project Orbiter required the combined efforts of the Navy and Army and, like the Soviet Sputnik program, it aimed to launch a satellite into orbit before the conclusion of the International Geophysical Year. The program wasn't formally started until

1957, after *Sputnik 1* had already made headlines (and simultaneously crushed U.S. visions of technological superiority to the Soviet Union), but its designers lost little time in creating a rocket suitable for such a mission. The Space Race had begun in earnest, with the U.S. in dire need of catching up.

## Racing to create an American satellite

Before the U.S. could develop a satellite to rival *Sputnik 1,* it needed to create a rocket that could launch said satellite. Enter Juno 1, a modified Jupiter-C rocket that included an added fourth stage and was completed in 1958. Of course, in the three months it took the U.S. to prepare Juno 1, the Soviet Union was proving its success in space with the launch of a second Sputnik satellite (which was built in just a few weeks).

*Explorer 1* was a much smaller satellite than *Sputnik 1,* because it was limited by the size of the available American launch vehicles. (The Soviet Sputnik rockets were modified from military applications that had a much larger physical launch capability.) *Explorer 1* was cylindrical rather than spherical (see Figure 6-2) and measured just 80 inches (2 meters) long by 6.25 inches (15.9 centimeters) in diameter; its mass was about one-sixth of *Sputnik 1*'s mass. *Explorer 1*'s mission centered on making several key measurements, including reading temperatures, gauging cosmic dust impacts, and detecting cosmic rays. Antennae transmitted the spacecraft's radio signals back to Earth.

The engineering for *Explorer 1* came from the Jet Propulsion Laboratory at the California Institute of Technology. A group of scientists at the State University of Iowa, led by Dr. James Van Allen, designed and developed the satellite's instrumentation.

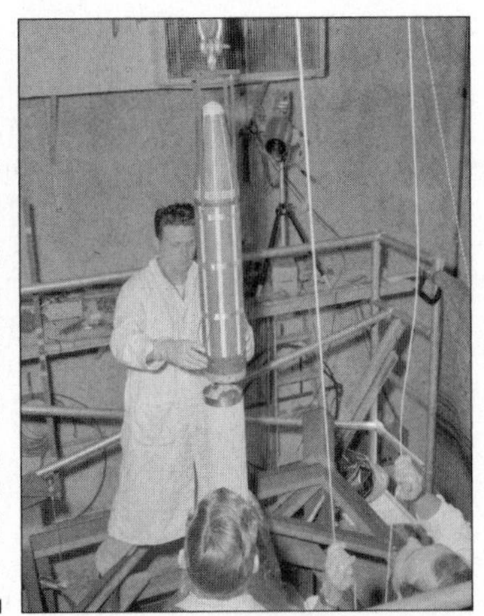

**Figure 6-2:** The *Explorer 1* satellite, which was smaller than *Sputnik 1,* had a cylindrical design, much like a small rocket.

*Courtesy of Jet Propulsion Laboratory*

## Fast facts about Explorer 1

Following are the basic facts about the Explorer 1 mission:

- **Launch date and site:** February 1, 1958; Cape Canaveral Air Force Station, Florida
- **Spacecraft name and mass at launch:** *Explorer 1;* 30 lb (14 kg)
- **Launch vehicle:** Juno 1
- **Number of Earth orbits:** 56,000
- **Length of time in Earth orbit:** 111 days

## Exploring the results

*Explorer 1* achieved orbit on February 1, 1958, formally cementing America's participation in the Space Race. Its orbit ranged from 1,563 miles (2,515 kilometers) above the Earth to 220 miles (354 kilometers), and it managed to complete about 12 orbits around the Earth per day. The satellite's last transmission to Earth was sent on May 23, 1958, and it remained in space until 1970 when it reentered the Earth's atmosphere, burning up in the process.

Based on the data received from *Explorer 1,* scientists discovered much about the strong radiation in the Earth's magnetic field. At certain altitudes, the sensors aboard *Explorer 1* sent back strange readings, and scientists were able to confirm these findings in later Explorer missions. This data led to the impressive discovery of the Van Allen radiation belt. Fewer cosmic rays were detected than scientists had expected, but they attributed that in part to the presence of the radiation belt.

More Explorer spacecraft followed the success of *Explorer 1;* they carried out a range of scientific missions and stand among the longest-running spacecraft programs in the U.S.

Yet perhaps the most-important result of *Explorer 1*'s success was the formation of the National Aeronautics and Space Administration, more commonly known today as NASA, in 1958. The agency's original mission was "to provide for research into the problems of flight within and outside the Earth's atmosphere and for other purposes." Since 1958, NASA has been responsible for worldwide innovations in satellites, probes, orbiting telescopes, space stations, and observatories — not to mention human spaceflight!

# Animals in Space

Humans aren't the only living creatures to take flight and travel in space. During the early days of the Space Race, various animals took on the role of astronaut for different purposes. Initially, animals were sent on spacecraft to ensure that any creature could survive a trip into Earth's orbit. (Scientists at the time figured it was more acceptable to run such experiments using animals rather than humans.) Other experiments focused on the effects of very low gravity on oxygen-breathing creatures, as well as the general act of being sent into space.

Animal space travel had humble beginnings. The first animals to travel to the edge of space were actually fruit flies that hitched a ride aboard a 1947 U.S. V-2 rocket to see how they fared under radiation. Fruit flies have since taken flight in many other space adventures, yielding valuable scientific data. Later years saw many other fish, insects, and mammals taking flight in the name of science.

In the following sections, we describe two major groups of animals sent into space: dogs from the Soviet Union and primates from the United States.

Animals have been integral components of space exploration throughout its history. They've provided invaluable data that has contributed to humans' ability to better understand space and, of course, to better prepare their human counterparts for extended space travel.

## The Soviet Union: Dogs in space

Years before the launch of *Sputnik 1* (which we describe earlier in this chapter), a 1951 rocket launch carried the world's first two Soviet space dogs, Tsygan and Dezik, to the very edge of space. Their suborbital flight reached a maximum height of 62 miles (100 kilometers) before their capsule returned to Earth where the dogs were retrieved successfully, both alive. Many more dogs traveled on later suborbital flights launched by the Soviet Union.

Perhaps the most-famous Soviet dog in space is Laika, who in 1957 traveled on the *Sputnik 2* spacecraft. Laika was actually a stray who was trained for space travel and became the first dog launched into orbit around the Earth. She rode in a specially designed harness that allowed her to change positions but not fully turn around inside the cabin.

Laika made it safely into orbit, but she perished a few hours into the flight. Numerous theories abound as to why and how she died. The most likely explanation is that even though the capsule was equipped with cooling fans and food, humidity and temperature levels inside it grew too high because of a malfunction with the spacecraft's thermal insulation and controls. Laika probably died from stress and overheating; however, her capsule was never intended to return her to Earth. *Sputnik 2* burned up as it reentered the Earth's atmosphere about five months after launch.

## It's a jungle out there . . . in space

Many countries have launched a wide variety of animals into space. Here's a sampling:

- France sent rats, mice, cats, and monkeys between 1961 and 1967.
- China launched mice and rats in the mid-1960s.
- The Soviet Union sent a tortoise and a variety of other life forms into deep space in 1968.

A wider variety of animals made it into space in later years. The 1990s saw the launching of desert beetles, brine shrimp, Japanese tree frogs, newts, crickets, snails, sea urchins, and a host of other creatures. Each species was chosen for its potential to teach humans something new.

Another significant Soviet canine spaceflight was that of Belka and Strelka. Like Laika, these female dogs were also strays selected mainly for their temperaments (calmer females were the preferred choice for canine cosmonauts). They practiced wearing spacesuits, entered simulators and centrifuges to get accustomed to the feeling of riding into space, and were exposed to long periods of confinement to prepare them for the tiny cabins. In 1960, aboard *Sputnik 5,* Belka and Strelka became the first dogs to survive an orbital launch into space and return home safe and alive. Other animals (mice, flies, and a rabbit) traveled aboard the same spacecraft, and they all survived the flight as well. Thanks to this accomplishment, the Soviet space program was off to a flying start with acquiring the knowledge it needed to support human space travel.

# The United States: Primates in space

In the true spirit of Space Race competition, the United States was developing its own program for sending animals into space. The rationale was much the same as for the Soviet dog studies (described in the preceding section): Engineers were eager to find out how living creatures would react to the pressures of microgravity and space travel, but they were far less eager to use humans as the first test subjects. (Generally speaking, scientists took as many humane precautions as they could to protect the animals' safety, including anesthetizing them before takeoff.)

The first noninsect to board a U.S. spacecraft was Albert, a rhesus monkey who was launched on a suborbital flight in 1948 via a V2 rocket that made it up to 39 miles (63 kilometers). Albert perished due to suffocation and was followed in 1949 by Albert II, whose flight reached a height of 83 miles (134 kilometers), making him the first monkey in space. Albert II died on impact, but he did survive the flight — and that was critical for engineers figuring out the basics of how to stay alive while in space. The next couple of Alberts (III and IV) also died on the return trip. A squirrel monkey named Gordo survived his 1958 launch on the Jupiter AM-13 but died on reentry because of a defect with the rocket's parachute.

The first primate-space-travel success story belongs to a pair of monkeys named Able and Baker. In 1959, they were launched on the Jupiter AM-18 rocket, made it more than 50 miles (80 kilometers) into space, and returned home alive. Able perished shortly thereafter due to surgery required to remove his medical electrodes, but the fact remained that living humanlike creatures had made it into space — and home again! NASA's work continued.

Ham the Chimpanzee made waves for being the first primate launched in Project Mercury (which we describe in detail in Chapter 7). In 1961, Ham was rocketed into space, even managing to survive a pressure loss in the capsule due to the effectiveness of his spacesuit. Thanks to his training, Ham proved that primates could do basic tasks in space (such as pushing levers, a task he practiced on Earth and replicated while in space), and his work paved the way for Alan Shepard's Freedom mission later that same year. (Check out an image of Ham in the color section.)

Another significant chimp was Enos, who trained at the Holloman Air Force Base in New Mexico with Ham. Enos became the world's first chimp to orbit the Earth during his successful Mercury-Atlas 5 launch in 1961. What Ham did for Shepard, Enos did for John Glenn: With the success of Mercury-Atlas 5, Glenn was able to successfully orbit the Earth for the first time on the 1962 Mercury-Atlas 6 flight.

# Chapter 7

# What a Ride! The First People in Space

### In This Chapter

- Being first in space with Vostok and Yuri Gagarin
- Orbiting in Mercury with John Glenn
- Spacewalking with Voskhod
- Docking with Gemini
- Surging ahead with Soyuz

Already playing catch-up after the Soviet Union launched the first satellite, *Sputnik 1,* in 1957 (see Chapter 6), the United States fell even farther behind in the Space Race when the Soviet Union launched the first person into space in 1961. Yuri Gagarin's revolutionary flight was a triumph for the Soviets. The Americans didn't give up, though. Gagarin's flight was quickly followed up by the first space trips by astronauts from the newly formed NASA — Alan Shepard took a suborbital flight in 1961, and John Glenn became the first American to orbit the Earth in 1962. The first people to travel in space became instant worldwide celebrities, and the Space Race continued with the American Gemini program and the Soviet Voskhod and Soyuz programs.

## The Vostok Spacecraft and Its Pioneers (1961–1963)

The Soviet Union's first foray into human spaceflight came in the form of the Vostok spacecraft, which was actually intended for use both in human spaceflight and as part of the Soviet spy network. The ship itself had two parts: The spherical upper portion was the capsule in which the cosmonaut and instruments traveled, and the cylindrical bottom portion housed the engine and its propellant.

In the following sections, we introduce you to two pioneers of the Vostok program: Yuri Gagarin, the first person in space, and Valentina Tereshkova, the first woman in space.

## Yuri Gagarin, the first human in space

The first spacecraft of the Vostok program was actually a prototype built in 1960 that didn't carry any cosmonauts; it did, however, make several automatic missions that helped prepare the designers for the finishing touches on the Vostok spacecraft. The first human-carrying spacecraft was *Lastochka* (a Vostok 3KA design). Its first manned mission, named Vostok 1, took place on April 12, 1961, and allowed cosmonaut Yuri Gagarin to take the honor of being the first human in space.

After launching from the Baikonur Cosmodrome (located in modern-day Kazakhstan, this facility was the main launch site of the then-Soviet and now-Russian space program), Gagarin completed one orbit around the Earth during his 108-minute first experience in space. Gagarin's spacecraft was put on automatic control during the flight because space engineers weren't sure how a human would react to the conditions of weightlessness.

The Soviets didn't exactly have a sophisticated plan for the spacecraft to deliver its passenger safely back to the ground upon the mission's completion. Gagarin actually parachuted to a safe landing on the ground from a height of about 23,000 feet (7,000 meters) above the Earth. The Vostok capsule itself also landed by parachute, but engineers thought that having the cosmonaut stay with the capsule posed too great of a physical hazard for him due to the anticipated rough landing.

Gagarin's parachute landing was kept secret for many years, because the international agency responsible for certifying the world's first orbital spaceflight required the astronaut to land along with the spacecraft. Soviet officials claimed for years that Gagarin did indeed land with his capsule — while still allowing cosmonauts on all future Vostok flights to parachute to the ground.

Gagarin's flight made him an instant hero in the Soviet Union. He was lauded many times over and quickly became the poster boy for the Space Race and the Communist Party. Gagarin was promoted several times in the Soviet Air Force, but due to his hero-level status, he was actually prohibited from taking on other spaceflight ventures that the government considered too risky. With Gagarin's trip into orbit, the Soviet Union had claimed another important "first" in the Space Race, leaving the U.S. to play catch-up once again.

## Fast facts about Vostok 1

Following are the basic facts about the Vostok 1 mission:

- **Launch date and site:** April 12, 1961; Baikonur Cosmodrome, USSR
- **Spacecraft name and mass at launch:** *Lastochka;* 10,400 lb (4,725 kg)
- **Launch vehicle:** Vostok 8K72K rocket
- **Size of crew:** 1 cosmonaut
- **Number of Earth orbits:** 1
- **Mission duration:** 1 hour, 48 minutes

# Valentina Tereshkova, the first woman in space

All in all, the Soviets sent six manned Vostok flights into space before the program ended in 1963. The last such mission, Vostok 6, wasn't manned by a man at all! Instead, a female cosmonaut named Valentina Tereshkova crewed the mission, making her the first woman in space. (At the time, the Soviet Union had a separate cosmonaut corps for women, and Tereshkova was one of only five women to make the grade.)

Tereshkova had a somewhat unusual background, having attended school largely through the mail, but her hobby of parachute jumping led her to apply for cosmonaut training. This specialty also made her a desirable candidate, because parachuting was a required skill for evacuation from Vostok spacecraft.

Tereshkova completed the Vostok 6 mission successfully, minus these minor onboard mishaps:

- Several communication lapses occurred between Tereshkova and Mission Control.
- She reported being unable to complete all scheduled experiments because some had been positioned beyond her reach.
- Although she was nauseous for most of the trip, Tereshkova's debriefing notes described the flight as largely pleasant and the food as awful (a far cry from the food astronauts get today, which you can read about in Chapter 4).

Tereshkova's flight was likely the product of the Soviet Union's desire for another first-place claim in the Space Race. Yet regardless of how she got there, her achievement opened the door for the success of future female astronauts.

## Fast facts about Vostok 6

Following are the basic facts about the Vostok 6 mission:

- **Launch date and site:** June 19, 1963; Baikonur Cosmodrome, USSR
- **Spacecraft name and mass at launch:** *Chayka;* 10,400 lb (4,710 kg)
- **Launch vehicle:** Vostok 8K72K rocket
- **Size of crew:** 1 cosmonaut
- **Number of Earth orbits:** 48
- **Mission duration:** 2 days, 22 hours, 50 minutes

# Project Mercury and Its Lofty Goals (1959–1963)

The United States, though bested by the Soviet Union in the race to place the first human into orbit, wasn't without a feisty comeback. NASA (which was formally created in 1958 thanks to the success of *Explorer 1,* as explained in Chapter 6) helped develop the first American plan for human spaceflight: *Project Mercury,* which ran from 1959 through 1963. Named in honor of the Roman god of speed, Project Mercury had very specific goals: Humans would orbit the Earth and return home safely, gathering valuable data about how mankind could live, breathe, and work in space in the process. (The program also included a number of unmanned missions, as well as a few that sent animals into space; see Chapter 6 for more details.)

A variety of governmental agencies were involved in creating the standards for selecting the first lucky batch of U.S. astronauts. The president at the time, Dwight D. Eisenhower, decreed that these first astronauts should come from the military. The candidates had to meet a range of physical and health conditions and be able to prove they had experience flying aircraft. Personal interviews were conducted, as well as a slew of tests and exams. After being subjected to extremely stressful conditions, both physically and psychologically, the aptitudes of the potential space pilots were narrowed down, and eventually NASA's first seven astronauts were selected.

The Mercury 7 astronauts were

- Scott Carpenter
- L. Gordon Cooper, Jr.
- John Glenn, Jr.

- Virgil "Gus" Grissom
- Walter Schirra, Jr.
- Alan Shepard, Jr.
- Donald Slayton

We describe the spacecraft and the missions of the Mercury 7 in the following sections.

## A little snug: The Mercury spacecraft

The spacecraft for the Mercury missions were quite small, due to the size of the available launch vehicles. The inhabitable interior space was only about 60 cubic feet (1.7 cubic meters), and much of that was stuffed with instrumentation and controls. The capsule itself measured about 82 inches (2 meters) by 74 inches (1.9 meters.) A 19-foot (5.8-meters) escape tower topped the capsule.

Due to the size constraints, each Mercury mission was crewed by just one astronaut. The capsule was equipped with three *posigrade rockets,* small rockets that fired in the direction of motion in order to break the capsule away from its launch vehicle. Three different rockets were used as part of the Mercury program: Redstone, Atlas, and Little Joe.

Each Mercury capsule also had three solid-fuel retrorockets designed to bring the spacecraft back to Earth; only one was necessary, so the system had some built-in redundancy to account for potential mechanical failure. The capsules were designed to withstand substantial reentry temperatures and provided an elaborate system of backup controls in case the astronaut inside became incapacitated. As opposed to the Soviet Union's Vostok capsules, which landed on solid ground, the Mercury capsules (as well as later NASA spacecraft constructed under the Gemini and Apollo programs) were designed to splash down in the ocean, with the crew and capsule retrieved via helicopter. You can see an image of a splashdown in the color section.

## The missions of Alan Shepard, John Glenn, and the rest of the Mercury 7

Table 7-1 provides a few basic facts about the Mercury 7's manned missions, all of which launched from Cape Canaveral Air Force Station in Florida.

**Table 7-1 The Manned Mercury Missions**

| *Mission Name* | *Launch Date* | *Spacecraft Name and Mass at Launch* | *Launch Vehicle* | *Size of Crew* | *Number of Earth Orbits* | *Mission Duration* |
|---|---|---|---|---|---|---|
| Mercury-Redstone 3 | May 5, 1961 | *Freedom 7;* 2,855 lb (1,295 kg) | Redstone rocket | 1 (Alan Shepard) | 0 (sub-orbital) | 15 mins, 22 secs |
| Mercury-Redstone 4 | July 21, 1961 | *Liberty Bell 7;* 2,840 lb (1,286 kg) | Redstone rocket | 1 (Virgil "Gus" Grissom) | 0 (sub-orbital) | 15 mins, 37 secs |
| Mercury-Atlas 6 | February 20, 1962 | *Friendship 7;* 2,700 lb (1,225 kg) | Atlas rocket | 1 (John Glenn) | 3 | 4 hrs, 55 mins, 23 secs |
| Mercury-Atlas 7 | May 24, 1962 | *Aurora 7;* 2,976 lb (1,350 kg) | Atlas rocket | 1 (Scott Carpenter) | 3 | 4 hrs, 56 mins, 5 secs |
| Mercury-Atlas 8 | October 3, 1962 | *Sigma 7;* 3,020 lb (1,370 kg) | Atlas rocket | 1 (Walter Schirra) | 6 | 9 hrs, 13 mins, 11 secs |
| Mercury-Atlas 9 | May15, 1963 | *Faith 7;* 3,000 lb (1,360 kg) | Atlas rocket | 1 (L. Gordon Cooper) | 22 | 34 hrs, 19 mins, 49 secs |

Donald Slayton was the only one of the original Mercury 7 astronauts not to get his own mission. He was scheduled to fly on Mercury-Atlas 7 but was removed from the flight list after doctors found an irregular heartbeat during a training exercise in a centrifuge that simulated the high gravitational forces astronauts encounter during takeoff and landing. Slayton served as the director of NASA's flight crew operations until he finally obtained medical clearance and got to fly on the Apollo-Soyuz flight (described in Chapter 12).

You may have noticed that all the capsule names have the number 7 after them. Surely they weren't each the seventh attempt at getting the capsule right? Nope. The astronauts were actually given the honor of naming their capsules, and all of them chose to use the number 7 to show their support for the entire Mercury 7 team.

Two Mercury missions are particularly notable:

## The Mercury 13: Women who qualified for space

Although history reveals that the first American astronauts were all men, that info doesn't quite tell the whole story. Dr. William Lovelace, the person responsible for developing the tests that helped select the Mercury 7, also solicited female recruits for the testing phase. The first such candidate, a female pilot named Geraldyn "Jerrie" Cobb, was invited to undergo the testing in 1960, and she passed the same three phases as the men.

With these results, more women were invited to take physiological tests, some of which were quite strange. The women were examined, poked, prodded, and filmed. Their stomach acids and other bodily fluids were tested, as were their reflexes, breathing, and balance. The group made consistent progress, until the training of the final 13 candidates was halted when the Naval School of Aviation Medicine in Pensacola, Florida refused to allow them to continue testing because the program was considered unofficial. Even though the Civil Rights Act of 1964 had made gender discrimination illegal, NASA claimed the women were ineligible because they weren't allowed to attend Air Force training schools at the time, and one of the astronaut requirements was to be a graduate of a military jet test pilot program. NASA refused to grant exemptions for comparable experience, as they had in the cases of some male astronauts, despite the fact that some of the women already had the required engineering degrees *and* civilian test pilot experience.

Despite cosmonaut Valentina Tereshkova making waves with her flight in 1963 (see the related section earlier in this chapter for details), the minds of NASA's powers-that-be remained unchanged, and the so-called Mercury 13 weren't able to proceed with astronaut training. As a result, the first American female astronaut, Sally Ride, didn't make her way into space until 1983.

The Mercury 13 were

- Myrtle Cagle
- Geraldyn "Jerrie" Cobb
- Jan Dietrich
- Marion Dietrich
- Wally Funk
- Janey Hart
- Jean Hixson
- Gene Nora Jessen
- Irene Leverton
- Sarah Ratley
- Bernice Steadman
- Jerri Truhill
- Rhea Woltman

- **Alan Shepard, the first American in space:** The astronaut with the honor of being the first American in space is none other than Alan Shepard. A former flight test pilot in the U.S. Navy, Shepard was invited to "audition" for NASA's new Project Mercury. He soon passed all the tests and joined the astronaut crew. Shepard made history when, in 1961 aboard the *Freedom 7* spacecraft, he became the first American (and only the second human) to venture into space on a suborbital flight. Shepard's achievement was dwarfed only by Yuri Gagarin's flight, which occurred a month earlier. Even though Gagarin orbited the Earth and Shepard only reached the edge of space, Shepard was still celebrated as a hero in the U.S. for being the first American in space.

- **John Glenn, the first American to orbit the Earth:** Another famous American astronaut, John Glenn, broke new ground by becoming the first American to orbit the Earth. Glenn joined NASA after a career in the U.S. Marine Corps. He made history in 1962 when he voyaged around the Earth on the *Friendship 7* spacecraft. Glenn led a long and storied career with NASA and as a long-serving U.S. senator. In addition to his Project Mercury–related work, Glenn became the world's oldest astronaut when he took a ride on the Space Shuttle *Discovery* in 1998.

# Voskhod 2 and the First Spacewalk (1965)

Another first in the Space Race went to the Soviet Union with the 1965 Voskhod 2 mission. Led by cosmonauts Alexei Leonov (pilot) and Pavel Belyayev (commander), *Voskhod 2* was a modified version of the original Vostok spacecraft. The major differences included an extra solid rocket booster for backup, additional seating, and an exterior airlock. This airlock was required for the prime goal of the mission, which was met when Leonov made the world's first *spacewalk.* Technically called an *extravehicular activity* (EVA), the ten-minute spacewalk marked the first time that a human had left the safety of a spacecraft and braved the vacuum of space, protected only by a spacesuit.

Although this mission made history with Leonov's spacewalk, it wasn't without its problems:

- The spacewalk itself was a success, but Leonov narrowly avoided disaster when his spacesuit became unexpectedly rigid in the vacuum of space and he had difficulty fitting back through the airlock to reenter the capsule.
- Following the spacewalk, the post-spacewalk hatch seal wasn't good, leading to an excess of oxygen in the cabin, which raised the risk of a fire (but none occurred).
- With the mission almost over, the systems responsible for the ship's reentry (specifically, the retrorockets) failed, leading to the cosmonauts having to make a hard, manual landing in a snowy Siberian forest far from the waiting recovery crews. The men were stranded in the woods overnight while rescue crews raced their way. (Fortunately, they landed in their capsule rather than parachuting as the Vostok crews had done, so at least they had minimal shelter and survival gear.)

### Fast facts about Voskhod 2

Following are the basic facts about the Voskhod 2 mission:

- **Launch date and site:** March 18, 1965; Baikonur Cosmodrome, USSR
- **Spacecraft name and mass at launch:** *Voskhod 2* (call sign "Almaz"); 2,570 lb (5,680 kg)
- **Launch vehicle:** Voskhod rocket
- **Size of crew:** 2 cosmonauts
- **Number of Earth orbits:** 17
- **Mission duration:** 1 day, 2 hours, 2 minutes, 17 seconds

The overall Voskhod 2 mission had enough problems that the rest of the Voskhod series of missions was canceled.

# The Gemini Program Proves Its Case (1965–1966)

With the successes of Project Mercury (described earlier in this chapter) firmly establishing the United States as a force to be reckoned with in the Space Race, American politicians, engineers, scientists, and the general public wanted more. Simply sending one astronaut into a brief orbit was no longer satisfactory. People across the nation clamored for longer flights, bigger crews, and more-advanced spacecraft that could raise the bar for space travel and rival the accomplishments of the Soviet Union. These cries were answered in NASA's second manned spaceflight program, *Project Gemini.*

Formally undertaken in 1962, shortly before the completion of its predecessor (Project Mercury), Project Gemini was named after the constellation Gemini, whose name means *twins* in Latin. This association with twins was meant to signify the main innovation of the Gemini spacecraft: It could carry two crew members, not just one.

Project Gemini had well-defined goals that were simple enough for any citizen to understand:

- Develop manned spacecraft that could sustain a spaceflight lasting up to two weeks.
- Meet and dock with other vehicles in space.

- Land on the ground in a controlled glide instead of parachuting into the water. (This goal was replaced during the design phase in 1964 with the more-familiar splashdown water landing, but the precision control of the landing site was still demonstrated.)

The first two overall Project Gemini goals were met successfully, as you can see in the following sections. Although none of the missions landed on solid ground, Gemini XI proved that computer-controlled landings were possible, and the project achieved around 1,000 man-hours of experience in space.

Table 7-2 provides at-a-glance info for the manned Gemini flights, all of which launched from Cape Canaveral Air Force Station in Florida.

**Table 7-2 Manned Gemini Flights**

| *Mission Name* | *Launch Date* | *Spacecraft Name and Mass at Launch* | *Launch Vehicle* | *Size of Crew* | *Number of Earth Orbits* | *Mission Duration* |
|---|---|---|---|---|---|---|
| Gemini 3* | March 23, 1965 | "Molly Brown;" 7,136 lb (3,237 kg) | Titan II rocket | 2 (Virgil "Gus" Grissom, John Young) | 3 | 4 hrs, 52 mins, 31 secs |
| Gemini IV | June 3, 1965 | *Gemini 4;* 7,880 lb (3,570 kg) | Titan II rocket | 2 (James A. McDivitt, Edward H. White) | 62 | 4 days, 1 hr, 56 mins, 12 secs |
| Gemini V | August 21, 1965 | *Gemini 5;* 7,950 lb (3,600 kg) | Titan II rocket | 2 (L. Gordon Cooper, Charles Conrad) | 120 | 7 days, 22 hrs, 55 mins, 14 secs |
| Gemini VII | December 4, 1965 | *Gemini 7;* 8,080 lb (3,660 kg) | Titan II rocket | 2 (Frank Borman, James Lovell) | 206 | 13 days, 18 hrs, 35 mins, 1 sec |
| Gemini VI-A | December 15, 1965 | *Gemini 6A;* 7,820 lb (3,540 kg) | Titan II rocket | 2 (Walter Schirra, Thomas P. Stafford) | 16 | 1 day, 1 hr, 51 mins, 24 secs |

| *Mission Name* | *Launch Date* | *Spacecraft Name and Mass at Launch* | *Launch Vehicle* | *Size of Crew* | *Number of Earth Orbits* | *Mission Duration* |
|---|---|---|---|---|---|---|
| Gemini VIII | March 16, 1966 | *Gemini 8;* 8,350 lb (3,790 kg) | Titan II rocket | 2 (Neil Armstrong, David R. Scott) | 6 | 10 hrs, 41 mins, 26 secs |
| Gemini IX-A | June 3, 1966 | *Gemini 9A;* 8,300 lb (3,750 kg) | Titan II rocket | 2 (Thomas P. Stafford, Eugene A. Cernan) | 47 | 3 days, 21 hrs |
| Gemini X | July 18, 1966 | *Gemini 10;* 8,295 lb (3,760 kg) | Titan II rocket | 2 (John Young, Michael Collins) | 43 | 2 days, 22 hrs, 46 mins, 39 secs |
| Gemini XI | September 12, 1966 | *Gemini 11;* 8,370 lb (3,800 kg) | Titan II rocket | 2 (Charles Conrad, Richard F. Gordon) | 44 | 2 days, 23 hrs, 17 mins, 8 secs |
| Gemini XII | November 11, 1966 | Gemini 12; 8,290 lb (3,760 kg) | Titan II rocket | 2 (James Lovell, Edwin "Buzz" Aldrin) | 59 | 3 days, 22 hrs, 34 mins, 31 secs |

** This flight didn't use a Roman numeral.*

## Two-person crews and longer flights

The Gemini spacecraft was based on the Mercury space capsule, but it was enlarged enough to support a two-person crew. The capsule measured 19 feet (5.8 meters) long with a diameter of 10 feet (3 meters), and its interior was redesigned for the astronauts' ease of movement. The ship was sent into space by the Titan II rocket.

The first manned launch of a Gemini spacecraft came with the Gemini 3 mission, which launched from Cape Canaveral Air Force Station in Florida on March 23, 1965. This mission was crewed by two NASA astronauts: Commander Virgil "Gus" Grissom and Pilot John Young. Grissom nicknamed the capsule "Molly Brown," from the popular Broadway play *The Unsinkable Molly Brown,* in a tongue-in-cheek reference to his sunken Mercury space capsule (see Chapter 5 for a description of this early disaster in the U.S. space program).

Alan Shepard was originally scheduled to command the Gemini 3 mission, but an inner ear problem kept him firmly on the ground.

The main goal of the Gemini 3 voyage was to prove that a two-member crew could survive a trip into space. Side goals were to practice maneuvering in space and come down in a more-controlled landing than had been seen in the days of Project Mercury. A variety of experiments and tests were planned, and medical data from the two crew members were collected.

Unlike the Mercury space capsules, which allowed the astronauts to make only slight adjustments to the spacecraft's orientation, Gemini astronauts could change the orbit of their space capsules. The Gemini 3 astronauts performed the first *orbital maneuver* during their flight, meaning they changed the size and period of their orbit midflight by firing maneuvering thrusters at carefully calculated orientations and durations. The Gemini 3 crew successfully changed the shape of its orbit from elliptical to circular and later decreased the orbit's altitude above the Earth. This sort of control over a spacecraft was unprecedented up to this point in NASA history.

Although the Mercury and early Gemini missions relied on batteries for power, NASA engineers knew that longer-duration missions — like those required for a trip to the Moon — needed long-lasting power. The later Gemini missions employed *fuel cells* for their onboard power needs. These electrochemical devices use hydrogen as a fuel source and oxygen as an oxidizer, and give off water as waste. Fuel cells, as well as improvements in spacecraft control that allowed conservation of fuel for orbital maneuvers, allowed for new endurance records, such as *Gemini 7*'s 13 day-mission.

## Rendezvous and docking in space

One of the most-significant historical achievements of Project Gemini was its ability to prove that rendezvous and docking activities could take place in space. These techniques were considered critical forbearers for a lunar mission, which was one of the ultimate goals of the U.S. space program. The target for these spacecraft docks was the *Agena* spacecraft, a separate unmanned spacecraft designed by NASA to provide a place for other space vehicles to practice docking in space.

The first planned Gemini space rendezvous would've taken place during the Gemini V mission in 1965, but the experimental fuel cells used on this flight provided a few glitches that depleted the ship's power reserves and made a docking experiment impossible. Gemini VI also had a planned rendezvous, but the entire mission was scrapped when the *Agena* target vehicle exploded on launch.

## Hold the mustard?

In addition to its more academic and scientific role in making history, the Gemini 3 mission also introduced the world's first space deli. Astronaut John Young snuck a corned beef sandwich onboard the flight. Aside from the risk that the floating bread crumbs posed to the spacecraft's equipment and controls, astronauts bringing along covert sandwiches distracted them from the more-important scientific goal of testing the rehydration and preparation of NASA's space food. Needless to say, future astronauts were warned rather strenuously about the dangers of space takeout.

After a quick change in plans, *Gemini 7* was launched before *Gemini 6A* in order to become the latter's new target vehicle. The ships successfully rendezvoused on December 15, 1965, flying to within 1 foot (30 centimeters) of each other while in orbit around the Earth. Figure 7-1 shows the view of *Gemini 7* from *Gemini 6A.*

Docking experiments were right around the corner. *Gemini 8* was launched on March 16, 1966, and it successfully rendezvoused with a new *Agena* target vehicle that had been successfully launched ahead of time. The docking proceeded without a hitch, and the two vehicles were joined together in space. (An extended spacewalk was planned for this same mission but had to be canceled due to a stuck thruster on the *Gemini 8.*)

**Figure 7-1:** *Gemini 7* (as seen from *Gemini 6A*) in orbit around the Earth.

*Courtesy of NASA*

While docked to the *Agena,* the Gemini spacecraft began to spin slowly. Following procedure, the astronauts undocked, only to find that their ship was now spinning at a rate of one spin per minute. After a few harrowing minutes, Neil Armstrong, in his first spaceflight, was able to recover control of the spacecraft. The ship then made an emergency return to Earth; the total *Gemini 8* flight lasted just 10 hours and 41 minutes. Armstrong's quick thinking and levelheadedness during this emergency likely helped in his selection as the first moonwalker of Project Apollo (which we cover in detail in Chapter 9).

# Soyuz: The Space Program That Keeps on Running (1967–Present)

The Soviet Union had its own response to the achievements of the United States' Mercury and Gemini programs: the *Soyuz program.* Begun in the 1960s, this Soviet space program involved both a Soyuz launch vehicle and a Soyuz spacecraft. It was originally conceived to test orbital docking and multicrew missions, with the intent of leading up to a trip to the Moon. The Soyuz program was meant to be a technologically advanced replacement for its problem-plagued Voskhod predecessor (which we cover earlier in this chapter). The Soviet Union's space program fell farther and farther behind, however, due to technical and political reasons, while NASA's Project Gemini (described in the previous section) put the U.S. firmly in the lead in the Space Race.

The original Soyuz launch vehicle was a three-stage kerosene and liquid oxygen propellant rocket (see Chapter 3 for more about how rockets work). Its design was simple but proved to be cheap and reliable over the years of its use. The first flight of a Soyuz rocket was in 1966; modified versions are still flying today. One of the main innovations of the Soyuz rocket is that it can be assembled on its side and then raised to a vertical position for launch (check out Figure 7-2). The Soyuz rocket is used to launch commercial payloads and send the Soyuz spacecraft into orbit.

The major counterpart to the Soyuz rocket is the Soyuz spacecraft, a three-part ship consisting of an orbital module, a reentry module, and a service module (see Figure 7-3). Overall dimensions are 8.9 feet (2.7 meters) wide by 23.6 feet (7.2 meters) long; solar panels on the service module, when fully extended, increase the total width of the spacecraft to 34.8 feet (10.6 meters). It can carry a crew of up to three cosmonauts. The three main portions of the Soyuz spacecraft were designed to fit together and come apart as components:

- The *reentry module,* which contains the inhabitable space that the cosmonauts require, is the only portion that has to be able to come back to Earth.

- The *orbital,* or habitation, *module* provides extra living space while in orbit.
- The *service module* houses the engines and other support systems, as well as solar panels to provide extra power in orbit.

**Figure 7-2:** A Soyuz spacecraft and its launch vehicle being erected at the launchpad.

*Courtesy of NASA/Bill Ingalls*

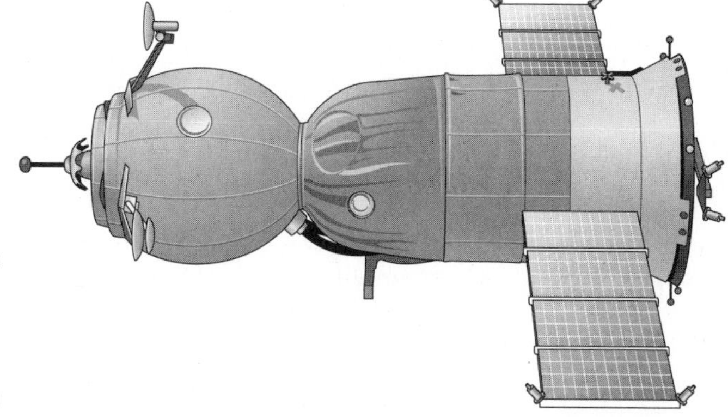

**Figure 7-3:** A Soyuz spacecraft.

This design approach, in which the orbital and service modules are jettisoned on the ship's return to Earth and burn up in the atmosphere on their way home, has a huge advantage for space travel. Because these modules aren't intended to protect human life during a fiery descent, they don't require the same shielding, or means of slowing themselves down, that the reentry module does. By limiting these comparatively expensive design components to just one module, the Soyuz spacecraft is able to expand its inhabitable space while keeping itself as light as possible. The Soyuz capsule itself lands on the ground, using a system of parachutes (to slow its descent) and braking engines fired just before landing (to soften the impact of landing).

*Soyuz 1,* which launched on April 23, 1967, was the first manned Soyuz flight. Cosmonaut Vladimir Komarov survived the launch and daylong excursion, but he perished upon landing. This disaster seriously delayed progress toward a Soviet mission to the Moon. (See Chapter 5 for more about the Soyuz 1 mission and other space disasters.)

The first Soyuz success with manned spaceflight was the Soyuz 3 mission, which was launched on October 26, 1968, and led by Georgi Beregovoi. Its main goal was to attempt a rendezvous and docking with the unmanned *Soyuz 2* spacecraft. Although *Soyuz 3* is significant for being the first successful manned flight of the Soyuz program because Beregovoi survived landing, the ship failed to dock with the *Soyuz 2* (although it did manage a successful rendezvous).

Perhaps the most-remarkable aspect of the Soyuz program is its expandability and longevity. Since the late 1960s, Soyuz spacecraft have been modified and reconfigured for a huge range of space applications. Some have never left the planning stages, whereas others came to fruition, as you find out in the following list:

- **Original Soyuz manned spacecraft:** Designed for orbital maneuvering around the Earth (1967–1971)
- **Soyuz 7K-L1:** Planned for lunar exploration and designed to be capable of navigating around the Moon (failed test flights 1967–1970)
- **Military Soyuz:** Planned for transporting crew and supplies to military space outposts (never flew)
- **Soyuz 7K-T and 7K-TM:** Used to transport crew to Soviet civilian and military space stations (1973–1981); docked with Apollo in 1975 (see Chapter 12)
- **Soyuz T:** Modified to allow crew to wear pressure suits and to accommodate longer flights (1976–1986)
- **Soyuz TM:** Upgraded radio communications and the maneuvering system; brought crews to Mir and the International Space Station (1986–2003; see Chapter 13)
- **Soyuz TMA:** Currently travels to the International Space Station and can hold taller astronauts (2003–present; see Chapter 19)

# Chapter 8

# Sending Robots to the Moon

***In This Chapter***

- Heading to the Moon with Luna
- Landing hard with Ranger
- Mapping the Moon with Lunar Orbiter
- Setting down and setting up shop with Lunar Surveyor

Although human visitation of the Moon was a major goal for both participants in the Space Race, the Americans and Soviets recognized that *robotic missions* (also known as unmanned missions) were a necessary first step. And what steps they were! Some of the early achievements were especially spectacular, such as the Soviet Union's automated soil sample return (which has yet to be replicated from another moon or planet) and the robotic Moon Rover *Lunokhod* (not replicated until the 1997 Mars Pathfinder Rover *Sojourner*). Major achievements such as these propelled space research to new heights and previously unachievable goals.

## First to the Moon with Luna (1959–1976)

The Soviet Union's first foray into robotic missions to the Moon took the form of the *Luna Program.* Forty-five missions were attempted but only about one-third of the launches resulted in successful flights between 1959 and 1976. The rest of the flights either didn't launch or didn't accomplish their missions. Despite the relatively large number of failures and no-shows, the Soviet Luna Program accomplished many significant achievements in space exploration.

## Early Luna missions: Setting the stage (1959–1968)

The Luna spacecraft were redesigned over the course of the Luna Program. The earliest versions were intended to simply crash-land on the Moon, whereas later models were designed as soft landers and lunar orbiters. In addition to their primary lunar missions, the Luna spacecraft were equipped with meteoroid and radiation detectors, as well as equipment to take a range of other measurements during space travel.

Highlights of the early Luna missions include the following:

- *Luna 1,* launched in 1959, had a spherical design reminiscent of the *Sputnik 1* satellite that had been put into orbit around Earth just two years before (flip to Chapter 6 for details on *Sputnik 1*). *Luna 1* missed its intended impact on the Moon due to a problem with the controls in the launch vehicle, but it was the first spacecraft in the world to go into orbit around the Sun and do a lunar flyby.
- *Luna 2,* launched later in 1959, was the first spacecraft to impact the Moon. A major scientific accomplishment of this *hard landing* (in which a spacecraft impacts the surface of a planet or satellite at a high speed without braking with parachutes or retrorockets) was acquiring the understanding that the Moon lacked a significant magnetic field.
- *Luna 3,* which also launched in 1959, introduced photography to space history. It was equipped with cameras that photographed the far side of the Moon. The photographs weren't incredibly clear, but they did show distinct mountainous sites in addition to darker areas. This new information fueled the imaginations of scientists, skeptics, and the general public alike.

- *Luna 9,* launched in 1966, was the first spacecraft to successfully land on the Moon. It was designed as a cylindrical spacecraft with a lander on top; this basic design was used for *Luna 4, 5, 6, 7, 8, 9,* and *13* (launched between 1963 and 1966). After touchdown, the lander could then right itself and open a series of petals that housed telecommunications equipment and cameras. The *Luna 9* cameras took a series of photographs (later assembled into a panoramic view) that became an instant part of the world's historical record (see Figure 8-1).
- Subsequent Luna incarnations (*Luna 10, 11, 12,* and *14,* launched in 1966, 1966, 1966, and 1968, respectively) focused on achieving lunar orbits and photographing the Moon's surface with more detail, as well as making more-advanced measurements and studying the Moon's surface and atmosphere.

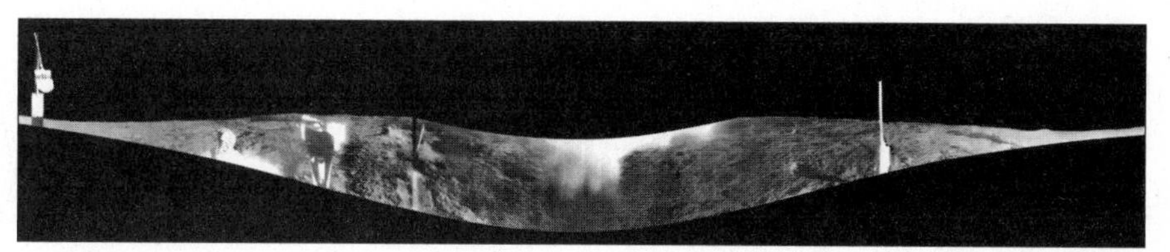

**Figure 8-1:** A panoramic view of the Moon's surface taken from Luna 9

Courtesy of Ted Styk

# Luna 16, 20, and 24: Sending samples to Earth (1970 and 1976)

The next wave of Luna spacecraft had the lofty goal of sending a little bit of the Moon back home to Earth. These Luna missions had landers onboard that were capable of gathering a soil sample from the Moon and launching a capsule containing that sample back to Earth. These landers were able to drill down into the Moon's surface, fill the hollow arm of the drill with soil, and send that arm home in the return capsule.

*Luna 15* was launched in July 1969 — just three days before NASA's Apollo 11 mission sent the first astronauts to the Moon (see Chapter 9 for details). By this point, the Soviets had all but conceded the Space Race to the Americans, but they saw *Luna 15*'s ability to return a lunar soil sample to Earth before the Apollo 11 team returned as one last chance for glory. However, they lost contact with the spacecraft during its final descent to the lunar surface (where Neil Armstrong and Buzz Aldrin had already made their footprints); *Luna 15* most likely crashed into the side of a mountain on its way down.

*Luna 16,* which was launched on September 12, 1970, successfully drilled for Moon dirt, landing and working near the Moon's Mare Fecunditatis (Sea of Fertility) area. Samples of the lunar surface were successfully returned to Earth 12 days after the spacecraft launched. At this point, American astronauts had already brought back soil and rock samples from both the Apollo 11 and Apollo 12 missions, but this automated sample return was still a technological coup for the Soviets. Another automated sample return mission in 1970, *Luna 20,* successfully returned a sample of the ancient lunar highlands. A final automated sample return mission in 1976 made *Luna 24* the last spacecraft to make a soft landing on the Moon.

The Luna soil samples were the first fully robotic sample return missions, a record that stood for decades. They also greatly increased the diversity of lunar sites from which samples were collected. Scientists back on Earth used the samples to learn about the geologic history of the Moon and study the differences in composition between the old lunar highlands and the younger volcanic plains.

### Photographing the Moon on the Zond missions

To prepare for a planned Moon mission by Soviet cosmonauts, the Soviet Union also launched a series of Zond missions in the late 1960s and 1970. These robotic spacecraft were meant to test future hardware to be used by cosmonauts, but a series of four Zond missions also took some spectacular images of both the Moon and the Earth from orbit.

- *Zond 5,* launched in 1968, went around the Moon and then landed safely on Earth. It took images of the Earth but failed to capture pictures of the Moon.
- Later in 1968, *Zond 6* carried an automatic film camera and took more than 150 pictures. Unfortunately, the film canister broke open when the capsule crash-landed on its return to Earth. A few pictures were recovered.
- *Zond 7,* launched in 1969, successfully took images of the Earth and Moon on both color and black-and-white film; these images were returned to Earth without incident.
- In 1970, *Zond 8* took very high-resolution images on very large film. Believe it or not, these photos rival today's digital images in quality.

## Lunokhod 1 and 2 aboard Luna 17 and 21: Driving robots (1970 and 1973)

NASA's recent Mars Pathfinder and Mars Exploration Rovers (which we highlight in Chapters 15 and 20), although infinitely impressive, weren't the first robotic rovers to have conducted science on another planet. The Soviet Lunokhod rover used cameras and other equipment to take photographs and test the soil on the Moon. Solar panels unfolded to provide enough electricity to run the rover and its instruments. *Lunokhod 1* (launched aboard *Luna 17* in 1970) traveled the Moon for about 11 months. Its follow-up, *Lunokhod 2* (launched aboard *Luna 21* in 1973) only survived about four months following landing, but it drove more than 23 miles (37 kilometers) on the lunar surface. For comparison, as of early 2009, the Mars Exploration Rover *Opportunity* has traveled more than 8 miles (13 kilometers) over the five years since it landed.

The two Lunokhod rovers were driven by Earth-bound controllers who used real-time television camera images onboard the spacecraft to operate them remotely. This technique works on the Moon, which is close enough to the Earth that signals take just about a second to travel from one body to the other. However, rovers on Mars must be completely autonomous, because the time delay there varies between 3 and 22 minutes, depending on the positions of the two planets in their orbits.

# The Ranger Program: Hard Lunar Landings (1961–1965)

In the true spirit of the Space Race, the Soviet Union's Luna missions compelled the United States to simultaneously develop its own series of robotic lunar missions that could photograph the Moon's surface. The spacecraft of the *Ranger Program* were intended to impact the Moon, taking pictures along the way with their six cameras (each one was intended to supply a different photograph through its unique lenses and settings).

The first six Ranger spacecraft in the program (launched between August 1961 and January 1964) all failed to return useful pictures for various reasons, such as the spacecraft's inability to escape Earth's orbit, the ship's loss of contact with Earth, or camera failures. One spacecraft, *Ranger 3,* even missed the Moon completely!

The loss of six spacecraft in a row had to be discouraging for the Ranger team and the newly formed NASA. The failures were probably the result of a combination of different factors, including the required sterilization of the spacecraft before launch for planetary protection reasons (see Chapter 11 for more on this topic). One particular issue was discovered just before *Ranger 6* was to launch: Tiny gold-plated diodes, which were used in many of the Ranger spacecraft's systems, turned out to produce gold flakes that could peel off and float around in zero gravity, producing unwanted connections that caused the diodes to short-circuit. The diodes were replaced in subsequent Ranger missions, including Ranger 6, which was successful except for a failed camera switch due to a lightning strike during launch.

Fortunately for the U.S. space program, the next few Ranger missions were much more successful. *Ranger 7* (also launched in 1964) made it into orbit as planned and was able to return more than 4,000 photos before crash-landing on the Moon's surface. *Ranger 8* (launched in 1965) sent home more than 7,000 photos. Last in the Ranger series, *Ranger 9* launched on March 21, 1965, and completed its chief mission of impacting the Moon and sending 5,800 high-resolution photographs back home.

# The Lunar Orbiter Project: Prepping the Way for Human Spaceflight (1966–1967)

Although manned spaceflight was always the ultimate goal of NASA's Moon program, everyone understood that it was necessary to first know a lot more about the Moon's surface. Toward this end, NASA promoted the *Lunar*

*Orbiter Project,* a series of five missions launched between 1966 and 1967. The project's goal was to use high-resolution photos and other data to create a map of the Moon's surface, paying particular attention to the areas that NASA considered prime candidates for a future astronaut mission (see Figure 8-2).

Five robotic missions occurred as part of the Lunar Orbiter series:

- *Lunar Orbiter 1* (launched August 10, 1966)
- *Lunar Orbiter 2* (launched November 6, 1966)
- *Lunar Orbiter 3* (launched February 5, 1967)
- *Lunar Orbiter 4* (launched May 4, 1967)
- *Lunar Orbiter 5* (launched August 1, 1967)

In the following sections, we describe how the missions not only helped scientists map the Moon but also made a surprise achievement: capturing the first image of Earth in its entirety.

**Figure 8-2:** The Moon's many regions were mapped with the aid of NASA's Lunar Orbiter missions, which helped select future landing sites.

# Mapping the Moon from a distance

Unlike some other space mission programs, all of NASA's Lunar Orbiter missions were a startling success. They managed to map out about 99 percent of the Moon's surface, with a minimum resolution of 197 feet (60 meters) per pixel; at its best, the resolution was around 3 feet (1 meter) per pixel. This achievement was a truly amazing feat because scientists could visualize the entire surface of the Moon for the very first time.

The Lunar Orbiter spacecraft measured about 5.4 feet (1.65 meters) in length by 4.9 feet (1.5 meters) in diameter and had a three-part, mechanical-looking design. The base portion contained the battery and other necessary equipment, including the photography gear. The middle module contained propellant, tanks, and sensors; and the upper deck provided the heat shield. Solar panels extended from the base to power the equipment and charge the spacecraft's battery.

The imaging system aboard the Lunar Orbiter spacecraft was very advanced. It included a dual-lens camera for both high-resolution and wider, lower-resolution photos, along with a module specifically for processing the used film. Other instruments included a readout scanner and a film-handling apparatus. Because the spacecraft itself was in motion, the engineers designed the film to move accordingly so that the resulting photos wouldn't appear blurry. After the images were exposed, the film was developed, scanned, and converted into signals that the spacecraft's antennae transmitted back home to Earth.

## Seeing a whole Earth

Thanks to a slight approved detour by *Lunar Orbiter 1,* the Lunar Orbiter Project gave the world its first picture of the Earth in its entirety. You can check out the inspirational image in Figure 8-3.

This historic photo, snapped on August 23, 1966, was taken about 236,000 miles (380,000 kilometers) from the Earth. Perhaps for the first time, it hit home that Planet Earth was but one cog in the universe. It also became clear that, at least from a distance, the Earth was one united entity.

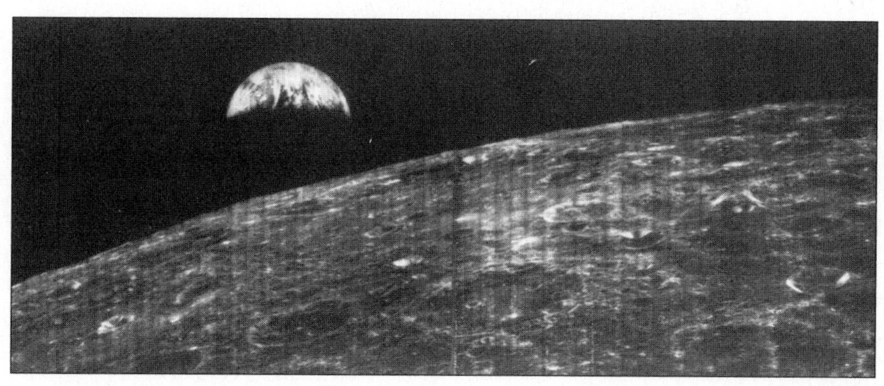

**Figure 8-3:** The Earth was first photographed as a whole, discrete object from *Lunar Orbiter 1.*

Courtesy of NASA

### Presenting the master robotic photographer of the 20th century

The photographs taken by the Lunar Orbiters in 1966 and 1967 were made long before the advent of digital photography. Unlikely as it may seem today, these photos were exposed onto 70-millimeter film! The Lunar Orbiter's imaging system, which scanned film, was a compromise between the low-resolution television camera systems on spacecraft such as the United States' Ranger and designs that returned physical film to Earth in a capsule (such as the Soviet Union's Zond spacecraft).

The volume of information returned by the various Lunar Orbiters, as archived in high-resolution photographs, was actually far greater than that returned by many modern digital spacecraft. The U.S. Geological Survey is currently engaged in a long-term project to digitize all the Lunar Orbiter data to make it available to modern image-processing techniques.

## The Lunar Surveyor Program: Landing on the Moon (1966–1968)

In the mid- to late 1960s, both competitors in the Space Race kept the ultimate goal of sending astronauts to the Moon in sight. The United States' next effort after the Lunar Orbiter Project was the *Lunar Surveyor Program,* a NASA project that ran between 1966 and 1968 and landed the first American spacecraft on the Moon. Seven robotic spacecraft were sent to the Moon during that time, primarily to test techniques and scout out landing spots for the Apollo missions (which we describe in Chapters 9 and 10):

- *Surveyor 1* (launched May 30, 1966; see an image of this launch in the color section)
- *Surveyor 2* (launched September 20, 1966)
- *Surveyor 3* (launched April 17, 1967)
- *Surveyor 4* (launched July 14, 1967)
- *Surveyor 5* (launched September 3, 1967)
- *Surveyor 6* (launched November 7, 1967)
- *Surveyor 7* (launched January 7, 1968)

The Lunar Surveyor Program originally included orbiters and landers, but the number and types of spacecraft used were scaled back due to cost and time constraints. After all, the Space Race really *was* a race. Both the Soviet Union and the U.S. wanted to be the first to make major achievements in space. Thus, time was of the essence in ensuring their respective places in history.

The mandate to focus the Lunar Surveyor Program on lunar landings was sanctioned at the highest level of the U.S. government by President John F. Kennedy, which helped focus the program's goals but also limited the amount of science the missions could accomplish — for good reason. By doing a faster series of test and data-gathering missions, more of the actual science could be done by the people who were slated to land on the Moon during Project Apollo.

Like the many other programs that preceded it, the Lunar Surveyor Program laid the groundwork for the human missions that were yet to come. Without this important background work, mission planners wouldn't have been able to target landing sites, understand and prepare for the properties of the lunar surface, optimize the instruments and techniques to be used, and safely send humans to the Moon.

## Testing soft-landing techniques

The main goal of the Lunar Surveyor missions was to practice and perfect a *soft landing* (where landing speed is controlled by retrorockets that allow the spacecraft to land gently) on the lunar surface. This goal was in direct preparation for Project Apollo, which would place the first Americans on the Moon's surface just a few years later. Even though these spacecraft were capable of landing without crashing, they didn't have the capability to launch themselves all the way back home.

Although most of the Lunar Surveyor flights were successful, some glitches did occur. *Surveyor 4,* for example, made a perfect flight to the Moon but lost contact with Earth during its final descent. *Surveyor 2* also had a good flight until it crash-landed on the Moon due to engine failure. The other five flights landed successfully.

## Capturing the details of the Moon's surface

The Lunar Surveyors were equipped with television cameras for photographing details of the lunar surface. These cameras could pan around to take multidirectional photos and photograph the lunar surface from different heights. Each spacecraft also had equipment to measure the surface temperature of the Moon and test the radar reflectivity and strength of its surface. In fact, more than 100 sensors were stashed aboard *Surveyor 1.*

A scoop added late in the project provided the capability for later Lunar Surveyors to dig a trench in the Moon's surface, which the cameras could then photograph so that scientists back home could study the lunar soil's composition. Without these critical bits of info, engineers wouldn't have known which parts of the Moon's surface would be capable of supporting the landing of an Apollo spacecraft.

One of the most-significant discoveries of the Lunar Surveyor missions was that the Moon's surface was firmer than expected, a fact that boded well for a future Apollo landing because previously scientists had worried that so-called "Moon dust" would prevent an easy landing.

## Making the Apollo 12 connection

A special connection existed between the Lunar Surveyor Program and Project Apollo: The Apollo 12 mission was intentionally targeted to land near the *Surveyor 3* spacecraft. Astronauts studied *Surveyor 3* to view the effects of more than two and a half years on the lunar surface. They detached the spacecraft's camera and a few other components to bring back to Earth for further analysis (see Figure 8-4).

The astronauts found that the components of Surveyor 3 were in surprisingly good shape after their stint in the harsh lunar environment. Perhaps the most widely reported result was the claim that a terrestrial bacterium, *Streptococcus mitis,* was found inside the *Surveyor 3* camera and thus must have survived exposure to harsh lunar conditions. If true, that would make the case for stringent planetary protection constraints to prevent contamination of possible nonterrestrial ecosystems. More-recent analysis, however, has suggested that the camera was probably contaminated with the bacterium sometime after its retrieval.

**Figure 8-4:** An Apollo 12 astronaut with the *Surveyor 3* spacecraft.

*Courtesy of NASA/Alan L. Bean*

# Chapter 9

# Man on the Moon: Project Apollo

### In This Chapter

- Examining the politics of exploring the Moon
- Checking out the components of a typical Apollo mission
- Conducting key tests with robotic and manned missions
- Walking on the Moon

Landing on the Moon was a far-away dream for engineers, NASA scientists, politicians, and just about everyone else in the world during the early 1960s. Thanks to the breakthroughs and achievements of the United States' Project Apollo, those dreams became reality just a few short years later — a reality witnessed by anyone with access to a television on July 20, 1969. The landing of astronauts on the Moon, as part of the Apollo 11 mission, is arguably one of the greatest human achievements ever. If you want to discover how this mission was possible, and find out about the many initial missions required to achieve it, then this chapter is for you.

## In the Beginning: The Politics of Manned Lunar Exploration

The major impetus of Project Apollo was political. In order to place America squarely in the lead during the Space Race of the 1960s, the highest levels of American government had to get involved, and President John F. Kennedy rose to the challenge. The Apollo program's stated goals — to meet the "national interests" of the United States in space as well as to establish its dominance beyond Earth — echoed this winner-take-all sentiment. Of equal importance, and the real means of achieving those first two goals, was executing significant scientific work on the Moon and developing the ability for people to work and function there.

In the following sections, we describe Kennedy's desires for the space program and the work NASA did immediately following the announcement of Kennedy's goal.

## President John F. Kennedy's goal: Winning the race to the Moon

Thanks to the success of the Soviet Union's *Sputnik 1* satellite (described in Chapter 6) and Vostok 1 mission (see Chapter 7), as well as the United States' Project Mercury (see Chapter 7), which had shown the world that both robotic and manned missions could orbit the Earth, the Space Race was becoming an all-out sprint. The Americans were clearly in second place in the early 1960s because the Soviets had more experience with space travel and possessed larger rockets. However, no country had yet won the ultimate prize of placing a human being on the Moon, and Dwight D. Eisenhower's administration in the 1950s hadn't put a heavy emphasis on this particular goal.

All that changed with the election of President John F. Kennedy in 1960. Although initially not a huge supporter of the space program, Kennedy seized the political and technological moment to initiate a program that would bring the nation together. In May 1961, in response to Yuri Gagarin's successful space flight the previous month (and perhaps as an antidote to the recent Bay of Pigs debacle), Kennedy committed intellectual and financial resources to putting an American on the Moon before the end of the decade. At this point in time, the U.S. had only launched one astronaut, Alan Shepard, on a suborbital flight just a few weeks earlier! (See Chapter 7 for more details on Gagarin's and Shepard's flights.)

In an instantly famous speech made at Texas's Rice University on September 12, 1962, Kennedy both summarized his position on how he saw America fitting into the race to the Moon and justified the expenses and sacrifices involved:

> "We choose to go to the Moon. We choose to go to the Moon in this decade and do the other things, not only because they are easy, but because they are hard, because that goal will serve to organize and measure the best of our energies and skills, because that challenge is one that we are willing to accept, one we are unwilling to postpone, and one which we intend to win, and the others, too."

To hear an audio file of Kennedy's entire speech (and to find more info on Kennedy's goals for the American space program), check out `www.jfklibrary.org`. Hold your mouse over the Historical Resources box and click on the JFK in History link. From there, simply click on the Space Program link.

Kennedy became one of the nation's biggest supporters of the space program, and his job was, as he clearly recognized, not easy. A significant portion of the U.S. budget was devoted to Project Apollo, and Kennedy had to make the case for why such a vast expenditure was needed. In his 1962 address at Rice University, Kennedy reminded the audience, and the nation, of the country's position behind the Soviet Union in the Space Race. He painted a picture of American ingenuity and hard work and laid out a bold goal of sending astronauts to the Moon within just eight years.

## NASA's first task from the president: Coming up with an efficient flight design

The scientists and engineers at NASA had to figure out how to bring President Kennedy's dreams to life. They started by designing a series of flights that would fulfill the president's goal of landing a human on the Moon in such a short time period, while at the same time protecting their investment in both the astronauts and equipment. Several different strategic approaches for Project Apollo were introduced, but ultimately NASA officials determined that a *lunar orbit rendezvous,* or LOR, was the most-efficient way to land a person on the Moon quickly.

The basic concept behind a LOR involved modules, as you can see from this description of the mission's three-part design (we include a diagram showing a typical mission to the Moon in Chapter 3):

1. A launch rocket would first hurl the spacecraft into space. After the spacecraft reached low Earth orbit, the rocket would reignite the third stage, sending the Lunar and Command/Service Modules on a trajectory that would reach the Moon a few days later.
2. Upon entering lunar orbit, two of the three crew members would enter the Lunar Module and detach it from the rest of the spacecraft before piloting it down to the Moon's surface.
3. The remaining crew member would stay with the Command/Service Module in lunar orbit and wait for the lunar explorers to return by way of the Lunar Module's ascent engine. After the Lunar Module arrived back safely, it would dock with the Command/Service Module. At this point, the crew members would jettison the Lunar Module and head back to Earth.

This ingenious mission design minimized the amount of rocket propellant required by only bringing along the parts of the spacecraft that were essential for each task.

The simplest mission concept to get to the Moon and back was the *direct ascent,* whereby a single spacecraft would be sent to the Moon and returned home in one piece. Although this method was initially the favored one, the size and weight of the required rocket boosters for both launches was ultimately prohibitive. Competing designs for a trip to the Moon included a lunar surface rendezvous, where two spacecraft (one manned, one unmanned) would launch into space, and the unmanned spacecraft would supply the manned one with additional fuel to return home. Another plan, the Earth orbit rendezvous, would've launched two rockets, one with a lunar capsule and one with additional fuel and power for propulsion; the two halves would've docked in orbit before continuing on to the Moon. These latter two suggestions involved sending more weight into space than the LOR, in addition to requiring some potentially risky docking maneuvers.

# The Components of an Apollo Mission

A typical Apollo mission had three major components, as shown in Figure 9-1:

- The Saturn V rocket
- The Command/Service Module
- The Lunar Module

This three-part division was one of the selling points of the lunar orbit rendezvous (LOR) method that we describe in the preceding section. It offered advantages in both weight and development time over other mission modes. The next few sections break down the details of the three Apollo components.

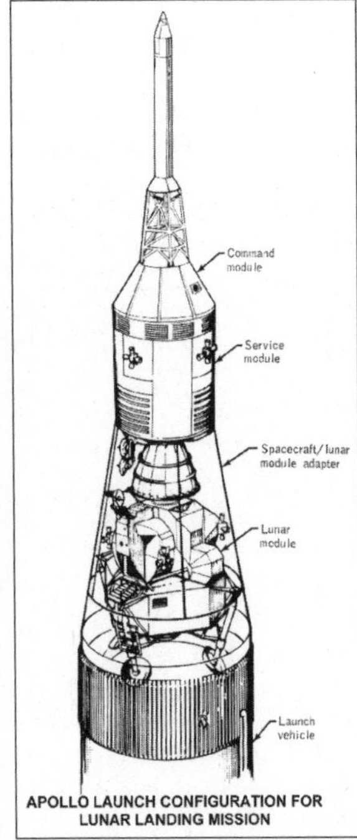


**Figure 9-1:** For the Apollo Moon landings, the Command/Service Module and the Lunar Module were stacked on top of the Saturn V launch vehicle.

Courtesy of NASA

## The Saturn V rocket: Blasting off toward the Moon

Propelling a three-astronaut lunar spacecraft such as the Apollo into space required a powerful rocket. Fortunately, the Saturn V rocket was up to the task. This launch vehicle was a three-stage rocket in which each stage contained a different number of engines, had a different burn time, and supplied different amounts of *thrust* (the force necessary for an object to take flight, defy gravity, and move through space). The huge Saturn V rocket is still the largest rocket ever used successfully, both in terms of size and in how much *payload* (cargo) it could get into orbit.

The Stage 1 module of a three-stage rocket like the Saturn V is the most powerful and is required to lift the rocket and other modules off the ground and into orbit. When its propellant has been used up, it disengages from the rest of the rocket and burns up in the atmosphere. Stage 2 provides the second burn that raises the rocket higher; it also disconnects and burns up when its job is done. Stage 3 ends up in the same lunar transfer orbit as the Command/Service Module (CSM). After the Lunar Module is removed from the third stage, the CSM heads for the Moon, and the third stage moves on to a solar orbit. (See Chapter 3 for more on how rockets work.)

## The Command/Service Module: Sustaining the mission

The Command/Service Module (CSM) of the Apollo spacecraft was a two-part entity that served to transport the astronauts, along with the Lunar Module, to the point where they could make their final lunar descent.

The CSM consisted of a Command Module plus a Service Module. The Command Module was designed to hold the crew plus their equipment; the Service Module contained the power supply, fuel, and storage. Because it housed the crew, the Command Module was intended to return home, whereas the Service Module was designed to be jettisoned and burned up in the atmosphere upon reentry. This design minimized the size of the capsule that had to be able to return safely to Earth, reducing mass and costs.

Smaller than you may've expected (see Figure 9-2), the Command Module measured just 127 inches (3.2 meters) by about 154 inches (3.9 meters). The Service Module was quite a bit larger — 295 inches (7.5 meters) long by 154 inches (3.9 meters) wide at the base — because it had to hold enough power and propellant for both mission activities and the flight home.

**Figure 9-2:** The Apollo Command Module with astronauts during a trial run.

Courtesy of NASA

The Command Module had a heat shield that was designed to partially burn off *(ablate)* as the capsule reentered the Earth's atmosphere. Though temperatures at the surface of the capsule reached 5,000 degrees Fahrenheit (2,750 degrees Celsius), the crew was kept safe inside the fireball. The atmosphere slowed down the capsule, which then deployed three drogue parachutes to further reduce speeds for a safe ocean landing (otherwise known as a *splashdown*).

## The Lunar Module: Designed to land

Perhaps the most-memorable component of the Apollo spacecraft, at least in the minds of everyone who saw it on television, the Lunar Module (also called the LEM, short for *lunar excursion module*) had just two purposes: land on the Moon and take the astronauts back to the Command/Service Module when their work was done. It was never intended to make it back to Earth by itself, and it could hold only two of the three Apollo crew members.

Designed in two stages (one for the ascent engine and one for the descent engine), the Lunar Module was a four-legged contraption that was actually one of the most-successful components of the entire Apollo program. Why? Because it was the only one to never fail in a way that compromised a mission.

The ascent portion of the Lunar Module contained the crew-related stuff: cabin, life-support, instruments, hatches, and docking equipment. The descent portion housed the landing gear, a ladder from which the astronauts could conduct their moonwalks, and cargo areas for batteries and fuel stores. Both parts had engines, antennae, radar equipment, and fuel. The descent stage was left behind on the Moon, whereas the ascent stage returned the crew to the orbiting Command/Service Module. (Check out an image of the Lunar Module in Figure 9-3.)

The first Apollo flight to carry a Lunar Module was the Apollo 9 mission, and the first successful lunar landing occurred during the Apollo 11 mission (we cover both missions in detail later in this chapter). Although the Lunar Module may look spindly and unstable, remember it was designed to operate solely in the low-gravity environment of the Moon — it just didn't need the reinforcements required for a similar vehicle on Earth, where the force of gravity is six times higher.

**Figure 9-3:** The Apollo 9 Lunar Module ascent stage on a trial run in Earth orbit, as photographed from the Command Module.

*Courtesy of NASA/David Scott*

# Laying the Groundwork for the Manned Apollo Missions

Making history is rarely a small affair. Project Apollo required significant resources, investments, and costs in order to reach its goals. Fortunately, all that time and money paid off: NASA made huge advances in the areas of computers and robotics as engineers worked to develop the first flight computer to aid in navigation and implemented complicated automated flights to test various components of the Apollo spacecraft and Saturn V rocket without astronauts onboard.

NASA's budget has expanded and shrunk over time, due in part to changes in emphasis on the space program. In the mid-1960s, the NASA budget rose to an all-time high, in no small part thanks to Project Apollo. Economists estimate that between 2 percent and 5 percent of every tax dollar went toward the space program during these years, and many employees (34,000 belonging to NASA plus more than 350,000 contractors) had the Apollo program to thank for their livelihood. No one ever said that putting a man on the Moon would come cheap. In total, about $25 billion in 1969 dollars was spent on Project Apollo (that's about $135 billion in 2005 dollars!).

## Developing the first flight computer

One important innovation of the Apollo missions was the first flight computer, the Apollo Guidance Computer (AGC). Designed by the Massachusetts Institute of Technology in the early 1960s, the AGC allowed the astronauts to control the flight of their spacecraft and gather and interpret navigational data in real time. Although cutting-edge at the time, the AGC seems extremely basic today with its small display, numeric keyboard, and translation table that was necessary for interpreting the computer's numerical codes output. It was the first computer to use *integrated circuits* (an early form of the microchips found in almost all of today's electronic devices), and it had an extremely low amount of memory (just 36K of RAM).

The AGC led the way for the development of desktop calculators, and versions capable of performing advanced calculations became commercially available during the 1960s. By the mid-1970s, hand-held calculators with both scientific and programmable functions were available; one such calculator was actually tested aboard the Apollo-Soyuz mission in 1975 (see Chapter 12 for details on this mission).

## Launching robotic flights: Apollo 4, 5, and 6

Although the manned flights are the ones that everyone thinks about when discussing Project Apollo, many *robotic* (unmanned) test flights went into making those landings possible. These missions laid the foundation for those important human flights and tested techniques that the astronauts would need to rely on later.

## The missing Apollos?

What about Apollo 1, 2, and 3? No, these missions didn't disappear into the dark side of the Moon. NASA began testing its Apollo spacecraft and Saturn launch vehicles with a different numbering scheme. The first robotic flights took place as early as 1966, with the AS-201 (Apollo/Saturn 201) flight, which tested a precursor Saturn 1 launch vehicle. AS-202, later in 1966, was a reentry test of the Command/Service Module, again on a Saturn 1 launch vehicle. AS-203, also in 1966, tested the Saturn IVB rocket. AS-204, in 1967, was scheduled to be the first manned flight, but the crew tragically perished in a fire during a prelaunch test (as you find out in Chapter 5). This mission was renamed Apollo 1 afterward in honor of the three astronauts who lost their lives. Missions then continued with the robotic AS-501 mission, which was renamed Apollo 4.

The first flight tests involved full-scale mockups of the Command/Service and Lunar Modules. This approach allowed the real (and costly) capsules to be spared and the Saturn V rocket to be tested properly.

Here's what the three robotic Apollo flights accomplished:

- The Apollo 4 flight was the first test of both the Saturn V rocket and the launch site (Launch Complex 39 at Florida's Kennedy Space Center) built for it. It was designed to test the launch and reentry of the lunar spacecraft. The thrust of the rocket, around 7.5 million pounds, was enough to rock the NASA buildings that were close to four miles away! This unexpected effect led NASA engineers to design special techniques to dampen the shock waves, such as pumping ocean water onto the launch site.
- Apollo 5 provided an opportunity to test the Lunar Module. After reaching Earth orbit, the Lunar Module separated from the launch vehicle so that mission controllers could test both the descent and ascent propulsion systems under real space conditions. Despite a slight glitch that caused the descent propulsion system to shut down prematurely, the 11 hours of testing showed that all the systems on the Lunar Module worked successfully.
- Apollo 6, the last robotic mission of the series, was the "qualification flight" that certified the Saturn V rocket for active duty. It would've tested return entry scenarios for the Command/Service Module, but these tests were scrubbed due to engine problems. A range of other problems plagued this test flight, but fortunately these issues didn't bode poorly for the future Apollo missions.

## Fast facts about Apollo 7

Following are the basic facts about the Apollo 7 mission:

- **Launch date and site:** October 11, 1968; Cape Canaveral Air Force Station, Florida
- **Spacecraft name and mass at launch:** *Apollo CSM-101;* 32,350 lb (14,674 kg)
- **Launch vehicle:** Saturn 1B
- **Size of crew:** 3 astronauts
- **Number of Earth orbits:** 163
- **Mission duration:** 10 days, 20 hours, 9 minutes, 3 seconds

# Apollo 7-10: Preparing for the Moon with Manned Missions (1968–1969)

Readying the spacecraft, astronauts, and techniques for a Moon landing required significant preparatory work, which is where Apollo 7 through Apollo 10 came in. Each manned flight was designed to test certain elements of a Moon mission, all of which built on each other and ultimately led to the success of Apollo 11. The next few sections illustrate how these early manned Apollo missions made history in their own right.

## Apollo 7: Testing the Command/Service Module (1968)

The significance of the first manned Apollo mission, Apollo 7, was simply that it succeeded. The purpose was to test the functioning of the Command/Service Module, as well as the effectiveness of the newly designed flotation system for use during splashdown.

Because the Apollo 7 mission employed only the Command/Service Module (and not the Lunar Module), it launched with a smaller rocket, specifically the Saturn 1B. In addition to orbiting the Earth for a sustained period of time, this mission marked the first time a full three-member Apollo crew performed a mission in space. The crew members on this historic flight were Commander Walter Schirra, Command Module Pilot Donn Eisele, and Lunar Module Pilot Water Cunningham. Ironically, this same crew served as the backup for Apollo 1, a mission that proved deadly for all the astronauts aboard (see Chapter 5).

## Fast facts about Apollo 8

Following are the basic facts about the Apollo 8 mission:

- **Launch date and site:** December 21, 1968; Kennedy Space Center, Florida
- **Spacecraft name and mass at launch:** *Apollo CSM-103;* 63,565 lb (28,833 kg)
- **Launch vehicle:** Saturn V SA-503
- **Size of crew:** 3 astronauts
- **Number of lunar orbits:** 10
- **Mission duration:** 6 days, 3 hours, 42 seconds

Apollo 7 was definitely an endurance mission for the crew, which spent nearly 11 days in Earth orbit to make sure the spacecraft's systems would operate long enough to get to the Moon and back. Unfortunately, the crew members weren't all that comfortable during their trip — all three men developed colds near the beginning of the mission. Stuffy noses are bad enough on the ground, but in microgravity, there's nothing to help all that mucus drain out of your sinuses. The discomfort made the crew fairly grumpy, and they resisted requests from Mission Control to add events to their timeline (which ended up causing the crew members to be passed over for future Apollo flights). The astronauts took a lot of decongestant tablets and spent a great deal of time blowing their noses to clear things out. As landing finally approached, one worry was that they wouldn't be able to blow their noses while wearing pressure suits with helmets for reentry. If the sinus pressure grew too high, the astronauts risked rupturing their eardrums. Fortunately, decongestants helped them reach the ground safely!

# Apollo 8: Orbiting the Moon (1968)

Apollo 8 earned its place in history as the first U.S. mission to orbit the Moon. Its core purpose was to see whether manned lunar orbit was possible, but it was also responsible for testing out the Saturn V, an enormous rocket that would be the launch vehicle of choice for many future missions. Apollo 8 also tested out the functioning of the Command/Service Module and its life-support systems and gathered Moon data, including photographs and information about the lunar landscape that directly benefitted later Apollo missions.

The Apollo 8 astronauts (Commander Frank Borman, Command Module Pilot James Lovell, and Lunar Module Pilot William Anders) had the honor of being the first humans to view the whole Earth from space, as well as from the Moon's far side. As the spacecraft rounded the Moon for the fourth time, the astronauts suddenly saw Earth come into view over the barren lunar surface. The image they captured (which we include in the color section) came to be an icon of both the Apollo missions, proving just how far they'd come, and the budding environmental movement, which adopted a view of Earth from space as a clear demonstration of how beautiful and fragile the planet appears. A live Christmas Eve television broadcast from Apollo 8 helped raise interest in the space program to even higher levels and paved the way for a successful Moon landing the following year.

## Apollo 9: Docking in Earth orbit (1969)

Apollo 9 went down in space history as the first successful trial of a lunar orbit rendezvous, NASA's chosen flight design for missions to the Moon (which we describe earlier in this chapter). The astronauts completed a successful separation, rendezvous, and docking of the Lunar and Command/ Service Modules in orbit around Earth, indicating that the ships were nearly ready for prime time.

The Apollo 9 crew (Commander Jim McDivitt, Command Module Pilot Dave Scott, and Lunar Module Pilot Rusty Schweickart) also executed a *spacewalk* (an *extravehicular activity,* or EVA, performed in space) to test the effectiveness of the newly designed Apollo spacesuit. The suit carried its own life-support system so that astronauts exploring the Moon's surface wouldn't have to be tethered to the spacecraft.

### Fast facts about Apollo 9

Following are the basic facts about the Apollo 9 mission:

- **Launch date and site:** March 3, 1969; Kennedy Space Center, Florida
- **Command/Service Module name and mass at launch:** *Apollo CSM-104* (call sign "Gumdrop"); 59,086 lb (26,801 kg)
- **Lunar Module name and mass at launch:** *Apollo LM 3* (call sign "Spider"); 32,132 lb (14,575 kg)
- **Launch vehicle:** Saturn V SA-504
- **Size of crew:** 3 astronauts
- **Number of Earth orbits:** 152
- **Mission duration:** 10 days, 1 hour, 54 minutes

## Fast facts about Apollo 10

Following are the basic facts about the Apollo 10 mission:

- **Launch date and site:** May 18, 1969; Kennedy Space Center, Florida
- **Command/Service Module name and mass at launch:** *Apollo CSM-106* (call sign "Charlie Brown"); 63,560 lb (28,830 kg)
- **Lunar Module name and mass at launch:** *Apollo LM 4* (call sign "Snoopy"); 30,735 lb (13,941 kg)
- **Launch vehicle:** Saturn V SA-505
- **Size of crew:** 3 astronauts
- **Number of lunar orbits:** 31
- **Mission duration:** 8 days, 3 minutes, 23 seconds

# Apollo 10: Setting the stage for the first lunar landing (1969)

Any big production requires a dress rehearsal, and that's precisely what Apollo 10 was. NASA considered this mission a dry run for Apollo 11 and so sent the complete Apollo spacecraft (Command/Service Module plus Lunar Module) and crew into space to test the Lunar Module in the environment for which it was really intended — lunar orbit. Commander Thomas Stafford, Command Module Pilot John Young, and Lunar Module Pilot Eugene Cernan crewed this particular flight and did everything that would be done during Apollo 11, with the exception of landing on the Moon.

Once in lunar orbit, the astronauts used thrusters to separate the Command/Service and Lunar Modules. The Lunar Module was programmed to make low-orbit passes by the surface of the Moon, just as Apollo 11 would need to do; at its lowest altitude, the spacecraft was less than six miles away from the Moon's surface. Numerous photos and television images of the Moon's surface were taken from both modules. The astronauts ran a full test of the instrumentation and all equipment aboard the spacecraft, and they redocked successfully about eight hours after they first separated. Most of the tests went according to plan, and engineers felt confident that they were ready for the biggest test of all.

Although Apollo 10 accomplished its mission objectives, there were a few moments of peril when the Lunar Module went out of control while ascending from its closest approach to the Moon due to an incorrectly set switch. The spacecraft suffered huge gyrations, and the crew managed to get control back just two seconds before a crash landing would've been inevitable. The understandably coarse language used by the crew during this close call was amusingly referenced in a banner that greeted them upon their return; it read "The flight of Apollo 10 — for adult audiences only."

### Snoopy and Charlie Brown in space

Although astronauts were perfectly capable of making names for themselves, sometimes the actual spacecraft needed a little help. The Apollo 10 mission found that help in the form of Charles Shultz, cartoonist of the immensely popular *Peanuts* strip, which was published in newspapers worldwide from 1952 to 1999.

Shultz left an indelible footprint in the form of books and comic strips, but he also impacted space exploration via the Apollo 10 mission. The Apollo crew members were allowed to assign nicknames to their spacecraft, and they chose *Peanuts* comic names for this particular flight. The Command/Service Module was dubbed "Charlie Brown," and the Lunar Module earned the moniker "Snoopy." Schultz then went on to create special *Peanuts* artwork for the mission.

# Apollo 11: Doing the Moonwalk (1969)

Despite numerous setbacks and intense competition from the Soviet Union, the United States won the most coveted accolade in the Space Race, if not the history of humanity, on July 20, 1969. On that day, Project Apollo placed an American on the surface of the Moon. The details that went into accomplishing this feat are no less remarkable than the results themselves.

## Landing on the Moon

Apollo 11 was staffed by the typical crew of three astronauts. Neil Armstrong served as commander, Michael Collins piloted the Command Module (call sign "Columbia"), and Edwin "Buzz" Aldrin piloted the Lunar Module (call sign "Eagle"). The spacecraft was launched using a Saturn V rocket, and the launch proceeded as planned. After first entering Earth orbit and then being rocketed into a lunar trajectory, the Command/Service Module separated from the third rocket stage and docked with the Lunar Module; this step was necessary because the Lunar Module was sent into space in the Lunar Module Adapter, a special structure on one of the rocket stages that helped protect the fragile Lunar Module during launch.

When the spacecraft entered lunar orbit, it headed for the anticipated landing site on the Moon's Sea of Tranquility (see Chapter 8 for a diagram of the Moon's regions). The Lunar Module undocked from the Command/Service Module, leaving Collins alone in lunar orbit. Because large boulders covered the intended landing site, Armstrong and Aldrin had to manually keep the Lunar Module moving in order to find a smoother place to land. They eventually landed the spacecraft with little fuel left to spare. Soon afterward, Armstrong spoke the first words to be sent from the Moon:

"Houston, Tranquility Base here. The Eagle has landed." After their landing was formally acknowledged (and celebrated!) on Earth, the astronauts prepared for their *moonwalk* (which is simply an *extravehicular activity* [EVA] done on the Moon).

## "One giant leap for mankind"

After a quick meal on the Moon, Armstrong and Aldrin prepared to take their first steps on the lunar surface. Neil Armstrong exited the Lunar Module first, descended the nine steps of the exit ladder, took that first historic step on the Moon, and uttered the words America longed to hear: "That's one small step for man, one giant leap for mankind."

Technically, Armstrong should've included the article "a" to make his line "That's one small step for *a* man . . .," but in the excitement of landing on the Moon, grammar mistakes likely weren't the first thing on his mind!

These first steps occurred about six and a half hours after the Lunar Module landed on the Moon. Aldrin soon followed Armstrong, and the two astronauts spent the next several hours photographing, drilling into the lunar surface for samples, and taking careful notes on everything around them. They photographed the soil, their own footsteps, and the Lunar Module itself; the latter photographs eventually helped NASA engineers evaluate both the spacecraft and its landing orientation.

An external television camera broadcast the astronauts' historic steps via live TV. The images were somewhat degraded because they had to pass through a second television monitor before being transmitted to radio telescopes on Earth, but they were good enough for people around the world to feel like they were really witnessing history (see Figure 9-4).

**Figure 9-4:** Neil Armstrong at the Lunar Module.

Courtesy of NASA

## Fast facts about Apollo 11

Following are the basic facts about the Apollo 11 mission:

- **Launch date and site:** July 16, 1969; Kennedy Space Center, Florida
- **Command/Service Module name and mass at launch:** *Apollo CSM-107* (call sign "Columbia"); 66,844 lb (30,320 kg)
- **Lunar Module name and mass at launch:** *Apollo LM 5* (call sign "Eagle"); 36,262 lb (16,448 kg)
- **Launch vehicle:** Saturn V SA-506
- **Size of crew:** 3 astronauts
- **Number of lunar orbits:** 30
- **Lunar landing date:** July 20, 1969
- **Mission duration:** 8 days, 3 hours, 18 minutes, 35 seconds

## *Working and walking in Moon dust*

The Americans' lunar landing wasn't just about beating the Soviets; it was also about testing the effects of the Moon's low gravity on motion and harvesting soil and rock samples to bring back to Earth. The astronauts tested a variety of motions on the Moon. They walked and hopped to demonstrate how they were able to balance, particularly while wearing their bulky spacesuits, in the Moon's low gravity. Both Aldrin and Armstrong adopted a sort of long-strided glide as the most-efficient way to move, and both noted that they needed to anticipate their movements because the Moon's surface dust was more slippery than anticipated. (You can check out an image of Aldrin walking on the Moon in the color section.)

The lunar dust was also quite abundant. It flew up as Aldrin and Armstrong walked about and gathered 47 pounds of lunar soil and rocks through collection tubes and core samples. The tubes had to be hammered into the Moon's surface, which was no small feat while wearing a bulky spacesuit and gloves. Other samples were taken using long-handled shovels.

The sampling and other work took longer than expected, which meant the astronauts wound up doing less sampling than originally planned. Mission Control back on Earth carefully monitored the astronauts' core temperatures and other physiological factors and let the men know when they were working too hard so they could reserve enough energy to make it back into the Lunar Module for the return journey.

## Planting the flag and heading home

In the ultimate display of American pride, Armstrong and Aldrin planted a United States flag into the lunar surface. Because there's no wind on the Moon to blow out the flag, the astronauts mounted it to a small rod that would (in theory) help keep the flag extended on the pole. In reality, the rod failed to extend all the way, giving the flag a slightly rumpled look that lends the appearance of fluttering (a look that was intentionally emulated on future Moon landings; see Figure 9-5). The astronauts also spoke via radio to President Richard Nixon during what was perhaps the most memorable (and probably expensive) telephone call to have been made from the White House. They also left behind a plaque, drawings of Earth, an olive branch, and other terrestrial memorabilia.

When it was time to head back, Aldrin reentered the Lunar Module first, followed by Armstrong. Along with them came camera film and boxes full of rock and soil samples. The astronauts disconnected themselves from their spacesuit life-support systems and hooked into the Lunar Module's life-support system instead. In an effort to make the Lunar Module lighter for the return trip, they jettisoned their life-support backpacks, parts of their spacesuits, and anything else that could be spared. The Lunar Module was repressurized for the return trip while the astronauts slept. After a quick repair to a circuit breaker, the ascent engine was fired, and the spacecraft returned to lunar orbit, leaving the descent stage behind on the lunar surface. Its rendezvous with the Command/Service Module was a success, and the ascent portion of the Lunar Module was jettisoned as planned.

**Figure 9-5:** Planting the first flag on the Moon.

*Courtesy of NASA*

The astronauts completed their mission, splashing down into the Pacific Ocean within five miles of the *USS Hornet,* their homeward-bound recovery ship. They were all quarantined for a period of time but were released within a few weeks after scientists realized they hadn't contracted any special lunar diseases. From then on the newly minted worldwide celebrities were free to engage in parades, publicity tours, and other celebrations of their achievement.

# Chapter 10

# Continued Lunar Exploration after Apollo 11

### *In This Chapter*

- Landing on the Moon again with Apollo 12
- Picking up the research and sample collection slack with Apollo 14
- Expanding American lunar exploration with Apollo 15, 16, and 17
- Looking at Soviet technology for manned Moon missions

With the incredible success of Apollo 11 (covered in Chapter 9), the U.S. space program had the support it needed to continue with the rest of the intended Apollo missions. Although putting the first man on the Moon in 1969 was an unparalleled achievement, other surprises were yet to come from the astronauts and their missions. After the frenzied pace leading up to Apollo 11, NASA was able to step back and space out the missions more, both to save money and to allow time for innovations and lessons learned from previous missions. As a result, the missions following Apollo 11 were rich in scientific exploration, as you discover in this chapter.

The news post–Apollo 11 wasn't all good, however. Project Apollo represented a huge investment of American resources, both in terms of funding and brainpower. Although the country certainly reaped tangible benefits in the form of spinoff technology and the creation of new technical industries, much of the driving force supporting this vast expenditure evaporated after the political goal of landing on the Moon first was accomplished. Science, always in second place to political reality, was allowed a few experiments in between flag plantings and live television broadcasts, but the political triumph of the Moon landings was always clear. Because the goal was "land on the Moon" rather than "do science on the Moon," after the initial Moon landings were accomplished, there was little reason to go back. Support for the Apollo program dried up, and the planned missions after Apollo 17 were canceled. Still, as NASA has discovered with the Space Shuttle program, there's only so much you can do in Earth orbit. NASA once again plans to return astronauts to the Moon for both political and scientific reasons.

# Apollo 12: Achieving A Second Successful Lunar Landing (1969)

Apollo 11 was a hard act to follow — after all, it put the first men on the Moon — but the Apollo 12 mission proved it was up to the task. Not only did the Americans put another pair of men on the Moon, but they did so just four months after their first successful attempt. This time the astronauts were to explore a different part of the Moon: the Oceanus Procellarum, a vast lunar plain that borders the near side of the Moon and was found to be covered with ancient lava flows. (Apollo 11 landed several miles west of the Sea of Tranquility, an area with varied topography that results from the close quarters of several other basins.)

Apollo 12 was led by Commander Charles Conrad. Richard Gordon piloted the Command/Service Module, and Alan Bean piloted the Lunar Module. Like its predecessor, the mission launched from a Saturn V rocket and followed the standard lunar orbit rendezvous flight design (described in Chapter 9) that called for the Lunar Module to detach from the Command/Service Module upon entering the Moon's atmosphere.

The Apollo 11 Lunar Module missed its mark due to geologic conditions, but Apollo 12 landed right on target, only about 600 feet away from the goal (which was, not coincidentally, near the site of the Surveyor 3 mission that we describe in Chapter 8). Because of this accurate landing, Conrad and Bean were able to take extensive photographs of the landing area. These images proved useful in planning the landing areas of later missions in high terrain areas.

Based on the experience of the first lunar astronauts, the Apollo 12 astronauts were better able to budget their time to tackle all of their assignments before the clock ran out. Two *moonwalks* (officially known as *extravehicular activities,* or EVAs, done on the Moon's surface) were conducted as part of the lunar mission:

- The first moonwalk, broadcast in real time from a television camera mounted on the Lunar Module, involved the astronauts collecting soil samples and setting up an experiment to test the composition of noble gases in the solar wind.
- The second moonwalk undertook a *geologic traverse.* In non–NASA speak, that's a long (4,300-foot [1,310-meter] to be exact) moonwalk with a designated scientific purpose. During this trek, Conrad and Bean collected several different Moon soil and rock samples that were brought to Earth for analysis.

## Fast facts about Apollo 12

Following are the basic facts about the Apollo 12 mission:

- **Launch date and site:** November 14, 1969; Kennedy Space Center, Florida
- **Command/Service Module name and mass at launch:** *Apollo CSM-108* (call sign "Yankee Clipper"); 63,577 lb (28,838 kg)
- **Lunar Module name and mass at launch:** *Apollo LM 6* (call sign "Intrepid"); 33,587 lb (15,235 kg)
- **Launch vehicle:** Saturn V SA-507
- **Size of crew:** 3 astronauts
- **Number of lunar orbits:** 45
- **Lunar landing date:** November 19, 1969
- **Mission duration:** 10 days, 4 hours, 36 minutes, 24 seconds

As part of its mission, Apollo 12 made a visit to the site of the *Surveyor 3* spacecraft. The Apollo 12 astronauts were able to bring several pieces of it back home, including the TV camera. The parts' main research value for scientists back on Earth was that they showed how metal and other elements were preserved after sitting on the Moon's surface for two and a half years. Initially, scientists also thought that a strain of terrestrial bacteria had survived the long exposure on the lunar surface, but more recently, experts think that the bacteria came from contamination back on Earth after the camera was retrieved. We include an image of the Apollo 12 astronauts visiting *Surveyor 3* in Chapter 8.

# Apollo 14: Making Up for Lost Research (1971)

Thanks to the failure of Apollo 13 to land on the Moon (as explained in Chapter 5), the next Apollo mission was under huge amounts of pressure to pick up the research and sample collection slack. With a few design changes, Apollo 14 fulfilled its new goals admirably and was able to take a wealth of photos, gather numerous samples, and conduct several interesting experiments.

Crewed by Commander Alan Shepard, Command Module Pilot Stuart Roosa, and Lunar Module Pilot Edgar Mitchell, the spacecraft launched from the now-standard Saturn V rocket on January 31, 1971. Despite a few noncritical glitches with the docking equipment, the Lunar Module landed in the Fra Mauro Formation, which had been the intended landing site for Apollo 13. Shepard and Mitchell descended to the Moon's surface and took numerous photographs of the landing site, the moonwalks, and all experiments.

## Fast facts about Apollo 14

Following are the basic facts about the Apollo 14 mission:

- **Launch date and site:** January 31, 1971; Kennedy Space Center, Florida
- **Command/Service Module name and mass at launch:** *Apollo CSM-110* (call sign "Kitty Hawk"); 64,463 lb (29,240 kg)
- **Lunar Module name and mass at launch:** *Apollo LM 8* (call sign "Antares"); 33,651 lb (15,264 kg)
- **Launch vehicle:** Saturn V SA-509
- **Size of crew:** 3 astronauts
- **Number of lunar orbits:** 34
- **Lunar landing date:** February 5, 1971
- **Mission duration:** 9 days, 1 minute, 58 seconds

Two moonwalks were included on the Apollo 14 agenda; together, they totaled more than nine hours of time working outside the Lunar Module. Shepard and Mitchell collected rock and soil samples from 13 different sites near the Lunar Module, and they performed a number of experiments, including meteoroid experiments, microgravity analysis, and the transfer of liquids in microgravity. The material at the Fra Mauro Formation was particularly interesting because it was created as a result of a series of large meteor impacts. The rocks had been broken down into smaller pieces with each impact, and in many cases, new rocks were actually formed from the heat generated by the impacts. The impacts excavated craters, which brought up rocks from deep beneath the surface; the astronauts collected samples from such boulders. The resulting information opened up an entirely new field of lunar geologic research.

One of several innovations aboard Apollo 14 was the MET, or *modular equipment transporter* (see Figure 10-1). This two-wheeled cart not only allowed the astronauts to transport equipment (such as tools and cameras) but it also let them carry back more rock samples than they could've transported manually. With the help of the MET, Shepard and Mitchell were able to bring back the largest amount of Moon rocks yet — about 93 pounds (42 kilograms) worth.

## Playing golf on the Moon

Although Apollo 14 landed on the Moon well before Tiger Woods became a fixture in the professional golf world, one of the mission's most-famous moments was a hole in one. Each astronaut participating in the Apollo 14 mission was allowed to bring along a few personal items. Stuart Roosa chose to bring tree seeds (later germinated and planted around the U.S. as Apollo Moon trees), whereas Alan Shepard brought two golf balls.

A full golf club wasn't sanctioned for the trip, but being an astronaut and an innovator, Shepard made do with one of his rod-shaped metal geologic tool handles and attached a golf club head to one end. During a moonwalk, he swung for two separate shots on the surface of the Moon. The balls traveled a good distance — about 590 and 820 feet (180 and 250 meters). Gravity on the Moon is considerably lower than that on Earth, and Shepard was put at a disadvantage by having to swing in his spacesuit. A short time afterward, Mitchell threw an instrument pole javelin-style, which traveled almost as far as Shepard's first golf shot.

**Figure 10-1** Apollo 14's Commander Alan Shepard with the modular equipment transporter, or MET.

*Courtesy of NASA/Edgar Mitchell*

# Apollo 15: Testing the Limits of Lunar Exploration (1971)

After the U.S. proved that landing and walking on the Moon was possible, it decided to raise the bar for future Apollo missions, beginning with Apollo 15. Because it was rapidly becoming clear that the immense cost of lunar exploration couldn't be sustained for long, it was imperative that Apollo 15 and later missions do as much science as they could. NASA officials pumped up the goals for Apollo 15 (as well as Apollo 16 and 17) to test the limits of lunar exploration as the world knew them. Earlier Apollo missions were limited in terms of the type of terrain they could land on, the amount of lunar samples they could bring back, and the length of time they could work on the Moon's surface. Apollo 15 was charged with breaking those limitations in part by traveling farther away from the lunar landing site and bringing more equipment for conducting experiments and gathering data.

Toward this end, changes to the equipment setup were made before Apollo 15; one of the most important was the inclusion of a new scientific instrument module. The Lunar Module was also redesigned to allow for greater *payloads* (cargo; in this case, primarily lunar rocks and soil samples) and to provide the astronauts with the tools they needed to survive for longer periods of time on the surface. Perhaps the best-known innovation of Apollo 15, though, was the Lunar Roving Vehicle (LRV), a four-wheeled vehicle that helped the astronauts collect samples on the Moon (see Figure 10-2).

Apollo 15, 16, and 17 were dubbed the *J missions,* the culmination of an alphabetic series of missions that began with unmanned tests of the various Apollo modules (A and B missions); manned tests in Earth orbit (C, D, and E); a manned test in lunar orbit (F); and a manned lunar landing (G, which was Apollo 11). H-series missions included two moonwalks (Apollo 12 and 14), and J missions included a three-day stay on the surface of the Moon and the use of the LRV.

The Apollo 15 mission was crewed by Commander David Scott, Command Module Pilot Alfred Worden, and Lunar Module Pilot James Irwin. These astronauts had to be trained differently than their predecessors due to the more-aggressive nature of their terrain hikes on the Moon; they spent more time hiking outdoors in their spacesuits, and of course had to undergo a series of training activities with the new LRV.

While Scott and Irwin went down to the Moon for their part of the mission, Worden remained in the Command/Service Module and conducted a series of experiments. He used special equipment to study the Moon's surface from afar, and he took highly detailed panoramic photos. His other work included using a gamma ray spectrometer, mass altimeter, and other devices to help create a more-sophisticated map of the Moon's surface.

**Figure 10-2:** Apollo 15 was the first mission to use a Lunar Roving Vehicle.

Courtesy of NASA/James B. Irwin

## Landing on the rugged terrain of the Hadley-Apennine region

Previous Apollo landings had the luxury of landing on *lunar maria,* the flat, dark, volcanic plains that cover much of the near side of the Moon. Apollo 15 took a more rugged approach: It landed in the Hadley-Apennine region of the Moon, a landing site chosen partially for its varied terrain. Hadley-Apennine lay along the Imbrium Basin, and scientists believed valuable information could be gleaned from any samples taken along the basin's rim. They expected that the astronauts would be able to obtain samples here that had come from deeper within the Moon than they had in the Fra Mauro Formation (the Apollo 14 landing site). In addition, they thought that landing in Hadley-Apennine would allow the astronauts to get an up-close view of the Hadley Rille, a region that was suspected to be the site of former volcanic activity.

The Hadley-Apennine site proved to be a fruitful, and beautiful, landing site, with mountains and valleys rather than the flat plains of earlier missions. The varied terrain allowed the crew to collect samples from a wide variety of geologic features, including what had been dubbed the "Holy Grail" of the remaining lunar science objectives: the collection of a piece of ancient lunar crust.

## Driving the first Lunar Roving Vehicle

Apollo 15 couldn't just return with some Moon rocks and dust. That was old news. It had to come back to Earth with more lunar souvenirs than any of its predecessors. Enter the Lunar Roving Vehicle, or LRV (nicknamed "Moon Buggy"). The *LRV* was an electric, four-wheeled vehicle designed specifically to run in the Moon's low-gravity environment. Here were some of its key characteristics:

- Rugged enough to handle being driven over a range of terrain types
- Enough storage space to be able to carry back the astronauts' anticipated payload and then some
- Able to be folded up during flight
- Equipped with two aluminum seats, one for the mission commander and one for the pilot of the Lunar Module
- Armed with cameras and antennae to provide the world with a live broadcast of the astronauts' travels and research

During the journey to the Moon, the LRV had been folded compactly and stored in the descent-stage portion of the Lunar Module (for more detail on this and other parts of the Apollo spacecraft, see Chapter 9). After landing, the astronauts busted out pulleys, tapes, and reels to lower the LRV down to the Moon's surface so they could unfold it and prepare it for use. Fortunately, the folding components were designed to lock as they were opened, so Scott and Irwin didn't have to attempt any finely detailed assembly work with their gloved hands.

### Fast facts about Apollo 15

Following are the basic facts about the Apollo 15 mission:

- **Launch date and site:** July 26, 1971; Kennedy Space Center, Florida
- **Command/Service Module name and mass at launch:** *Apollo CSM-112* (call sign "Endeavour"); 63,560 lb (30,370 kg)
- **Lunar Module name and mass at launch:** *Apollo LM 10* (call sign "Falcon"); 36,222 lb (16,430 kg)
- **Launch vehicle:** Saturn V SA-510
- **Size of crew:** 3 astronauts
- **Number of lunar orbits:** 74
- **Lunar landing date:** July 30, 1971
- **Mission duration:** 12 days, 7 hours, 11 minutes, 53 seconds

### Gravity: It's the law

In the days before Isaac Newton formulated the mathematical basis for gravity, Galileo Galilei performed a series of experiments into how gravity worked. His research showed that objects fall at a constant acceleration that's independent of the mass of the object. One story, probably a made-up one, says that Galileo tested this theory by dropping cannonballs of the same size but different weights from the Leaning Tower of Pisa and observing that they hit the ground at the same time. Other factors such as air resistance and friction come into play on the Earth, meaning that the surface area of a falling object can change the speed at which it falls. For example, a feather floats slowly to the ground, whereas a rock falls quickly.

Because the Moon has no atmosphere, there's no air to slow down falling objects, making it a perfect place to test Galileo's theory. During the Apollo 15 mission, Commander David Scott brought a falcon feather (in honor of the Lunar Module's "Falcon" call sign) and a hammer. He proceeded to drop them both at the same time while on live television. Sure enough, accounting for slight differences in Scott's right- and left-hand synchronization, the two objects struck the Moon's surface simultaneously. Physicists everywhere breathed a sigh of relief!

Changes to the spacesuit gave the astronauts increased flexibility, particularly around the hips and waist. Being able to bend over more fully, something that hadn't been possible with the previous Apollo spacesuits, allowed them to sit down and ride in the LRV. This was, of course, critical to their being able to use it!

The astronauts found that the LRV's four-wheel-drive allowed them to maneuver over the surface of the Moon without too much trouble. The ride was quite bumpy though, so they were very glad to have seat belts. With the help of the LRV, Scott and Irwin were able to bring home about 170 pounds (76 kilograms) worth of lunar rocks and soil, including the *Genesis Rock,* an almost unaltered 4-billion-year-old sample of the lunar crust — the very type of rock mission designers had been hoping to find! The LRV traveled a total of about 17 miles (28 kilometers), stopping 12 times along the way to collect rock and soil samples.

## Embracing new life-support systems for extended space work

The longer surface missions NASA wanted to run weren't possible without enhanced life-support systems, so the Apollo 15 astronauts got to test out the changes NASA engineers had made, such as the extra batteries and other life-support necessities added to the Lunar Module in order to allow the astronauts to make extended moonwalks. In addition to the life-support system on the Lunar Module, the astronauts' spacesuits had been modified with improvements to the Portable Life-Support System. These enhancements allowed them to stay outside the Lunar Module for additional time. Unlike other Apollo missions, Scott and Irwin were able to complete three moonwalks during Apollo 15; that added up to about 18 hours of EVA time on the Moon's surface.

Altogether, the improvements to the life-support systems allowed Scott and Irwin to spend about 67 hours on the Moon's surface — more than twice the amount of time previous Apollo astronauts were able to devote. In between excursions in the LRV, the astronauts returned to the Lunar Module to perform additional scientific experiments, which included using a lunar surface drill to gather samples and place underground probes, as well as deploying the now-standard experiment for testing solar wind composition.

Scott and Irwin performed other tests on lunar soil and planted an American flag at their landing site. Before leaving the Moon's surface, they made sure to place the LRV's camera in a position that would allow it to broadcast a video of the Lunar Module's ascent back into orbit. (The LRV was never intended to make the flight back to Earth.)

# Apollo 16: Breaking Records (1972)

Apollo 16 was like the bigger, better version of Apollo 15. It may have been the tenth manned Apollo mission and the fifth lunar-landing mission, but it was the first one to land in the lunar highlands. Second in the J-series of missions (defined by their ability to bring home larger payloads, run more experiments on the Moon, and use the Lunar Roving Vehicle [LRV] that we describe earlier in this chapter), Apollo 16 also trumped the payload and lunar surface time records of its immediate predecessor.

The crew of Apollo 16 consisted of Commander John Young, Command Module Pilot Ken Mattingly, and Lunar Module Pilot Charles Duke, Jr. Oddly enough, all three astronauts had a connection to Apollo 13: Young and Duke were part of the mission's backup crew, and Mattingly was actually slated to pilot the Command/Service Module until he was exposed to German measles and had to be removed from the flight rotation.

Apollo 16's Lunar Module landed on the edge of the Moon's Descartes Mountain range, marking the first time that an Apollo mission succeeded in landing in the ancient lunar highlands. Consequently, the astronauts and their trusty LRV were able to explore a brand-new type of geology and learn about the composition of the highlands. To maximize the geologic potential of the mission, the target landing site was set between two impact craters: South Ray and North Ray Craters. (In Figure 10-3, Duke is standing at the edge of Plum Crater at the landing site.)

**Figure 10-3:** The LRV allowed the Apollo 16 astronauts to visit a variety of geologic sites, including this crater.

*Courtesy of NASA/John W. Young*

## Fast facts about Apollo 16

Following are the basic facts about the Apollo 16 mission:

- **Launch date and site:** April 16, 1972; Kennedy Space Center, Florida
- **Command/Service Module name and mass at launch:** *Apollo CSM-113* (call sign "Casper"); 67,009 lb (30,395 kg)
- **Lunar Module name and mass at launch:** *Apollo LM 11* (call sign "Orion"); 36,255 lb (16,445 kg)
- **Launch vehicle:** Saturn V SA-511
- **Size of crew:** 3 astronauts
- **Number of lunar orbits:** 64
- **Lunar landing date:** April 21, 1972
- **Mission duration:** 11 days, 1 hour, 51 minutes, 5 seconds

Apollo 16 managed to break the lunar surface record set by Apollo 15; the astronauts spent a total of 71 hours on the Moon. They traveled via the LRV about 17 miles (27 kilometers), roughly the same distance the Apollo 15 crew traveled, but they brought home more samples — 212 pounds (96 kilograms) worth to be exact — and stayed on the Moon's surface about two hours longer. Two additional hours may not seem like a lot, but many incremental improvements in spacesuit and life-support technology were required to make 'em possible.

Young and Duke conducted three moonwalks during their 71 hours on the Moon's surface, getting in about 20 hours of EVA time. After deploying the LRV and establishing a television broadcast, the astronauts set off on their *geologic traverses* (a long moonwalk with a dedicated scientific purpose, such as gathering samples or taking pictures). They also did what was becoming routine for lunar astronauts: deploying experiments packages, setting up a cosmic ray detector, digging for soil samples, and planting an American flag (see the color section for an image of Young jumping to salute the flag). Some of their experiments involved

- Testing the properties of Moon soil
- Collecting solar wind samples through the solar wind composition experiment
- Measuring cosmic rays with a cosmic ray detector
- Taking ultraviolet pictures using the Far Ultraviolet Camera/Spectrograph, which allowed astronauts to take pictures of the stars at wavelengths that aren't available on Earth

Apollo 16's experiments led to an important discovery about the lunar highlands, ancient battered regions of the Moon that look bright when seen from Earth. (The lunar maria are the other main geologic feature on the Moon; they look smooth and dark from Earth.) Based on Duke and Young's discovery of *breccias,* mixed-up rocks formed by impact rather than volcanic activity, in the Descartes region, scientists determined that the highlands were predominantly the result of impact cratering rather than volcanic activity. Whereas the dark lunar maria are thought to be ancient lava flows, even those surfaces have been battered by impacting objects from space, such as pieces of asteroids and comets. Because the Moon lacks an atmosphere to protect it from small meteors and has no wind or water to erode its surface, the lunar surface bears almost a perfect record of the many bits of material that have hit it, from tiny micrometeorites that mix up the soil to larger bodies that form big craters.

# Apollo 17: Presenting Project Apollo's Grand Finale (1972)

The last manned Moon landing of the Apollo program was Apollo 17, a mission that was designed to maximize the number of lunar samples collected and experiments run. Why? Because mankind was making its last journey to the Moon for the remainder of the 20th century. All future Apollo missions had been canceled due to a combination of budgetary concerns and a need to allocate resources to other projects. The equipment originally slated for Apollo 18, 19, and 20 was used for other missions, put on display, or scrapped.

Unlike the previous Apollo flights, Apollo 17 launched at night (see Figure 10-4). A failure in the countdown sequencer caused a slight delay, but after a few hours, the flight launched as planned. The spacecraft itself was similar to the previous Apollo spacecraft, with the exception of several small changes intended to correct minor problems noticed on Apollo 16.

**Figure 10-4:** Apollo 17 was the only Apollo mission to launch at night.

*Courtesy of NASA*

As with the earlier Apollo flights, the Command/Service Module (CSM) and Lunar Module launched together with the aid of the Saturn V rocket and then separated once in orbit. Commander Eugene Cernan and Lunar Module Pilot Harrison Schmitt descended to the lunar surface, and Command Module Pilot Ronald Evans stayed aboard the CSM to conduct experiments from space.

## Traveling around varied terrain

The landing site for Apollo 17 was the subject of some debate. NASA scientists wanted to be sure they chose wisely because it wasn't like they were going to get another chance. Some sites of interest, such as the Tycho Crater or the Tsiolkovsky Crater, were rejected because they presented too dangerous a landing site and would've required communications satellites due to their location. Other promising sites, such as the Mare Crisium highlands, were turned down because they were more accessible to Soviet spacecraft than American ones.

The ultimate selection was on the rim of the Mare Serenitatis. This area, known as the Taurus-Littrow region, provided a range of terrain types from which the astronauts could take samples, including mountains, boulders, older highlands, younger volcanic regions, and impact craters. It certainly didn't hurt that previous Apollo astronauts had spotted odd craters with dark haloes surrounding them, which some scientists thought could be volcanic. One such crater, Shorty, was an important target for the Apollo 17 crew.

Like its predecessors, Apollo 15 and 16, Apollo 17 made use of a Lunar Roving Vehicle (LRV) to increase the astronauts' ability to travel across the Moon's surface and collect heavy samples. Cernan and Schmitt deployed and tested the LRV, making sure the attached television camera was working so the vehicle could broadcast their work live back on Earth.

Together they gathered more than 700 samples taken with two main goals in mind: Scientists wanted examples of ancient highlands rocks, as well as samples of newer volcanic material. Some of the samples the astronauts brought back included the following:

- The volcanic samples taken from the Taurus-Littrow Valley floor were shown to be primarily made of mare basalt, which is created from molten lava.
- Lunar highland rock samples (much older than the more-recent mare basalt) were found in the mountains on either side of the landing area, and they helped show that the meteorite impact that created the Serenitatis basin was close to 4 billion years old. These 4.2- to 4.5-billion-year-old rocks were brought closer to the surface as a result of the impact.
- One particularly interesting sample was orange, glassy soil found near the dark-haloed crater Shorty, which turned out to be from an ancient volcanic eruption that was stirred up by the impact.

Over the course of three moonwalks, which lasted a total of approximately 22 hours, Cernan and Schmitt collected more than 243 pounds (110 kilograms) of lunar soil and rock samples. They were also involved in the solar system's first lunar fender bender, which occurred when the right rear fender fell off the LRV. The cause wasn't one of the left-behind LRVs from previous missions but rather an errant tool. The astronauts were able to make repairs using the objects they had available, and the moon rides proceeded without further problems.

Not one of the Apollo missions was long enough for any sort of in-depth geologic exploration, but they did greatly enhance scientists' understanding of the formation and history of the Moon. More importantly, the samples returned from known locations on the lunar surface could be backdated in Earth-based labs. This information was used to build up a chronology that has allowed estimates of surface ages to be made for the other planets in our solar system.

## Setting up complex science experiments

Because Apollo 17 was the last chance NASA scientists had to harvest scientific data from the Moon, they prepared a battery of complex experiments for the astronauts to perform. (The expanded experiments package actually caused Apollo 17's initial geologic traverse to be shorter than anticipated.) The experiments package included studies of subsurface gravity, heat flow, and seismic data. The purpose of these experiments was to discover more about lunar surface material, as well as what was going on deep underneath the Moon's surface. The setup for these deep experiments was more complex because additional skill and precision were necessary to drill for soil and rock samples. Apollo 17 had just the man for the job in the form of Schmitt.

An American geologist as well as an astronaut, Schmitt joined NASA in 1965 as part of an effort to involve geologists more in the space program. He worked directly with other astronauts to provide training for geologic field work that would be done on the Moon. When it became apparent that Apollo 17 would end the United States' forays to the Moon, Schmitt was selected for the final mission. As a trained field geologist, he was able to select rock samples with an educated eye (bringing home a treasure trove for geologists) and provide expert assistance with the complex experiments that needed to be run.

The Apollo 17 results showed, again, no evidence for recent volcanic activity on the Moon. Even the orange, glassy soil found near the Shorty crater turned out to have come from an ancient, long-buried volcanic deposit that was simply exposed by the more-recent impact that formed the crater.

The last human steps on the Moon were taken by Eugene Cernan on December 14, 1972. File that away for the next time you play Trivial Pursuit.

## Fast facts about Apollo 17

Following are the basic facts about the Apollo 17 mission:

- **Launch date and site:** December 7, 1972; Kennedy Space Center, Florida
- **Command/Service Module name and mass at launch:** *Apollo CSM-114* (call sign "America"); 63,560 lb (66,952 kg)
- **Lunar Module name and mass at launch:** *Apollo LM 12* (call sign "Challenger"); 67,759 lb (30,735 kg)
- **Launch vehicle:** Saturn V SA-512
- **Size of crew:** 3 astronauts
- **Number of lunar orbits:** 75
- **Lunar landing date:** December 11, 1972
- **Mission duration:** 12 days, 13 hours, 51 minutes, 59 seconds

# Scoping Out the Soviets' Lunar Exploration Preparations

Although the Soviet Union began the Space Race in the 1950s with a substantial lead, the U.S. gained a triumphant victory when it landed astronauts on the Moon in the 1960s and 1970s. The Soviets worked hard to compete with Project Apollo's successes, but the lunar program they devised was marred by accidents that slowed progress to such a point that they had no hope of catching up. President John F. Kennedy offered to perform joint Moon missions with the Soviet Union, but the Soviets mistrusted the American proposal. Instead, they worked on their own lunar exploration preparations. Yet ultimately, cosmonauts were excluded from lunar exploration and the exclusive club of moonwalkers.

Because the Soviets' lunar program was primarily politically motivated (much like its American counterpart; see Chapter 9 for details), after the political impetus to be "first" was no longer present, the Soviet lunar cosmonaut program collapsed. Instead, the nation quickly switched its political, and scientific, space goals to building space stations in orbit around the Earth, which it accomplished successfully in 1971 (see Chapter 12 for details).

## Preparing the N1 rocket

The N1 rocket was the Soviet Union's version of the Saturn V rocket (the launch vehicle of choice for each manned Apollo mission; see Chapters 3 and 9 for the scoop on the Saturn V). Development on the N1 began in 1959. Designed for carrying a manned spacecraft to the Moon and intended to have other military implications as well, the N1 (which you can see in Figure 10-5) was a five-stage rocket with a hefty mass of 6,029,643 pounds (2,735,000 kilograms). Measuring about 345 feet (105 meters) in height with a 56-foot (17-meter) diameter, it was hardly a small rocket. In fact, it was only slightly smaller than the Saturn V.

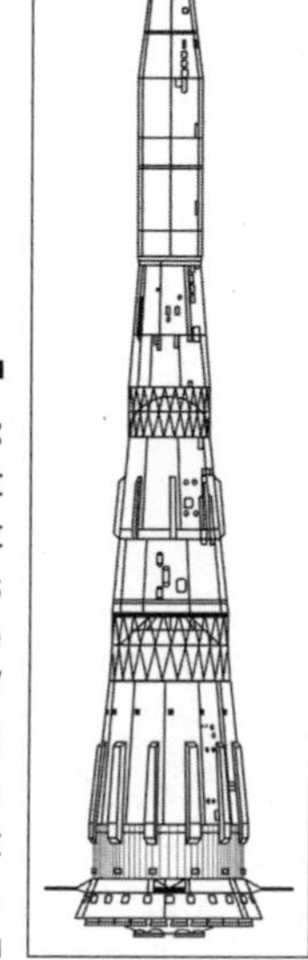

**Figure 10-5:** The Soviet N1 rocket was as tall as the Saturn V and had 30 engines on its first stage.

Courtesy of NASA

Although the N1 design was a powerful one, it didn't have unanimous support from Soviet rocket designers. Each of the major players at the time supported a different proposal. By 1961, the N1 project received only minimal funding, yet it was still slated for a test launch in 1965. Due to various political mishaps, technical difficulties, and the death of the N1's primary supporter, Sergei Korolev, the project never succeeded in its stated mission. Four attempts to launch the N1 occurred from 1969 through 1971, but they all ended in disaster.

## Building the LK lunar lander

In the event that a suitable rocket was developed, Soviet space designers did have a plan for sending cosmonauts to the Moon. Their 1964 design coalesced into the *LK lunar lander,* a single-crew spacecraft that, the Soviets hoped, would provide for a Moon landing before the Americans had a chance to win with Project Apollo (see Figure 10-6 for a comparison of the LK with the American Lunar Module).

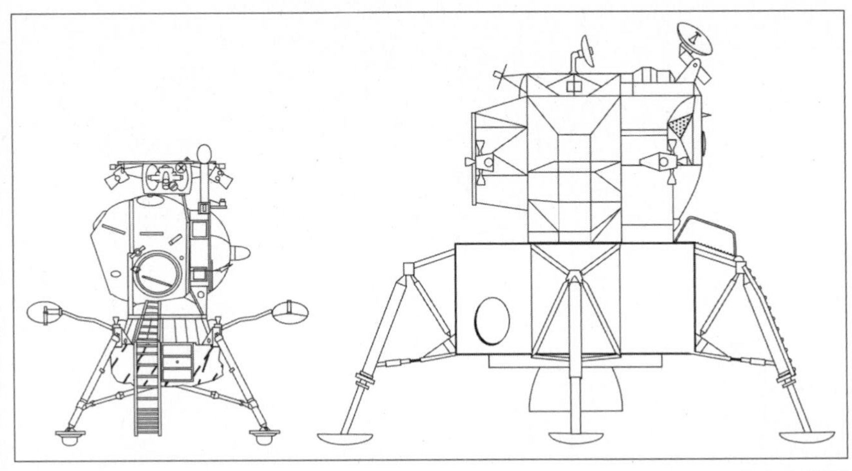

**Figure 10-6:** The Soviet Union developed a one-person lunar landing module, the LK (left, also called the L3), which was smaller than NASA's Lunar Module (right).

*Courtesy of NASA*

The LK differed from the American Lunar Module in several ways:

- Because the N1 rocket had a smaller *payload* (cargo) capacity than the Saturn V, the LK was smaller and lighter and could hold only one cosmonaut.
- There wasn't room for a docking tunnel or bay to connect the LK to its counterpart orbiter, so the cosmonaut would've had to do a *spacewalk* (an *extravehicular activity,* or EVA, performed in space) from one spacecraft to the other.
- The ascent and descent mechanisms were also handled differently. The Soviet LK used a single stage rocket for both landing on the lunar surface and lifting off of it, whereas NASA's Lunar Module used a separate landing stage and ascent stage.

Like Project Apollo's Lunar Module, though, the LK was based on a modular design. It had landing gear for a touchdown on the Moon, a rocket stage to provide for a soft lunar landing as well as liftoff back into orbit, a pressurized cabin module for the cosmonaut, and thrusters to keep the spacecraft oriented properly. It also contained flight controls, life-support systems, a power supply, and all the other essentials required for a Moon landing.

Unlike the N1, the LK program was considered quite successful. The lander was test-flown in low Earth orbit several times under the Cosmos 379, Cosmos 398, and Cosmos 434 missions. However, without a launch vehicle capable of carrying it into lunar orbit, the LK had nowhere to go. As a result, the Soviet lander never made it to the Moon.

# Chapter 11

# Going on a Grand Tour of the Solar System

### In This Chapter

- Probing Mars early on
- Venturing to Venus with Venera and Vega
- Meandering to Mercury, Venus, and Mars with Mariner
- Pioneering views of Jupiter and Saturn
- Landing on Mars with Viking
- Voyaging into space with Voyager

At the same time as the successes of the United States' Project Apollo, which ultimately sent astronauts to the Moon (see Chapters 9 and 10) were mounting, both the U.S. and the Soviet Union developed a large suite of robotic space probes that traveled to other planets. The Soviets had a series of failed missions to Mars, but the Americans sent missions that successfully flew past, orbited, and eventually landed on Mars. The Soviets had better luck with missions to Venus, including the first interplanetary landing in 1970. These early missions were exploration in the truest sense of the word — before NASA's Mariner 4 mission in 1964, for example, some scientists still thought that Mars might have canals and other signs of an ancient civilization.

More-sophisticated spacecraft followed. NASA's Viking missions to Mars in the 1970s provided a wealth of data from two orbiters and two landers. Not only that, but these landers actually performed the first search for life on another planet. NASA also launched two pairs of spacecraft on a grand tour of the outer solar system: Pioneer 10 and 11, and Voyager 1 and 2. Both pairs of spacecraft are currently on their way out of the solar system; they carry information that may one day identify their origins to a future alien civilization.

# Early Soviet Mars Probes: Mixed Results (1960–1973)

Although not tremendously successful, the Soviet Union made several attempts at sending probes to Mars from 1960 through 1973. The Soviets deserve credit for getting the ball rolling in terms of Mars exploration, but not one of these missions achieved its goal. Of course, with as new an enterprise as building interplanetary spacecraft, some failures can be anticipated . . . but in the case of Soviet Mars missions, bad luck appears to have been involved as well! In the following sections, we describe the few successful Mars missions and the many failures.

## There were some successes . . .

Despite a number of attempts in the 1960s, the Soviets didn't have any luck with an operational Mars probe until the 1970s. Even then, the few Soviet spacecraft that were modest successes couldn't match the data quality and quantity from contemporary NASA missions.

The successful Soviet Mars missions were:

- **Mars 2 and Mars 3:** Shown in Figure 11-1, these identical missions (1971) featured both a lander and an orbiter. Both orbiters succeeded and sent back a few observations of the Martian surface and atmosphere. However, transmitter problems, less-than-ideal orbits, and a global dust storm on Mars substantially reduced the quality and quantity of the data. The *Mars 2* lander crash-landed, and the *Mars 3* lander succeeded with a soft landing on the Martian surface, but returned just 20 seconds worth of data to Earth before mysteriously going silent. However, the working orbiters and one successful soft landing qualify Mars 2 and Mars 3 as some of the more successful Soviet Mars missions.
- **Mars 5:** The Mars 5 mission (1973) made it into orbit around Mars, and in a surprising turn of events for the Soviet Mars program, the lander's cameras worked and returned about 60 images during the nine days it survived on the planet. The other onboard instruments relayed information about the planet's surface temperature (a maximum of approximately 32 degrees Fahrenheit, or 0 degrees Celsius), atmosphere, ozone layer, and soil composition.

**Figure 11-1:** The *Mars 2* orbiter sent back limited observations.

Courtesy of NASA

## . . . But even more failures

The successful missions outlined in the preceding section were preceded and followed by a variety of failures. In addition to a series of flops belonging to the numbered Mars program, the Soviet Union launched several other unsuccessful missions to explore the Red Planet. Sending a spacecraft to another planet is a difficult task, and the Soviet rockets and boosters of the 1960s suffered from a proliferation of designs that weren't systematically tested. These failures weren't addressed until the 1970s, when existing designs were consolidated and thoroughly tested, a strategy that has resulted in some extremely reliable spacecraft and rocket designs, such as Soyuz (which we introduce in Chapter 7).

Following is a rundown of the failed Soviet Mars probes:

- *Marsnik 1* and *2,* both launched in October 1960, would've attempted Mars flybys, but both missions failed on launch.
  - *Marsnik 1* (called Korabl 4, or Mars 1960A) never made it into orbit at all, having self-destructed on launch.
  - *Marsnik 2* (called Korabl 5, or Mars 1960B) lifted off, but the third rocket stage wasn't successful in launching the spacecraft into Earth orbit.
- The Sputnik Mars missions constituted three more failed attempts.
  - Sputnik 22 was launched in 1962 with the intent of doing a Mars flyby, but the spacecraft broke apart in orbit.
  - Sputnik 23, also called Mars 1 (1962), launched successfully, but this mission lost communication with Earth before reaching Mars.

  - Sputnik 24 (also 1962) would've landed on Mars had the mission been successful; instead, the spacecraft broke up during flight.

- The failures unfortunately continued with the Zond 2 mission (1964), which aimed to conduct a Mars flyby. The *Zond 2* spacecraft was equipped with cameras and ultraviolet spectrometers, as well as an infrared spectrometer that would've helped identify the presence of methane. But *Zond 2* lost communication with Earth and was unable to return any data.
- *Mars 1969 A* and *B* (launched in 1969, of course) were matching probes designed to orbit the Red Planet, with each spacecraft containing a probe and an orbiter. The missions were somewhat covert and not formally introduced to the public at the time of their launch. Unfortunately, both missions failed on launch.
- Mars 1971C (also dubbed the Cosmos 419 mission and launched in 1971) was another attempt by the Soviet Union to take advantage of Mars's relative proximity to Earth. Although the Soviets hoped to beat the Americans in the race to Mars, this mission succumbed to a rocket malfunction and never reached its destination.
- *Mars 4* (1973) suffered from equipment failures that doomed its scientific productivity. The spacecraft did reach Mars, but a computer malfunction kept the spacecraft from entering orbit. Only one batch of images was returned from this mission.
- The Mars 6 mission (1973) reached Mars and actually did well enough to start sending data back home to Earth as the *Mars 6* lander descended through the atmosphere. Unfortunately, the transmitted data was mostly useless due to a computer chip problem, and all contact was lost just before the lander reached the Martian surface.
- To complete the comedy of errors, the Mars 7 mission (1973) arrived at Mars, but its landing probe was prematurely disconnected from the rest of the spacecraft. The probe (which was intended to land on the surface) missed the planet entirely.

# The Soviet Venera and Vega Programs: Reaching Venus Successfully (1961–1986)

The Soviet Union may've been unlucky when it came to exploring Mars (okay, that's an understatement; see for yourself in the previous section), but it experienced dramatic success with missions to Earth's other neighbor: Venus. The Venera program ran from 1961 through 1984 and boasted 10 landing probes that gathered data on Venus's surface, as well as 13 flyby or orbital probes that sent data home from the atmosphere surrounding Venus. The landers, in particular, were technological marvels that were able to withstand the extreme conditions at Venus's surface: high pressure, temperatures hot enough to melt lead, and a corrosive atmosphere. (Check out a Venera lander in Figure 11-2.) The subsequent Vega program sent two more Venera-type spacecraft to visit Venus, with a little flyby of Comet Halley thrown in for fun, from 1984 to 1986.

**Figure 11-2:** The Soviets used Venera landers (shown during testing) to explore Venus.

*Courtesy of NASA/National Space Science Data Center*

In the following sections, we focus on the details of several particularly notable missions: Venera 7, 9, 10, 13, and 14, as well as Vega 1 and 2.

## Venera 7 (1970): Landing accomplished

Venera 7 accomplished what no space mission had up to that point: It landed a capsule on another planet. (Remember that the Moon doesn't count as a planet!) Previous Venera missions had included atmospheric probes, which sent back data as they traveled through Venus's atmosphere, but these spacecraft didn't survive the trip to the surface. Launched in 1970, the *Venera 7* spacecraft made it to Venus and successfully jettisoned its landing capsule, which parachuted down to the planet's surface. Using a raised antenna, the capsule sent home radio signals with data regarding Venus's surface temperature for 23 minutes before communications ceased.

The transmitter actually got stuck and sent back the surface temperature over and over again. However, the Venera 7 mission still counts as the first time that data was sent back from the surface of another planet.

### Surveying the surface conditions of Venus

Probes to the Moon or to Mars were capable of surviving and transmitting data for weeks or months (or, in some cases, years), but the surface conditions on Venus were much less hospitable. Although Venus is similar in size to Earth, its atmosphere is so thick that the atmospheric pressure at the planet's surface is almost 100 times that at the Earth's surface. Venus's thick clouds trap the Sun's heat, resulting in a surface temperature of about 500 degrees Fahrenheit (260 degrees Celsius). Because of the extreme temperature and pressure conditions, the space probes that managed to reach Venus's surface lasted for only around an hour, so the amount of data they were able to return home was limited.

Ironically, although surface conditions on Venus are much harsher than on Mars, landing on the former is actually easier than landing on the latter. The thick atmosphere means that a series of parachutes can slow a spacecraft down from interplanetary speeds to a safe landing speed without having to rely on complex retrorockets to be fired before landing (like the ones needed on Mars due to the planet's thin atmosphere). Other landing schemes, such as airbags, have also been used on Mars (see Chapter 15), but in general, a Venus landing is technologically less complicated — at least for 1960s technology. Then again, no spacecraft has landed on Venus since *Vega 1* and *Vega 2* in 1985!

## *Venera 9* and *10* (1975) and *Venera 13* and *14* (1981): Taking historic pictures

The four Venera spacecraft covered in this section all succeeded in capturing images of Venus's surface. These photos were valuable because they were mankind's first views of the surface of a mysterious world whose surface can't be seen from space (unless you have radar eyes!). Though science-fiction-style steamy jungles had mostly been ruled out due to Venus's hot, hot, hot surface temperatures, scientists were surprised to see a world that was more Earth-like than they'd originally imagined.

Although its main goal was to take measurements of the Venusian atmosphere with a variety of instruments, *Venera 9,* launched in 1975, wound up taking the first pictures returned from the surface of another planet. The spacecraft had two cameras, one on each of its sides (although one didn't work because a lens cap failed to pop off correctly). The functioning camera used a periscope to peek through *Venera 9*'s hull so it could capture images without exposing itself to the extreme temperature and pressure conditions on the Venusian surface.

*Venera 9* captured relatively low-resolution images, because it was limited in the data it could send back during the estimated 30-minute lifetime of the lander. It returned multiple images that could be assembled into a panorama of Venus's surface (see Figure 11-3) that was intentionally angled to include both the horizon (visible at the top right and left corners) and the base of the spacecraft (bottom center). The images show a clearly recognizable landscape covered with rough rocks.

**Figure 11-3:** A panoramic shot of Venus's surface, as taken from *Venera 9.*

Courtesy of NASA

As an interesting aside, *Venera 9* was also the first spacecraft to successfully orbit Venus. Not bad for a Soviet space program that was struggling to reclaim its status in the Space Race after its many setbacks in Mars exploration

Later in 1975, *Venera 10* was similarly successful in taking images of Venus's surface, which revealed a smooth surface similar to a terrestrial lava flow. Complete lens cap failures on *Venera 11* and *12* caused a slight delay in the flow of photos from Venus until 1981 when *Venera 13* and *14* successfully returned color images of the planet's surface.

*Venera 13* and *14* used a more-sophisticated camera design that allowed them to send higher-resolution images back to Earth for a combined total of approximately three hours. The color images were made by taking versions of the same image using different color filters on the camera. Unlike their predecessors, both cameras worked on *Venera 13* and *14,* allowing a complete view of the terrain surrounding the landers. Turns out the surfaces at both of these landing sites are covered with smooth, platelike rocks that are likely lava flows. The sky is yellowish, and the surface appears brownish-yellow. (*Venera 13* and *14* landed relatively close to each other, slightly south of Venus's equator.)

## Venera 15 and 16 (1983): Mapping Venus

*Venera 15* and *16,* launched just five days apart in 1983, achieved what previous Venera missions couldn't: They mapped Venus's surface. Although the images sent back from previous Venera missions were both novel and useful, Venus's thick atmosphere prevented traditional cameras from seeing through to the surface. Thus, a global study of the planet's surface was impossible until radar instrumentation was added to the Venera spacecraft. Why is a global study so valuable? Well, global surface maps are essential for understanding the geologic history of planets and how features at different parts of a planet's surface relate to each other, when and how they were formed, and so on. Such maps can also be used to determine approximately how old the surface is, and how recently it was likely to have had active volcanoes or other activity.

*Twinning* of spacecraft, or launching a pair of identical spacecraft within a very close time frame, was typical in the Soviet space program (as well as the American space program), because the combined effort of the two spacecraft was able to provide much more coverage for image mapping than either would've been able to do alone. In the case of *Venera 15* and *16,* the two spacecraft were designed identically, and the surface-imaging instrumentation took up enough space that these missions didn't have entry probes. Each spacecraft contained dish antennae for transmitting a signal through the thick Venusian clouds and then picking up that signal as it bounced off the surface. They also used solar arrays to power the equipment.

## Vega 1 and 2 (1984–1986): Landers, balloons, and a comet flyby

The final chapter in the Soviet exploration of Venus came with the Vega 1 and 2 missions, launched in 1984 and arriving at Venus in 1985. These twin spacecraft took advantage of the United States' decision to cancel its planned probe to Comet Halley, which passed near the Earth in 1986. The Soviet Union stepped in to build a comet probe and soon realized that planetary alignments would allow a spacecraft to fly past Venus on its way to the famous comet. To take advantage of this phenomenon, the Soviet Union built two more copies of its now well-tested Venera spacecraft, with a few variations. The most innovative of these tweaks was the addition of a balloon, or *aerostat,* that was released during the spacecraft's descent into the Venusian atmosphere.

The 11-foot (3.4-meter) diameter balloon carried an attached gondola of instruments on a long tether. The two aerostats successfully measured Venus's atmospheric composition and tracked its wind currents, sending back data for 46 hours until their batteries ran out. During that time, *Vega 1* and *2* floated more than one-third of the distance around the planet at an altitude of about 33 miles (54 kilometers) above the surface. After releasing the landers and aerostats, the two Vega spacecraft used Venus's gravity to travel on to their rendezvous with Comet Halley in 1986. By mid-1986, they'd sent back more than a thousand pictures of the comet's nucleus.

# The Mariner Program: Visiting Mercury, Venus, and Mars (1962–1975)

Early planetary exploration didn't belong solely to the Soviet Union. American planetary research started with the *Mariner Program,* the United States' first long-running interplanetary spacecraft program, which was intended to start exploring the strange new worlds of Mercury, Mars, and Venus. This program consisted of ten space missions, seven of which were successes. Two additional missions (Mariner 11 and 12) were planned but ended up rolling into the start of the later Voyager Program (which we cover later in this chapter).

The Mariner spacecraft, which wound up becoming the basis for almost all of NASA's planetary exploration for decades to come, were based on either octagonal or hexagonal core units to which the instruments (cameras, antennae, and the like) were attached (see Figure 11-4). Like most other space missions, solar panels provided the main source of power for both the spacecraft and its instrumentation. Atlas rockets, either Centaur or Agena, were used to launch the Mariner spacecraft into space.

**Figure 11-4:** The Mariner spacecraft.

*Courtesy of NASA/JPL-Caltech*

In the following sections, we provide details of the Mariner missions of the 1960s and 1970s.

## Missions of the 1960s

Even though its focus was on putting a human on the Moon before the Soviets could, NASA also set its sights on exploring the rest of the solar system in the 1960s. The Moon was close enough to make human space travel there feasible, but the initial forays to other planets had to be done by robotic spacecraft.

The leadoff mission of the Mariner series, Mariner 1 (1962), would've done a flyby of Venus, but it was destroyed shortly after launch due to a problem with the launch rocket. *Mariner 2,* which was originally intended as a backup to the Mariner 1 mission, was a stunning success and became the first spacecraft in the world to fly past another planet. On December 14, 1962, *Mariner 2* came as close as 21,644 miles (34,833 kilometers) to the Venusian surface. Another achievement of this monumental flyby was the variety of important data it collected, such as cloud temperatures, Venusian surface temperatures (the planet measured about 752 degrees Fahrenheit [400 degrees Celsius] on the surface), and other data concerning Venus's atmosphere. Additionally, the sensors and instruments aboard *Mariner 2* discovered that Venus has no major magnetic field or radiation belt surrounding it.

After success at Venus, Mars was next on the agenda. The Mars-bound *Mariner 3* spacecraft (1964) was a flop due to launch failure, but *Mariner 4* (1964) successfully became the first spacecraft to fly by Mars at close range. This mission was the first time scientists got pictures of the surface of another planet, and they were very surprised to see an old, cratered surface that looked much more like the Moon than the Earth. *Mariner 4*'s images thus revolutionized society's view of Mars as a planet (although ironically *Mariner 4, 6,* and *7* all imaged old cratered parts of the Martian surface, missing the younger valleys and other interesting features that make Mars more Earth-like). The images taken by *Mariner 4* at its closest approach were surprisingly sharp; they even showed clouds in the thin Martian atmosphere, something scientists didn't expect to see.

*Mariner 5* (1967) was originally intended as a backup Mars mission, but after the success of *Mariner 4,* it was retrofitted and sent to Venus. It succeeded in taking measurements of Venus's atmosphere, refining the measurements made by *Mariner 2* and providing further clues about the planet's hot surface temperature and thick atmosphere.

An American tag team of octagonal spacecraft, *Mariner 6* and *7,* gained valuable information from their flybys of Mars in 1969. Launched about a month apart (on February 24 and March 27, respectively), the two spacecraft took about 200 pictures of Mars, some from their approach and others at closer range. These photos showed a number of details about the north and south poles of Mars, in addition to providing images of Mars's moon Phobos.

## Missions of the 1970s

Flyby missions are fine for initial reconnaissance, but to really understand a world you need detailed maps produced from observations made in orbit. The Mariner flyby missions of the previous decade had produced intriguing glimpses of Venus and Mars, but without global context, such snapshots aren't that meaningful to geologists. So orbiters were the logical next step in planetary exploration.

A pair of twin ships, *Mariner 8* and *9,* would've become the world's first Martian satellites. They were intended to continue the data collection begun by the Mariner 6 and 7 missions and perform a detailed mapping of the Martian surface. However, *Mariner 8* (launched in 1971) failed during launch and fell back into the ocean. The honor of being the first spacecraft to orbit another planet fell to *Mariner 9* alone.

*Mariner 9* entered orbit around Mars in November 1971. From there, it prepared to photograph the Martian surface and took atmospheric measurements using instruments much like those aboard its immediate predecessors. The actual taking of the photos was delayed due to a giant global dust storm on Mars that completely obscured the surface. Fortunately, engineers on Earth were able to wait for the dust to clear several months later before beginning the imaging program. Ultimately, *Mariner 9* was able to map more than 80 percent of the Martian surface from orbit. A full geologic picture emerged as a result, complete with images of valleys, huge volcanoes, canyons, and evidence of geologic features carved by both wind and water. This mission rekindled interest in Mars as a possible location for past, or current, life.

The final spacecraft in the Mariner Program, *Mariner 10* (1973) was the first spacecraft ever to take advantage of *gravity-assisted trajectory,* a technique that uses the gravitational attraction of one planet to propel a spacecraft toward its next goal in an effort to conserve both fuel and power. In this case, *Mariner 10* was slated to fly past Venus and then continue on to Mercury, where it could make multiple flybys. *Mariner 10* used its ultraviolet cameras to take detailed images of the structure of the Venusian clouds, but the real prize of the mission was Mercury. Three Mercury flybys, in 1974 and 1975, revealed an ancient, battered surface covered with impact craters, but with signs of old volcanic activity and some cracks and faults. *Mariner 10* was only able to map one hemisphere of Mercury; the other hemisphere remained unseen for more than 30 years until the MESSENGER mission (see Chapter 21).

# Pioneer 10 and 11: Exploring the Outer Solar System (1972–2003)

Americans weren't content with just exploring the Moon and the inner solar system. They longed to expand their pioneering spirit to the vast reaches of the outer solar system. Thus, the *Pioneer Project* was born. This widely focused series of missions, which started in the late 1950s and continued through 1978, served to increase people's knowledge both of space science and technological potential in the inner and outer solar system.

- *Pioneer 0–2* were failed attempts to reach the Moon.
- *Pioneer 3* and *4* were successful Moon exploration missions.
- *Pioneer 5–9* explored magnetic fields and solar activity.
- *Pioneer 12* entered orbit around Venus in 1978, mapped the planet's surface via radar, studied the atmosphere, and recorded other key measurements for the next 14 years.
- *Pioneer 13,* which launched in 1978, took specialized measurements using four different probes that were sent to different locations around Venus.

*Pioneer 10* and *11,* though, stand out from the crowd as the first spacecraft to make the trip to Saturn and Jupiter. These planets are referred to as *outer planets* or *outer solar system planets* because they lie outside the asteroid belt; the other outer planets are Uranus and Neptune. In the following sections, we describe the launch and journey of both spacecraft.

## Providing the first views of Jupiter and Saturn

*Pioneer 10,* launched on March 2, 1972, made history as the first space probe, or man-made object of any kind, ever to fly through the asteroid belt successfully. It returned valuable images and other data from Jupiter. *Pioneer 10* also found that Jupiter has a very strong magnetic field that produces high levels of radiation in its immediate vicinity. After its Jupiter flyby, *Pioneer 10*'s trajectory took it out of the solar system.

*Pioneer 11* was no less of a success. The twin to *Pioneer 10, Pioneer 11* launched on April 5, 1973. As it made its Jupiter flyby, the spacecraft came within 21,000 miles (34,000 kilometers) of the clouds surrounding Jupiter and was able to photograph the planet and Callisto, one of its primary moons. After its successful Jupiter flyby, *Pioneer 11* used Jupiter's gravity to change its trajectory and send itself on to Saturn. The spacecraft's Saturn flyby was equally successful, sending home images of Saturn, as well as its rings and moons.

Some of the instruments aboard *Pioneer 11* were designed to measure solar wind, cosmic rays, and magnetic fields — the same types of things studied by the Apollo astronauts on the Moon (as explained in Chapters 9 and 10). However, the Pioneer spacecraft also measured specific elements in the Jupiter and Saturn atmospheres, such as radio waves and *aurorae* (atmospheric interactions due to a planet's magnetic field and solar wind; the result looks like the *aurora borealis* [northern lights] observed here on Earth).

## Staying in touch for more than 20 years

Although it seems that nothing lasts beyond its warranty these days, spacecraft can be an exception. Both *Pioneer 10* and *Pioneer 11* stayed in contact with Earth for more than two decades.

- *Pioneer 10* was still making contact with Earth as recently as 2003, when it was 7.5 billion miles (12 billion kilometers) from Earth. The spacecraft's power has likely declined now to the point where sending signals home is no longer possible.
- *Pioneer 11*'s mission came to completion in 1995, because no further communications have been received since then. The likely reason for this communications shutdown is that the spacecraft's onboard power generator has expired.

The fact that these spacecraft remained in contact as long as they did is a testament to superior design and mission planning. The two spacecraft sent back valuable data about particles and fields in the far reaches of the solar system. In addition, the simple tracking of their orbits has yielded interesting information about chaos theory and orbital interactions in complex systems. And last but certainly not least, *Pioneer 10* was used as a test target for flight engineers learning how to talk to faraway spacecraft.

## Bearing plaques for alien encounters

Even though they've lost touch with Earth, *Pioneer 10* and *11* are still capable of communicating. As the first man-made objects to be sent out of the solar system, both spacecraft bear a plaque (shown in Figure 11-5) designed by leading scientists as a message for any future intelligent beings that might encounter them. The plaque includes a drawing of a naked man and woman next to a schematic of the spacecraft, for scale. It also includes a diagram of a hydrogen atom, markers to indicate the position of our Sun, and a map of the solar system with a special marker to point out Earth.

**Figure 11-5:** As the first terrestrial objects to leave the solar system, *Pioneer 10* and *11* have plaques to greet civilizations that might discover them.

Courtesy of NASA

Where might these plaques come in handy? Well, both spacecraft are on trajectories that will eventually take them to the stars:

- *Pioneer 10* is heading toward the star Aldebaran, in the constellation Taurus, and will take about 2 million years to get there.
- *Pioneer 11* is heading toward the constellation Aquila, a journey that'll take about 4 million years to complete.

# Viking 1 and 2: Embarking on a Mission to Mars (1975–1982)

With the knowledge that Mars is much more Earth-like than originally thought (thanks to the findings of *Mariner 9* in 1971; see the earlier section for details), NASA was inspired to create an ambitious Martian exploration program that was motivated as much by scientific goals as by political showmanship (the Soviet Union had yet to have a successful Mars landing). Martian features such as canyons and valleys that appeared to have been carved by liquid water, combined with the hot surface temperatures and noxious clouds of Venus that could in no way support life, rekindled interest in Mars as a possible location for life in the solar system. So began the United States' efforts to find evidence of life on Mars.

Taking a page from the successful Soviet strategy of twin spacecraft, described earlier in this chapter, NASA built *Viking 1* and *2*. The two spacecraft included both an orbiter, for global mapping, and a lander, for landing on another planet and searching for life. (Both the orbiter and lander go by *Viking 1* and *2* . . . so much for originality in the naming process, huh?)

The Viking Program didn't come out of nowhere. It built on the past achievements of the Mariner Program, as well as lessons learned from successful lunar landers and orbiters such as Surveyor and Lunar Orbiter (see Chapter 8 for details). The concept behind the mission was to send two spacecraft to Mars, separated by about a month. Each would land at a different site while their orbiters remained in space to complete their missions of imaging the planet and conducting other studies from space.

The Viking 1 mission departed on August 20, 1975, and arrived in Mars orbit nearly a year later. The orbiter and lander stayed united for about a month while they photographed the planet and looked for a landing site. Meanwhile, the Viking 2 mission launched on September 9, 1975, reaching Mars orbit in August 1976. The *Viking 1* lander detached from its orbiter on July 20, 1976, and reached the Mars surface at Chryse Planitia. The *Viking 2* lander wasn't far behind, landing at another area of Mars called Utopia Planitia on September 3, 1976.

## Orbiting and landing on Mars

The Viking orbiters and landers had different goals and designs, but both provided necessary functions:

- The orbiter for both *Viking 1* and *2* was primarily for transporting the corresponding lander to Mars, taking survey photos that would help determine a landing site, and relaying transmissions from the lander back to Earth while continuing to map the surface. The orbiter design was octagonal in shape, measuring about 8 feet (2.5 meters) wide. Solar panel wings provided power for the spacecraft on its voyage to Mars and while in orbit; initial propulsion was provided by a liquid-fueled rocket.
- The Viking mission lander was a six-sided, three-dimensional object with three legs that came out to rest on the ground. It was propelled by rocket thrusters and engines; landing radar, gyros, and an altimeter helped control its landing. It was powered by two *radioisotope thermoelectric generators,* a type of electrical generator that uses the radioactive decay of plutonium to provide energy. Communications back to the orbiter, which were relayed to Earth, took place via transmitters and antennae. The *Viking 1* and *2* landers used hydrazine fuel in their landing thrusters to slow their descent through the thin Martian atmosphere for a safe arrival on the surface.

The orbiters from the Viking 1 and 2 missions led productive lives: The *Viking 1* orbiter made more than 1,400 orbits around Mars over the course of four years, and the *Viking 2* orbiter completed about half that many orbits in two years. The images taken from these spacecraft gave scientists the most complete and detailed image maps of the Martian surface until the *Mars Global Surveyor* arrived at Mars 20 years later, in 1997 (see Chapter 15).

The Viking landers were also very long-lived: The *Viking 2* lander successfully sent back data for more than three and a half years until its battery failed in 1980, and the *Viking 1* lander operated for more than six years until 1982, when a faulty software update mistakenly deleted essential data used by the lander to point its antenna at Earth, shutting down communication.

## Searching for Martian life

The scientific goal of the Viking missions was clear: Study the biological, physical, chemical, and other properties of Mars's atmosphere and surface in an effort to better understand the planet and its potential to support life. Each Viking lander housed numerous instruments, including cameras, sample-collection devices, and temperature and wind sensors, to help it fulfill this goal. (You can see some of these instruments in Figure 11-6, taken on the surface of Mars.)

**Figure 11-6:** The *Viking 2* lander on Mars.

*Courtesy of NASA*

## Planetary protection

At the dawn of the Space Age, science fiction blended with science fact to produce a view of the solar system that included life on other planets. Although the first missions to the Moon, Mars, and Venus revealed harsh conditions that seemed unlikely to support life as we know it, scientists knew life could still exist elsewhere in the solar system. In 1966, the U.S. and several other nations signed a space exploration treaty, part of which requires that nations protect other planets and the Earth from contamination due to space exploration.

Planetary protection is designed into space missions for two reasons: to avoid contaminating the planets being explored and to prevent any byproduct of the mission from causing reciprocal harm to Earth. Keeping the other planets free from biological contaminants helps preserve them in pristine form for future research. Keeping the Earth free from potential extraterrestrial contaminants is, similarly, a clear preventative measure against the unknown. Although all missions have some form of planetary protection, it becomes most critical when astronauts or robotic landers bring samples back to Earth.

The first few Apollo crews to land on the Moon were actually kept in isolation following their return to Earth. The Apollo 11 astronauts, for example, climbed out of their floating Command/Service Module (CSM) after it splashed down and went straight into a decontamination raft. They were washed with a bleachlike cleanser and hustled into a quarantine unit for the next several weeks. Their decontamination raft was actually sunk so as to eliminate the potential spreading of any lunar microbes they may have picked up, and their CSM was washed with an antiseptic before it was hauled onto the *USS Hornet*. Fortunately for the astronauts, these procedures were substantially reduced after Apollo 14 when it became clear that returning astronauts didn't succumb to any sort of lunar flu.

The Viking landers, with their life-detection experiments, took extraordinary precautions to avoid contaminating the Martian surface. In many ways they set the gold standard for planetary protection. The hydrazine used in the landing rockets, for example, was purified in order to try to reduce the chance of contaminating the Martian surface as the chemical burned. As the landers separated and traveled into Martian orbit, they were covered with an aeroshell that prevented microbes and organisms from flying off into the Martian atmosphere. The landers were also deliberately exposed to very high temperatures inside this aeroshell in another effort to prevent the spread of Earth bugs into Martian space.

The lander cameras were able to image the surface using a high-tech version of smoke and mirrors. Movable mirrors were aimed at vertical strips of the landscape; photodetectors then measured the light reflected from those mirrors back into the cameras. The cameras were then moved for the next scan, and these strips were recombined later to form a complete image.

The *Viking 1* and *2* landers didn't just rely on imaging in their search for Martian life. Both landers had their own robotic arm with a soil scoop that could dig small trenches on the surface of Mars and bring the soil into the spacecraft for chemical analysis.

Following are some of the experiments the landers ran on the Martian soil:

- **The gas-exchange experiment** looked to see whether compounds in Martian soil would exhibit lifelike symptoms after being stimulated with organic material.
- **The labeled-release experiment** studied the surface after introducing radioactive gas to see whether live organisms would emit carbon dioxide gas.
- **The carbon-assimilation experiment** tested for the creation of carbon compounds.

None of these tests proved conclusively that there was life (as we know it, at least) on Mars. This fact doesn't mean the Viking missions weren't enormously successful. However, the intense excitement surrounding the search for life — and the failure to successfully detect it — meant that Mars exploration basically came to a standstill at the end of the Viking missions. After an expenditure of close to a billion dollars on the Viking Program, no successful missions were sent to Mars for more than 20 years (see Chapter 15 for the scoop on these later successes).

# Voyager 1 and 2: To Jupiter and Beyond (1977–Present)

Once named *Mariner 11* and *Mariner 12, Voyager 1* and *Voyager 2* are the main players in the most successful and longest-lasting missions of solar-system exploration. The two spacecraft, launched in 1977, were conceived of as a way to take advantage of a rare planetary alignment that would allow them to visit and study Jupiter and Saturn, as well as Uranus and Neptune if all went well. The particular orbital locations of these four planets provided a prime opportunity for the spacecraft to conserve both fuel and power by using the gravitational attraction of one planet to propel them toward their next goal, a technique known as *gravity-assisted trajectory.* Without this extra boost, the missions likely wouldn't be enjoying such long-lasting success.

The identically designed twin Voyager spacecraft were stabilized and controlled using gyroscopes. Data was managed through a digital tape recorder (an 8-track, no less!); it was sent and received via the spacecraft's 12-foot (3.7-meter) antenna. The wide- and narrow-angled cameras that made up the imaging subsystem were controlled via computer. Both Voyager spacecraft still carry a range of instruments designed to help them do their jobs, including high- and low-field magnetometers to measure the strength and direction of magnetic fields and spectrometers to measure the composition of planets and atmospheres. At the height of the missions, each of the 11 main instrument systems had a team on Earth dedicated to studying and compiling the results of its findings.

## *Voyager 1*: Whizzing by Jupiter and Saturn

Launched on September 5, 1977 — about two weeks after *Voyager 2* — *Voyager 1* arrived at Jupiter first to fulfill the initial part of its mission: observing Jupiter and its moons. Beginning in March 1979, the spacecraft took images and gathered other data during its month-long trip through Jupiter's system. One of its major discoveries was the existence of active volcanoes erupting on Jupiter's moon Io (see Figure 11-7).

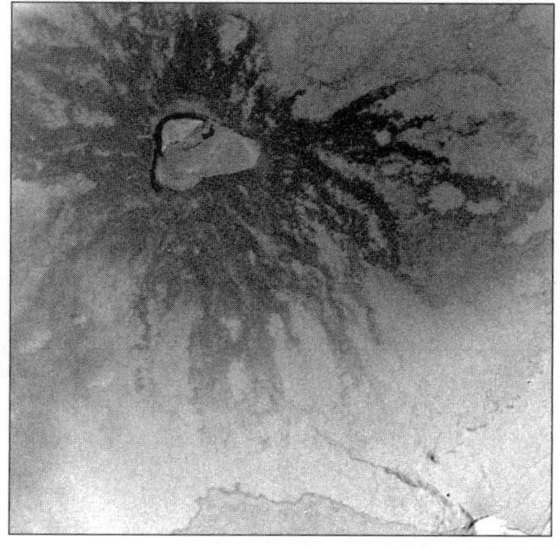

**Figure 11-7:** An image of volcanic features on Io taken from *Voyager 1.*

Courtesy of NASA/JPL

As *Voyager 1* used Jupiter's gravity to swing onward toward Saturn, it was able to take close observations of Saturn's moon Titan, as well as Saturn's rings. Its closest approach to Saturn occurred in November 1980. (You can see an image of Saturn taken by *Voyager 1* in the color section.)

The mysterious, large moon Titan was a prime target of the Voyager 1 mission, and the spacecraft trajectory was carefully designed to allow for a close flyby. Unfortunately, Titan's thick, hazy atmosphere completely blocked any views of the surface from orbit, and the much-anticipated pictures of Titan showed nothing more than a fuzzy orange ball.

Despite the Titan disappointment, the data *Voyager 1* collected over the years has contributed immensely to astronomers' understanding of interplanetary space. With its mission completed, *Voyager 1* began making its way out of the solar system; it currently can lay claim to the title of "farthest manmade object from the Earth."

## Voyager 2: Making discoveries everywhere

Despite losing the race to Jupiter (*Voyager 2* reached the planet three months after *Voyager 1*), *Voyager 2* has become known as one of the greatest and most-prolific exploratory space missions in history because it made significant discoveries everywhere it went. During the first leg of its journey, *Voyager 2* accomplished the following:

- It encountered and documented new rings of Jupiter. It also showed the Great Red Spot, a storm vortex located on the South equatorial belt of Jupiter. *Voyager 1* and various Earth-based astronomers had viewed the huge storm system before but never with such detail and accuracy.
- It took close-up pictures of Jupiter's moon Europa, revealing a network of cracks and ridges and a very young surface.
- It made measurements of Saturn's temperature (ranging from –333 degrees Fahrenheit to –202 degrees Fahrenheit [–203 degrees Celsius to –130 degrees Celsius]) and took some of the most-dynamic views of Saturn ever seen.

After *Voyager 2* completed exploring Saturn at close range, it proved to have sufficient fuel and functionality left to continue on what scientists had hoped would be the next leg of its journey: a visit to Uranus. Following the disappointment of *Voyager 1*'s hazy observations of the moon Titan (described in the preceding section), scientists fortunately were able to change *Voyager 2*'s trajectory before it reached the Saturn system. Instead of achieving another unsatisfying Titan flyby, the mission navigators at NASA redirected *Voyager 2* so it could continue its grand tour of the solar system past Uranus and Neptune. Fortunately *Voyager 2* had enough fuel to continue on its new trajectory, because there aren't any gas stations in the outer solar system!

After a mission continuation was authorized and funded, the spacecraft proceeded to gather valuable images and data along the way to Uranus. Upon its arrival there in August 1986, *Voyager 2* took photos; contributed new information about the planet's magnetic field, rotation, and radiation belts; and made the rings and moons of Uranus visible in clear detail for the first time. *Voyager 2* was eventually funded for a pass by Neptune as well, and the spacecraft came within close range of the planet in August 1989. Major discoveries from this part of the mission included images of the moon Triton and a large, long-lived storm in the atmosphere of Uranus dubbed the Great Dark Spot.

## Carrying the gold (-plated) record

As we note earlier in this chapter, the Pioneer spacecraft carry plaques to be read by future alien civilizations that might encounter the probes, but the Voyager spacecraft went one step further. Both *Voyager 1* and *Voyager*

The Andromeda Galaxy, which is the largest galaxy near the Milky Way, is made up of cool, dusty star formation regions (red) and young, bright stars (blue). See Chapter 2 for more about stars and galaxies.

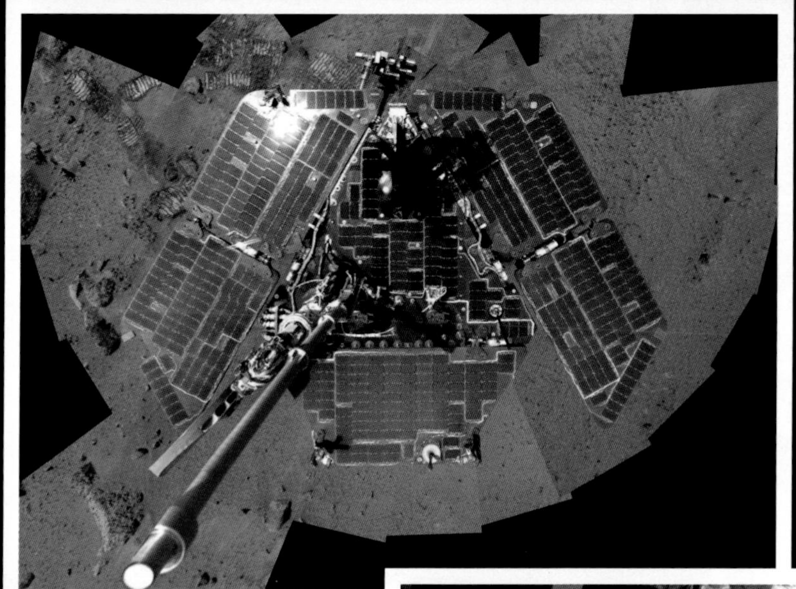

This top view of the Mars Exploration Rover Spirit (a self-portrait) shows solar panels coated in reddish Mars dust. Flip to Chapter 3 to see how solar panels power spacecraft.

NASA astronaut Linda Godwin performs a spacewalk, or extravehicular activity (EVA), outside the Space Shuttle Endeavour. Check out Chapter 4 for more on spacewalks and other astronaut duties.

COURTESY OF NASA

Ham the chimp, one of NASA's first space travelers, greets the commander of the recovery ship following his successful flight on a Mercury-Redstone rocket. See Chapter 6 to find out more about his flight.

Gemini VIII crew members Neil Armstrong and David Scott are in their capsule after splashdown following their successful mission. Check out Chapter 7 for more on Project Gemini.

COURTESY OF NASA

NASA's Surveyor 1 mission was launched on an Atlas-Centaur rocket. The Surveyor missions made soft landings on the Moon; see Chapter 8 for more about these robotic missions.

The first view of the Earth rising over the Moon, as seen by the Apollo 8 astronauts on their trip around the Moon. Check out Chapter 9 for more on Project Apollo.

Apollo 11 astronaut Buzz Aldrin walks on the Moon in this pictur taken by Neil Armstrong; see Chapter 9 for more about the Apollo 11 mission.

OURTESY OF NASA

Apollo 16 astronaut John Young jumps as he salutes flag; both the Lunar Module and the Lunar Roving ehicle are visible in the background. See Chapter 10 for more on this mission.

This mosaic of Saturn images, taken by the Voyager 1 spacecraft, also includes three of the planet's small satellites. Check out Chapter 11 for more on the Voyager missions.

NASA's Skylab space station (as seen from orbit) suffered damage during launch that led to the loss of one solar panel. See Chapter 12 for details on Skylab and the end of the Space Race.

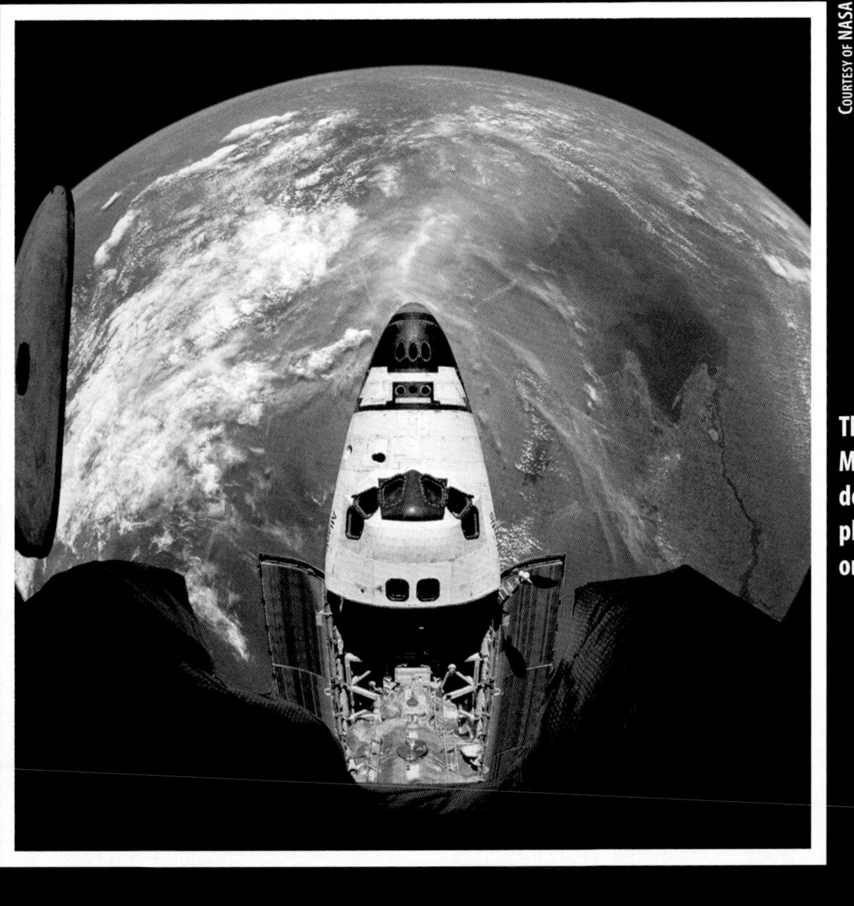

Courtesy of NASA

The Space Shuttle as seen from the Mir space station, with its payload bay doors open and the docking tunnel in place. Check out Chapter 13 for more on the Space Shuttle's missions to Mir.

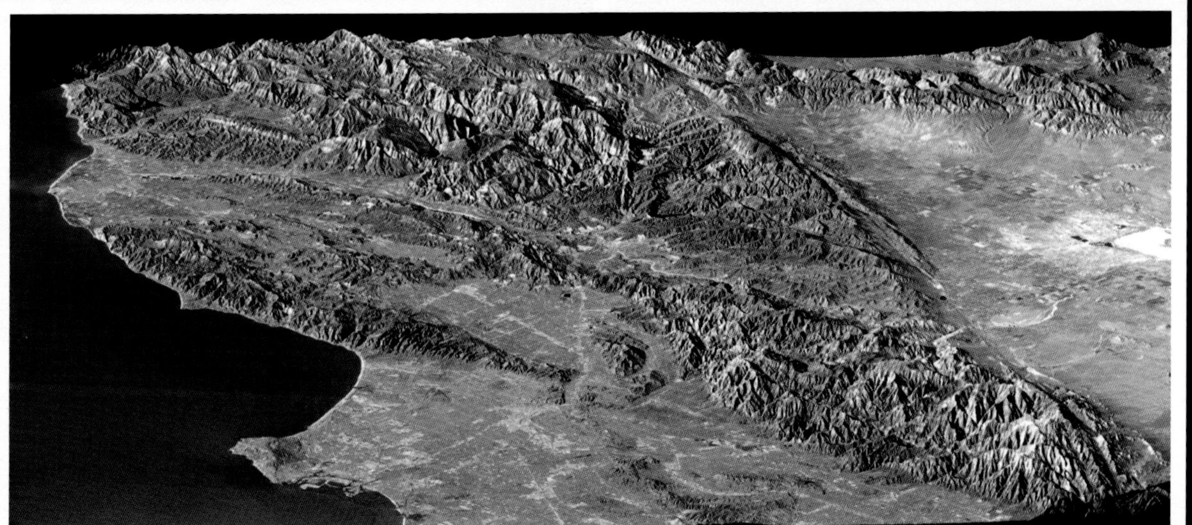

Courtesy of NASA/JPL/NIMA

This three-dimensional view of the Los Angeles area was made with data collected by the Shuttle Radar Topography Mission (SRTM), which used a huge radar-mapping instrument installed on the Space Shuttle. Check out Chapter 14 for all the details.

**A view of one hemisphere of the surface of Venus using radar imaging and topography data by the Magellan spacecraft. See Chapter 16 for more on the United States' Magellan mission.**

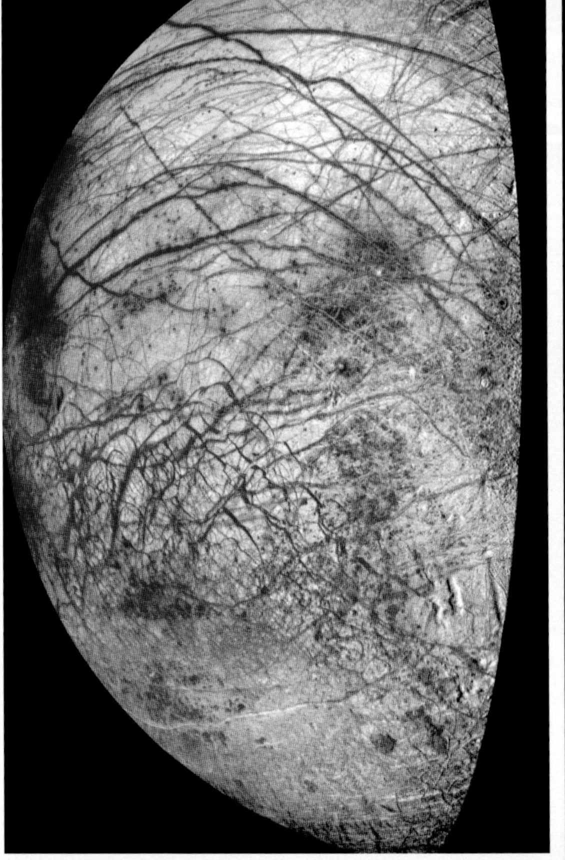

**An enhanced-color mosaic of images of Jupiter's moon Europa, taken by the Galileo spacecraft, shows cracks in the icy surface. Check out Chapter 17 for more on Galileo.**

Courtesy of NASA, ESA, and the Hubble Heritage Team (STScl/AURA)

An image of a planetary nebula taken by the Hubble Space Telescope. This slowly expanding cloud of dust and gas is all that's left after the central star exploded at the end of its lifetime. More information on the Hubble Space Telescope is in Chapter 18.

Courtesy of NASA

NASA astronaut Steve Smith uses a specially designed ratchet to repair and upgrade the Hubble Space Telescope during the second servicing mission in 1997. See Chapter 18 for the details.

The International Space Station as seen from the Space Shuttle Discovery in 2008. Flip to Chapter 19 for the scoop on the ISS.

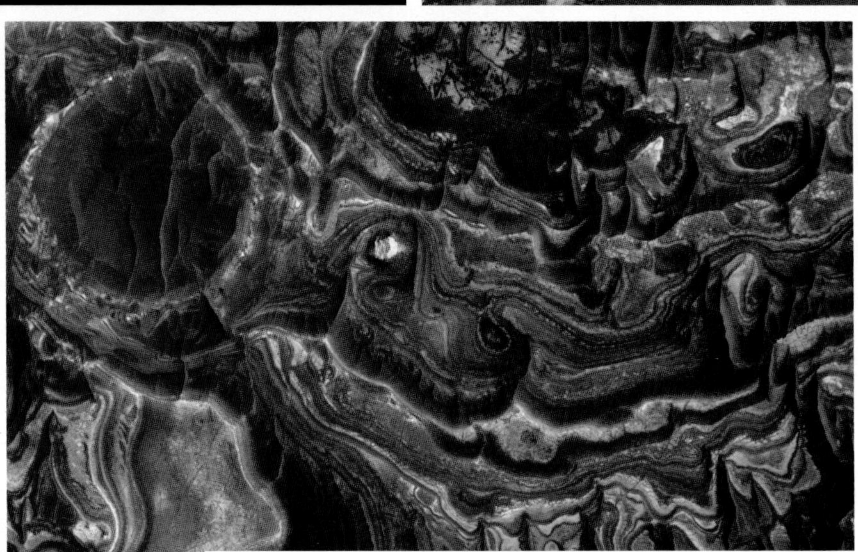



A high-resolution color view of layers on the surface of Mars taken by the HiRISE camera on NASA's Mars Reconnaissance Orbiter mission, described in Chapter 20.

A proposed mission to Saturn's moon Titan would include an orbiter, built by NASA, and a lander and balloon, built by ESA. Check out the details in Chapter 21.

*2* carry golden photograph records that can be played (see Figure 11-8). For those aliens whose music collections are entirely digital, handy instructions are included on the spacecraft itself for how to build a device to play back the record.

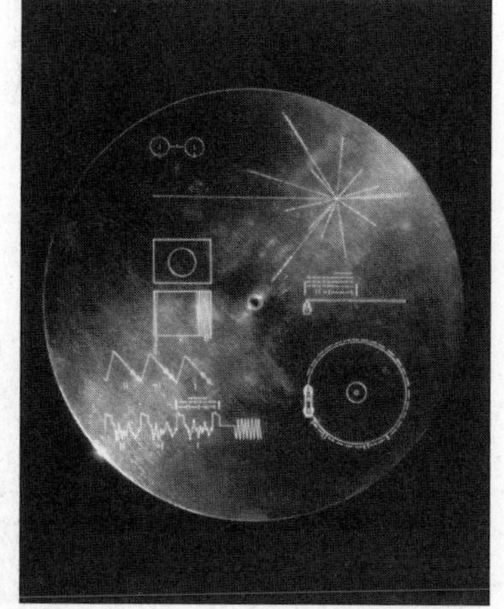

**Figure 11-8:** The Voyager spacecraft's gold-plated record.

*Courtesy of NASA*

The gold-plated copper records, filled with a wide variety of sounds and images intended to capture the range of human experience, are meant to serve as a time capsule of sorts. Scientists, led by the famed astronomer Carl Sagan, carefully debated the impression that each and every sound and image might give a future extraterrestrial civilization. Here's what they decided to include on the records:

- **Audio:** Voices in numerous languages, natural sounds, and music from many cultures
- **Visual:** Photographs of people, buildings, animals, and other natural and artificial scenes

The cover of the record includes reference points to help determine the location of the solar system in which Earth resides.

## Still checking in

The Voyager spacecraft were built and designed to last for about five years. Yet 30 years later, they're still sending data and information back to Earth (albeit not a lot these days), thereby contributing to the history of space research.

Both spacecraft are currently on their way out of the solar system. In 1998, *Voyager 1* passed *Pioneer 10*'s distance to become the man-made object that's most distant from the Sun. The meager information the twin spacecraft now send back to Earth focuses on interstellar studies, and 5 of the original 11 instrument teams are still supporting their group's data acquisition and analysis. Both spacecraft are investigating the very edge of the Sun's influence and the interaction of that region with the space between the stars. It'll be at least 40,000 years before either spacecraft makes a close encounter with another star system.

How long can the Voyager spacecraft continue producing data? Scientists estimate that the two spacecraft are good for at least another ten years, or until the probes finally run out of power. (Currently, power levels are still about 60 percent of what they were at launch time.) Another consideration, though, is the fact that the thermocouples used to convert that heat into electricity degrade over time, which could further decrease the amount of time that the power supplies will be useful. Eventually, the spacecraft will no longer have enough power to send a signal back to Earth, similar to what occurred with the Pioneer spacecraft. In the vacuum of space, however, they'll continue silently on their journey until something, or someone, gets in their way.

For the latest status of the Voyager spacecraft, check out their Web site: `voyager.jpl.nasa.gov`.

# Chapter 12

# The First Space Stations and the End of the Space Race

### *In This Chapter*

- Inhabiting space in the various Soviet Salyut stations
- Jumping into the space station game with the American Skylab program
- Cooperating internationally with Apollo-Soyuz

The Apollo Moon landings (described in Chapters 9 and 10) were a triumph for the United States, but such a level of accomplishment couldn't be sustained on either side of the Space Race after the politically motivated megagoal of being first to the Moon was achieved. Thus, both the U.S. and Soviet space programs set their sights on more-modest goals closer to home, in Earth orbit. The Americans scaled down Project Apollo to meet new orbital objectives, and the Soviets launched their successful Soyuz program, which took cosmonauts into orbit and ferried them to and from a succession of space stations. (Believe it or not, a version of the Soyuz spacecraft is currently servicing the International Space Station; see Chapter 7 for more about Soyuz.)

Following the limited-duration flights of the Apollo and Soyuz programs, both the U.S. and the Soviet Union developed early space stations that allowed humans to remain in space for longer periods of time. The Soviet Salyut station and American Skylab station both allowed humans to remain in Earth orbit for months at a time. (One goal of such long-duration flights was to begin preparation for a human mission to Mars, a feat that would involve months or years in flight and has yet to be achieved.)

A final politically motivated mission remained: In 1975, the U.S. and Soviet Union staged a joint mission, Apollo-Soyuz, during which spacecraft from the two nations actually docked in orbit. Widely considered the end of the Space Race, this incident ushered in a new era of reduced hostilities between the two superpowers. In this chapter, we fill you in on the details of this joint mission, as well as the two nations' respective space station programs.

# Shrouded in Secrecy: The Early Salyut Program (1971–1973)

After failing to reach the Moon before its American rivals, the Soviet Union set its sights on putting the first space station into orbit around Earth — a goal that ultimately led to the creation of the Salyut program. The Soviets launched nine separate space station modules, six of which met enough of their mission requirements to be considered successes.

Unlike today's International Space Station (which we cover in Chapter 19), all the Salyuts were completely separate, self-contained units that didn't dock with one another. Each Salyut module was designed as a stand-alone space station that fit one of two models:

- **Long-Duration Station (acronym DOS in Russian) units:** This model was made up of a transfer compartment, a main compartment, and an auxiliary compartment. The main compartment included scientific equipment such as telescopes and cameras. Six of the Salyut stations fell into this category.
- **Orbital Piloted Station (OPS) units:** Intended for military rather than civilian work, these models included an airlock, a larger work compartment, and a smaller living compartment. The work compartment was primarily designed around a large Earth observation camera that the cosmonauts used to take photos of various sensitive locations. They then developed the film and scanned it for return to Earth. Three of the Salyut stations were OPS units.

In the following sections, we describe the first few missions of the Salyut program, from 1971 to 1973. Flip to the later section "Building on Experience: Later Salyut Missions (1974–1991)" for the scoop on Salyut 3 through Salyut 7.

## Salyut 1: Launching and docking with Soyuz 11 (1971)

The first space station in the world, Salyut 1, was launched into orbit by the Soviets on April 19, 1971. Its purpose — other than beating the Americans into space with a station — was to begin exploring how humans could survive in space for long periods of time. In addition, Salyut 1 provided an excellent opportunity to photograph Earth from afar and study cosmic rays, gamma rays, and other astronomical phenomena that can't be observed from the ground. A DOS-type orbital station, Salyut 1 measured about 52 feet by 13 feet (16 meters by 4 meters) and was launched into space by a Soviet Proton launch vehicle.

## Fast facts about Salyut 1

Following are the basic facts about the Salyut 1 mission:

- **Launch date and site:** April 19, 1971; Baikonur Cosmodrome, USSR
- **Spacecraft name and mass at launch:** Salyut 1 (a DOS-type station); 40,620 lb (18,425 kg)
- **Launch vehicle:** Three-stage Proton rocket
- **Size of crew:** Launched unoccupied; visited by 1 crew of 3 cosmonauts (who spent a total of 24 days inside)
- **Number of Earth orbits:** 362
- **Length of time in Earth orbit:** 175 days

Salyut 1, like all space stations to come in the future, was launched without a crew onboard because the constraints to get humans safely into orbit are much stricter than for equipment. By launching the space station unoccupied, the designers didn't have to include space-wasters like seats that would only be used on launch, and the station could be blasted into orbit at gravitational accelerations that would be unsafe for human occupants.

Of course, a manned space station is fairly useless without a crew, so the Soviets sent three cosmonauts to Salyut 1 as part of the Soyuz 10 mission. The men successfully launched from Baikonur Cosmodrome on April 22, 1971, but they were unable to dock with the space station because of a hatch problem.

The next mission to try to visit Salyut 1, Soyuz 11, had more initial success. Crewed by Georgi Dobrovolski, Vladislav Volkov, and Viktor Patsayev, *Soyuz 11* made it to the space station and successfully docked with it on June 7, 1971. This event was a monumental achievement because it was the first time in history that a spacecraft had performed a docking with a space station.

Unfortunately, the Soyuz 11 mission ended tragically (see Chapter 5 for details). After Dobrovolski, Patsayev, and Volkov spent 24 days living on the world's first space station, their Soyuz capsule lost pressure during reentry, and all three astronauts were found dead upon landing. Salyut 1 itself reentered the Earth's atmosphere on October 11, 1971, burning up over the Pacific Ocean.

# Salyut 2 and Cosmos 557 (1973)

The second Salyut mission, named DOS-2, launched in July 1972 but never made it into orbit due to a faulty launch vehicle. The actual Salyut 2 mission was launched on April 4, 1973, and the space station was a military OPS unit rather than the more-common DOS type used for civilian purposes. Salyut 2 launched successfully, but the crewed mission was postponed and eventually canceled as the station began losing pressure and power. The Soviets abandoned the station, and it burned up in the Earth's atmosphere on May 28, 1973.

The last of the early Salyut missions, dubbed Cosmos 557, was another DOS-type station that would've further advanced Soviet space station capabilities. However, it also ended in failure. This space station launched on May 11, 1973, but it failed to reach a stable orbit and burned up in Earth's atmosphere a week later.

## Skylab: NASA's First Home in the Sky (1973–1979)

While the Soviet Union was successfully launching the world's first space station into orbit (see the earlier section on Salyut 1), the U.S. was developing its own version of a space station. Named Skylab, this orbiting space station may have come in second in this particular Space Race contest, but it achieved a number of important scientific goals. Its primary mission was to prove that astronauts could live in space for more than a few days; it was also intended to study the Earth and stars from a new perspective.

Another important programmatic reason for Skylab was to continue NASA's human spaceflight program during the period between the cancellation of Project Apollo and the launch of the next generation of space-transportation systems, the Space Shuttle, which was still almost a decade away at this point. (Check out Chapters 13 and 14 for the scoop on the Space Shuttle.) The space station even featured some Apollo hardware left over from the canceled Apollo 18, 19, and 20 flights.

The following sections highlight the launch of Skylab, the subsequent missions to it, and the space station's eventual return to Earth. (Be sure to flip to the color section for an image of Skylab.)

### Launching Skylab

Skylab's debut took place on May 14, 1973, when it launched into space on a two-stage version of NASA's huge Saturn V Moon rocket. The space station suffered damage during launch and ended up losing both a main solar panel and a cover designed to protect the laboratory and living areas from heat and micrometeorite damage. A second solar panel became stuck and didn't deploy correctly.

## Fast facts about Skylab 1

Following are the basic facts about the Skylab 1 mission:

- **Launch date and site:** May 14, 1973; Kennedy Space Center, Florida
- **Spacecraft name and mass at launch:** Skylab; 170,000 lb (77,088 kg)
- **Launch vehicle:** Saturn INT-21
- **Size of crew:** Launched unoccupied; visited by 3 crews of 3 astronauts each (who spent a combined total of 171 days inside)
- **Number of Earth orbits:** 34,981
- **Length of time in Earth orbit:** 2,249 days

Because the Saturn rockets had never experienced a launch failure, NASA scientists and engineers were stunned at the possible loss of their space station. Teams worked around the clock following the launch to devise ways to fix the damage, delaying the first crew launch by ten days to test out the procedures and fabricate replacement hardware. The astronauts of the first Skylab crew worked to repair the damage during their first *spacewalk* (the more-common term for an *extravehicular activity* [EVA] done in space) after arriving at the station in an Apollo capsule (see the next section for details).

Not a small affair, Skylab weighed closed to 100 tons when it left Earth and contained both living space and scientific instruments. It basically served as a floating laboratory stuffed with instruments that could take atmospheric and other scientific measurements. It also carried water, food, oxygen, and other life-supporting equipment for its future astronaut occupants.

# Living in the sky

Three separate crews of three astronauts each took up residence in Skylab during its six years in orbit, for a total length of 171 days. Perhaps even more impressive is that the crew of each of these Skylab missions set a series of increasing records for the longest amount of time in space. Each group was brought to Skylab and returned home via Apollo spacecraft launched on Saturn IB launch vehicles. (Figure 12-1 shows a crew inside Skylab.) The following sections break down the details of the three manned Skylab missions.

## Skylab 2

Skylab 2 (named that because the first launch of the Skylab program was unmanned) was the first manned mission to reach the United States' first orbiting space station. Commander Charles Conrad, Jr.; Pilot Paul Weitz; and Science Pilot Joseph Kerwin were charged with getting the station up and running after the problems at launch, testing the ability of astronauts to live and work in space for long periods, and performing a number of scientific experiments. This first crew launched on May 25, 1973.

## Fast facts about Skylab 2

Following are the basic facts about the Skylab 2 mission:

- **Launch date and site:** May 25, 1973; Kennedy Space Center, Florida
- **Spacecraft name and mass at launch:** *Skylab 2* (a modified Apollo CSM); 44,048 lb (19,980 kg)
- **Launch vehicle:** Saturn IB
- **Size of crew:** 3 astronauts
- **Number of Earth orbits:** 404
- **Mission duration:** 28 days, 49 hours, 49 minutes

When the astronauts arrived at the space station, they found that the temperature inside Skylab, which had lost its heat shield at launch, had risen to 126 degrees Fahrenheit (52 degrees Celsius). The men immediately set about deploying a replacement heat shield before the plastic inside the space station could melt. They were also able to perform a spacewalk to deploy a stuck solar panel, restoring minimum power levels to Skylab.

**Figure 12-1:** Astronauts inside Skylab.

*Courtesy of NASA*

## Fast facts about Skylab 3

Following are the basic facts about the Skylab 3 mission:

- **Launch date and site:** July 28, 1973; Kennedy Space Center, Florida
- **Spacecraft name and mass at launch:** *Skylab 3* (a modified Apollo CSM); 44,357 lb (20,120 kg)
- **Launch vehicle:** Saturn IB
- **Size of crew:** 3 astronauts
- **Number of Earth orbits:** 858
- **Mission duration:** 59 days, 11 hours, 9 minutes, 1 second

Of course, the Skylab 2 astronauts didn't just spend their time repairing the space station and going about their day-to-day business. They actually performed more than 300 experiments, including many on themselves in an effort to better understand how humans react to living in microgravity. Other experiments focused on studying the atmospheres of the Sun and Earth from space. In particular, the 25,000 pictures of the Sun taken by the Skylab 2 crew advanced studies of solar physics immensely with observations that couldn't be done from the ground.

When the Skylab 2 astronauts left the space station, they had spent a record 28 days in orbit, doubling the previous U.S. record. However, they left behind a little something on Skylab: When the next crew arrived, they were surprised to find dummies made by the first Skylab crew already hard at work, perched on the exercise bicycle, toilet, and medical station!

### Skylab 3

Skylab 3 launched one month after its predecessor. The mission further tested the abilities of astronauts to perform valuable scientific experiments in space and understand the medical effects of long periods in microgravity. The Skylab 3 astronauts (Commander Alan Bean, Pilot Jack Lousma, and Science Pilot Owen Garriott) installed a more-permanent sunshade on the station to protect it from overheating. They also performed a number of other experiments and medical tests and kept careful records regarding the changes in the human body after long stays in space.

The Skylab 3 crew also brought a number of experiments to the space station that had been designed by high school students around the country. One interesting student experiment brought two spiders to the station and monitored how the webs they spun in zero gravity differed from spider webs spun on Earth.

All together, the astronauts spent a total of 59 days in space, breaking all endurance records up to that point. Of course, the astronauts got so used to being in space that after their return to Earth, Lousma absent-mindedly let go of his razor while shaving, expecting it to float in place as it had aboard Skylab!

### Skylab 4

Skylab 4 featured the last crew to occupy Skylab. Commander Gerald Carr, Pilot William Pogue, and Science Pilot Edward Gibson launched to the space station on November 16, 1973, and stayed there through February 8, 1974, setting a world record with 84 days in space. Among their significant work was the observation of Comet Kohoutek, which was discovered in March 1973 and was kind enough to make an appearance for the Skylab 4 astronauts. Observations of the comet were made from inside Skylab, outside the spacecraft on two separate spacewalks, and with the onboard Apollo Telescope Mount.

The Apollo Telescope Mount was a solar observatory that went for a ride on Skylab. It was made from a reconfigured Apollo Lunar Module from one of the canceled Moon landings. The telescope included a range of ultraviolet instruments designed to be used for solar observation, along with equipment to perform other science experiments.

When the final Skylab crew came home, the space station was essentially abandoned in orbit. The original plan was to send another crew in 1979 via NASA's next-generation spacecraft, the Space Shuttle, in order to boost Skylab into a higher, safer orbit. However, the first Space Shuttle flight wasn't until 1981.

## Making a splash

Skylab was intended to remain in orbit for about eight years, which still would've made a Space Shuttle docking in 1981 a possibility, but the Sun didn't quite cooperate. The Earth's atmosphere was hotter than expected due to high levels of solar activity, which increased the drag on Skylab from the Earth's outer atmosphere and degraded the space station's orbit more quickly that initially expected.

Skylab ended up reentering Earth's atmosphere in 1979. Most of the station burned up as it passed through, but some pieces of the 100-ton space station survived the blaze and became scattered throughout the Indian Ocean and parts of Australia.

## Fast facts about Skylab 4

Following are the basic facts about the Skylab 4 mission:

- **Launch date and site:** November 16, 1973; Kennedy Space Center, Florida
- **Spacecraft name and mass at launch:** *Skylab 4* (a modified Apollo CSM); 44,092 lb (20,000 kg)
- **Launch vehicle:** Saturn IB
- **Size of crew:** 3 astronauts
- **Number of Earth orbits:** 1,214
- **Mission duration:** 84 days, 1 hour, 15 minutes, 30 seconds

The backup flight-quality Skylab (built in case a replacement station was ever needed) is safely housed at the National Air and Space Museum in Washington, D.C. Better yet, you can even walk through a portion of the station there! Check out `www.nasm.si.edu` to see the museum's collection of Skylab and other space artifacts online and consider visiting in person to get a first-hand view of space exploration.

# Building on Experience: Later Salyut Missions (1974–1991)

At the same time that NASA was conducting missions to its single Skylab space station, the Soviet Union continued to enhance its leading role in space station design and occupation by launching multiple Salyut stations. Each one was visited by more and more crews and remained in space for longer periods of time.

The Salyut program progressed through 1991, with each mission experiencing varying degrees of success. Although Salyut set several records, it was also plagued by problems that led to its eventual abandonment as a human-occupied space station.

## Salyut 3: Controlling its own orientation (1974–1975)

Unlike any of the previous Salyut space stations (described earlier in this chapter), Salyut 3 used special thrusters to regulate its orientation while in space, becoming the first spacecraft to keep stable while pointing at the same part of Earth. The overall dimensions of Salyut 3 (which is depicted in Figure 12-2) were about 48 feet by 13 feet (14.5 meters by 4 meters). The OPS-type station was powered by solar panels.

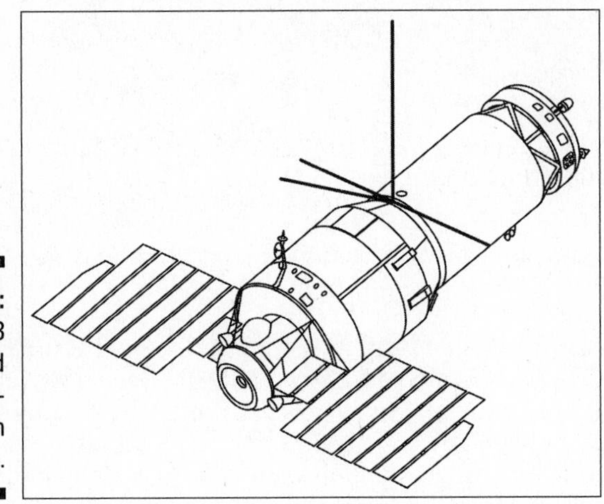

**Figure 12-2:** Salyut 3 regulated its own orientation in space.

*Courtesy of NASA*

Tensions were still high at this point in the Cold War, and although Salyut 3 was disguised as part of the civilian Salyut program, its true mission was to test military capabilities in space. Not only was Salyut 3 equipped with a high-powered camera that could resolve objects as small as 10 feet (3 meters) on the ground, but it also had a so-called self-defense cannon that was actually fired in space. Records from that time are sketchy, but later reports from cosmonauts indicate that the onboard gun, similar to an aircraft cannon, fired shells that destroyed a test satellite.

Crew members for Salyut 3 were transported to the space station via the Soyuz 14 mission. Commander Pavel Popovich and Flight Engineer Yuri Artyukin launched on July 3, 1974, and docked successfully with the space station. During their roughly two weeks of living inside Salyut 3, the cosmonauts tested a range of military reconnaissance features, as well as the feasibility of a newly designed exercise suit created especially for the mission.

## Fast facts about Salyut 3

Following are the basic facts about the Salyut 3 mission:

- **Launch date and site:** June 25, 1974; Baikonur Cosmodrome, USSR
- **Spacecraft name and mass at launch:** Salyut 3 (an OPS-type station); 40,785 lb (18,500 kg)
- **Launch vehicle:** Three-stage Proton rocket
- **Size of crew:** Launched unoccupied; visited by 1 crew of 2 cosmonauts (who spent a total of 15 days inside)
- **Number of Earth orbits:** 3,442
- **Length of time in Earth orbit:** 213 days

# Salyut 4: Three visits from Soyuz (1974–1977)

Salyut 4, a DOS-type station launched into space on December 26, 1974, was significant in that two manned Soyuz missions managed to make contact and dock:

- In January 1975, *Soyuz 17* brought Commander Aleksei Gubarev and Flight Engineer Oleg Makarov to the space station, where they stayed for a month. The main science achieved by this cosmonaut team focused on astronomical observations of the Sun and Earth. The station included one of the first X-ray cameras to make astronomical observations (which can't be done from Earth), and it discovered a number of celestial objects whose X-ray emissions varied with time.
- *Soyuz 18* brought two new crew members, Commander Pyotr Klimuk and Flight Engineer Vitali Sevastyanov, in May 1975. They orbited on the space station for 63 days, approaching the record set by the American Skylab astronauts the previous year. This crew continued the astronomical observations begun by the previous crew, took pictures of the Sun and Earth, and experimented with growing vegetables in space.

A third, unmanned Soyuz spacecraft docked with Salyut 4 for an additional few months. The goal of this test was to ensure that the space station could remain in orbit for that long with an attached module, in addition to proving the increased longevity that the Salyut missions hoped to be capable of.

### Fast facts about Salyut 4

Following are the basic facts about the Salyut 4 mission:

- **Launch date and site:** December 26, 1974; Baikonur Cosmodrome, USSR
- **Spacecraft name and mass at launch:** Salyut 4 (a DOS-type station); 40,785 lb (18,500 kg)
- **Launch vehicle:** Three-stage Proton rocket
- **Size of crew:** Launched unoccupied; visited by 2 crews of 2 cosmonauts each (who spent a combined total of 92 days inside)
- **Number of Earth orbits:** 12,444
- **Length of time in Earth orbit:** 770 days

## Salyut 5: The last military station (1976–1977)

Salyut 5 has the distinction of being the last OPS unit launched into orbit. The space station left Earth on June 22, 1976; carried out additional military tests; and entertained two crews that docked with it in space. Two cosmonauts joined Salyut 5 via the Soyuz 21 mission between July and August 1976, and the *Soyuz 24* spacecraft delivered two cosmonauts in February 1977.

The cosmonauts conducted a variety of military tests and experiments aboard the station, including making observations of a Soviet military exercise in Siberia to test the capabilities of the station's cameras to perform reconnaissance.

The *Soyuz 21* crew, who were scheduled to spend two months on the station, had to return early due to the illness of one cosmonaut. Vitali Zholobov developed both space sickness and psychological problems that could've been due to the combination of leaking gases on the station and the fact that the crew neglected their physical training and sleep schedules. The *Soyuz 24* crew, which followed, didn't find any problems with the air on Salyut 5, but the men completely changed the cabin air before beginning their stay.

## Fast facts about Salyut 5

Following are the basic facts about the Salyut 5 mission:

- **Launch date and site:** June 22, 1976; Baikonur Cosmodrome, USSR
- **Spacecraft name and mass at launch:** Salyut 5 (an OPS-type station); 41,888 lb (19,000 kg)
- **Launch vehicle:** Three-stage Proton rocket
- **Size of crew:** Launched unoccupied; visited by 2 crews of 2 cosmonauts each (who spent a combined total of 67 days inside)
- **Number of Earth orbits:** 6,666
- **Length of time in Earth orbit:** 412 days

# *Salyut 6: Undergoing many upgrades (1977–1982)*

Some of the first major changes of the Salyut program came in 1977 with the launch of the DOS-type station Salyut 6. This new design included a second docking port and an improved propulsion system. It also included an overhaul of the space station's telescopes and an upgrade to its living quarters. Salyut 6 could be resupplied via robotic supply ships, which could dock with the station automatically to bring aboard fuel and supplies for the crew.

Because of these changes, which allowed crews to change hands and supplies to be brought to the station more easily, Salyut 6 was a real second-generation space station built to be inhabited for years. It also had the most visitors of any space station to date. During its nearly five years in orbit, 16 different missions made their way to the Salyut 6. For the first time, cosmonauts from other countries were able to visit a Soviet space station as part of the Soviet Union's new Intercosmos program. In addition to the Soviet Union, visitors to Salyut 6 hailed from Czechoslovakia, Poland, the German Democratic Republic, Hungary, Vietnam, Cuba, Mongolia, and Romania. The dual docking ports allowed a long-term crew, with their Soyuz spacecraft docked to one port, to receive a shorter-term, visiting crew that could dock at the other port for a visit of a few days.

## Fast facts about Salyut 6

Following are the basic facts about the Salyut 6 mission:

- **Launch date and site:** September 29, 1977; Baikonur Cosmodrome, USSR
- **Spacecraft name and mass at launch:** Salyut 6 (a DOS-type station); 43,704 lb (19,824 kg)
- **Launch vehicle:** Three-stage Proton rocket
- **Size of crew:** Launched unoccupied; visited by 16 missions carrying 33 cosmonauts (who spent a combined total of 683 days inside)
- **Number of Earth orbits:** 28,024
- **Length of time in Earth orbit:** 1,764 days

## *Salyut 7: Expeditions for repair and additions (1982–1986)*

The last space station to be launched into orbit under the Salyut program was the DOS-type Salyut 7. The main goals of this final expedition were to test how long the station could remain operational and to experiment with a more modular design for adding future space stations (essentially laying the groundwork for the interlocking designs of future missions such as the International Space Station, covered in Chapter 19).

Part of the reason Salyut 7 was able to operate for so long was the expertise brought by its visitors, especially when it came to making essential repairs to systems such as a ruptured fuel line. Salyut 7 was ultimately visited by ten crews of cosmonauts, including some from India and France. It was also briefly home to female cosmonaut Svetlana Savitskaya, making her the second woman to travel in space and the first to perform a spacewalk. Salyut 7 endured in space for more than eight years, lasting until after its replacement, Mir, was launched. (Flip to Chapter 13 for details on Mir.)

At the end of its lifetime, Salyut 7 was placed in a high orbit to keep it safe until the planned Soviet Buran spacecraft could reach it. However, increased solar activity caused the station to reenter the Earth's atmosphere in 1991 and burn up over Argentina, creating a spectacular meteor shower with many fragments reaching the ground.

### Fast facts about Salyut 7

Following are the basic facts about the Salyut 7 mission:

- **Launch date and site:** April 19, 1982; Baikonur Cosmodrome, USSR
- **Spacecraft name and mass at launch:** Salyut 7 (a DOS-type station); 43,704 lb (19,824 kg)
- **Launch vehicle:** Three-stage Proton rocket
- **Size of crew:** Launched unoccupied; visited by 10 missions carrying 26 cosmonauts (who spent a combined total of 816 days inside)
- **Number of Earth orbits:** 51,917
- **Length of time in Earth orbit:** 3,216 days

# Apollo-Soyuz: Friends in Space (1975)

The Space Race didn't end with anything definitive like a closing ceremony or medal distribution. Instead, the intense atmosphere of competition between the United States and the Soviet Union gradually gave way to a wave of unprecedented cooperation. The first great manifestation of this new relationship came in the form of the Apollo-Soyuz mission in 1975 — the first time that space vehicles belonging to different nations docked in orbit around Earth.

## Mission planning

Apollo-Soyuz was considered a test flight that would prove the feasibility of a mission combining the American and Soviet space programs. Its official designation at NASA was *ASTP,* short for Apollo Soyuz Test Project. Science experiments and equipment were included, but they were really secondary to the goal of promoting peace and success between the Space Race superpowers.

The heart of the mission was the docking of an American Apollo capsule with a Soviet Soyuz spacecraft. The plan was to launch an Apollo spacecraft that was very much like the ones sent to the Moon along with a Soyuz spacecraft, plus a NASA-designed docking module that would allow astronauts and cosmonauts to move between the two spacecraft. In addition to promoting a working relationship between the U.S. and the Soviet Union, scientists wanted to test whether the two countries' space vehicles could play nicely together. If so, future internationally supported spaceflights and ventures might be possible.

## A docking and a handshake

The Apollo-Soyuz mission made history when Commander Thomas Stafford, an American astronaut, and Commander Alexei Leonov, a Soviet cosmonaut, performed the first-ever international handshake in space (as shown in Figure 12-3). Command Module Pilot Vance Brand and Docking Module Pilot Donald Slayton comprised the rest of the American crew; Flight Engineer Valeri Kubasov was the second half of the Soviet crew. It was a stellar crew on both sides, with years of experience and history (in fact, Leonov conducted the Soviet Union's first spacewalk from the *Voskhod 2* spacecraft in 1965; see Chapter 7 for more details).

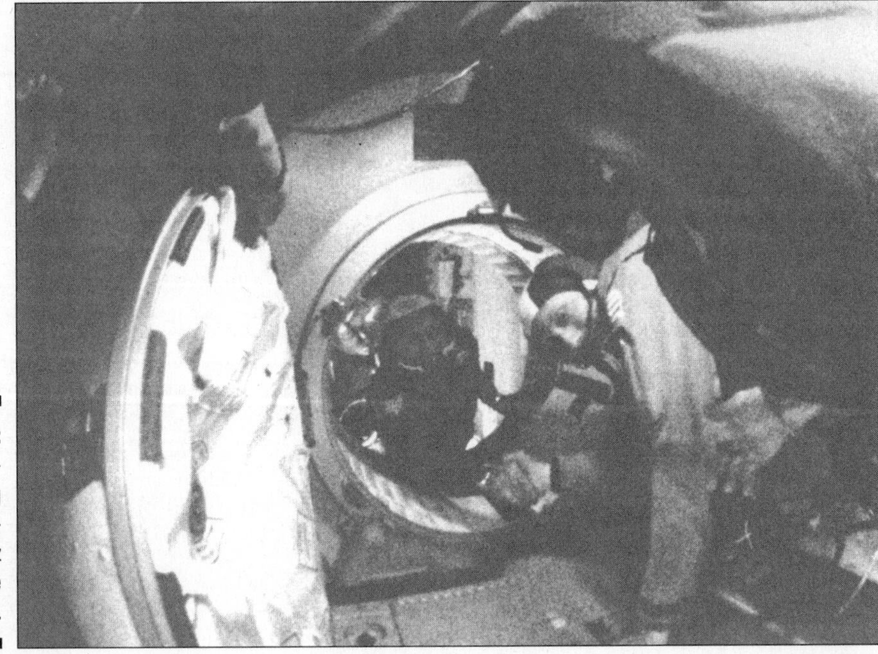

**Figure 12-3:** The history-making Apollo-Soyuz handshake in space.

*Courtesy of NASA*

After docking for the first time and completing the handshake, the crews spent about two days together, using the time to exchange gifts, eat meals (American and Soviet space food was about equally "delicious" in the 1970s), and float through the other country's spacecraft. They also practiced rendezvousing, docking, and undocking while in orbit.

## Fast facts about Apollo-Soyuz

Following are the basic facts about the historic Apollo-Soyuz mission:

| | *United States* | *Soviet Union* |
|---|---|---|
| **Mission name** | ASTP Apollo | Soyuz 19 |
| **Launch date and site** | July 15, 1975; Kennedy Space Center, FL | July 15, 1975; Baikonur Cosmodrome, USSR |
| **Spacecraft name and mass at launch** | Apollo CSM plus Docking Module; 37,000 lb (16,780 kg) | Soyuz spacecraft; 15,000 lb (6,790 kg) |
| **Launch vehicle** | Saturn IB | Soyuz rocket |
| **Size of crew** | 3 astronauts | 2 cosmonauts |
| **Number of Earth orbits** | 148 | 96 |
| **Mission duration** | 9 days, 1 hour, 28 minutes | 5 days, 22 hours, 30 minutes |

The Apollo and Soyuz spacecraft remained in orbit for several days before heading home. This would be the last flight of an Apollo mission and, as historians would later suggest, the “official” end of the Space Race. Despite a few technical difficulties during Apollo-Soyuz, by all accounts the joint mission was a magnificent success for both countries. For the first time, Soviets and Americans proved they could put aside their past differences and achieve monumental greatness — together.

# Part III
# Second-Generation Missions

The 5th Wave                                  By Rich Tennant

"Shoot, that's nothing. Watch me spin him!"

## In this part . . .

Part III begins with the development of a mature program of orbital science and robotic missions. By the 1980s, the American Space Shuttle provided innovative, reusable access to space, and the Soviet Union's Mir space station provided the first, albeit temporary, home for humans in space. Both the United States and the Soviet Union gained expertise with longer-duration stays in orbit and performed useful scientific and engineering tasks. Robotic exploration continued on Mars, which remained a difficult target, and within the inner and outer solar system.

## Chapter 13

# The Space Shuttle and Mir Space Station: The Dawn of Modern Space Exploration

### *In This Chapter*

- Developing the Space Shuttle program
- Launching, living on, and landing the Space Shuttle
- Checking out the Space Shuttle's first missions
- Visiting the Mir space station

Whereas the spacecraft of the 1960s and 1970s were primarily single-use and meant to last for just a few years, the 1980s introduced a switch to reusable spacecraft that signified a long-term commitment to space exploration.

- In 1981, the United States sent the first Space Shuttle on its way. With its reusable parts, the Space Shuttle was designed to provide low-cost, regular access to space.
- In 1986, the Soviet Union began constructing the Mir space station, meant for long-term, constant occupation. A symbol of international cooperation, Mir housed astronauts from many different nations who spent time doing scientific research and fixing systems on the station.

The new spirit of cooperation that allowed Soviet and American crews to train together and the Space Shuttle to dock with Mir truly set the stage for a new era of space exploration. We give you the scoop on the beginnings of the Space Shuttle program and the Mir space station in this chapter.

# The Birth of the American Space Shuttle Program

The first flights of the Space Shuttle program (which is technically called the *Space Transportation System,* or STS) began well after the end of Project Apollo in 1975. However, the design and engineering work that went into the Space Shuttle commenced long before that. As early as 1968, scientists were developing plans and specs for a fleet of vehicles that could deliver and retrieve *payloads* (in other words, cargo, such as experiments, equipment, and anything else that hitches a ride to or from orbit via the Space Shuttle) in space. The Space Shuttle's original goals were to

- Be reusable, thereby reducing some of the cost of space exploration
- Provide a means of encouraging commercial, military, and scientific achievements in space

The Space Shuttle program became a reality in 1972 when President Richard Nixon confirmed that the federal government would support the development of a reusable spacecraft that launched like a traditional rocket but landed on a runway like a glider. The concept for the Space Shuttle was a clear, three-part plan:

- A winged orbiter, which could be used repeatedly, would house the mission's crew. This orbiter would be 122 feet (37 meters) long, 78 feet (24 meters) wide from wing to wing, and 57 feet (17 meters) high.
- Reusable solid rocket boosters would help propel the spacecraft into space.
- An external single-use fuel tank would provide propellant for the orbiter's main engines on the trip to space.

By 1976, contractors had developed and started testing the *Enterprise,* the first Space Shuttle orbiter. It lacked engines and heat shields but provided enough of the core design to be suitable for glide-landing tests. After the concept of glide landing a spacecraft was proven, NASA was able to refine the orbiter's design and build *Columbia,* the first production Space Shuttle (Figure 13-1 shows the Shuttle's first launch).

**Figure 13-1:** The Space Shuttle, in all its technological glory.

*Courtesy of NASA*

The whole STS package consists of the orbiter plus the solid rocket boosters and external fuel tank; this configuration at launch is called the "stack." Yet the orbiter is really the guts of the Space Shuttle. It contains the pressurized crew cabin (which normally holds up to 7 crew members but can fit up to 10 in case of emergency), payload bay, engines, and orbital-maneuvering systems. The orbiter is organized into forward, mid, and aft fuselages, each with a different purpose (see Figure 13-2):

- The forward fuselage contains the crew cabin, cockpit, and operator station.
- The mid fuselage houses the payload bay.
- The aft (rear) fuselage contains the Shuttle's engines.

As the first partially reusable spacecraft, the Space Shuttle made "green" history. After all, space exploration is, in many ways, a history of spending billions of dollars on spacecraft that can never be used again. The original Space Shuttle plan, on the other hand, intended for each orbiter to survive ten launches apiece over the course of ten years, or about 100 total flights. In reality, the Space Shuttle fleet as a whole boasted nearly 125 flights as of late 2008, with each orbiter making between 10 *(Challenger)* and 35 *(Discovery)* flights over a 27-year period. (Of course, this greatly reduced flight frequency resulted in greatly increased preflight costs.)

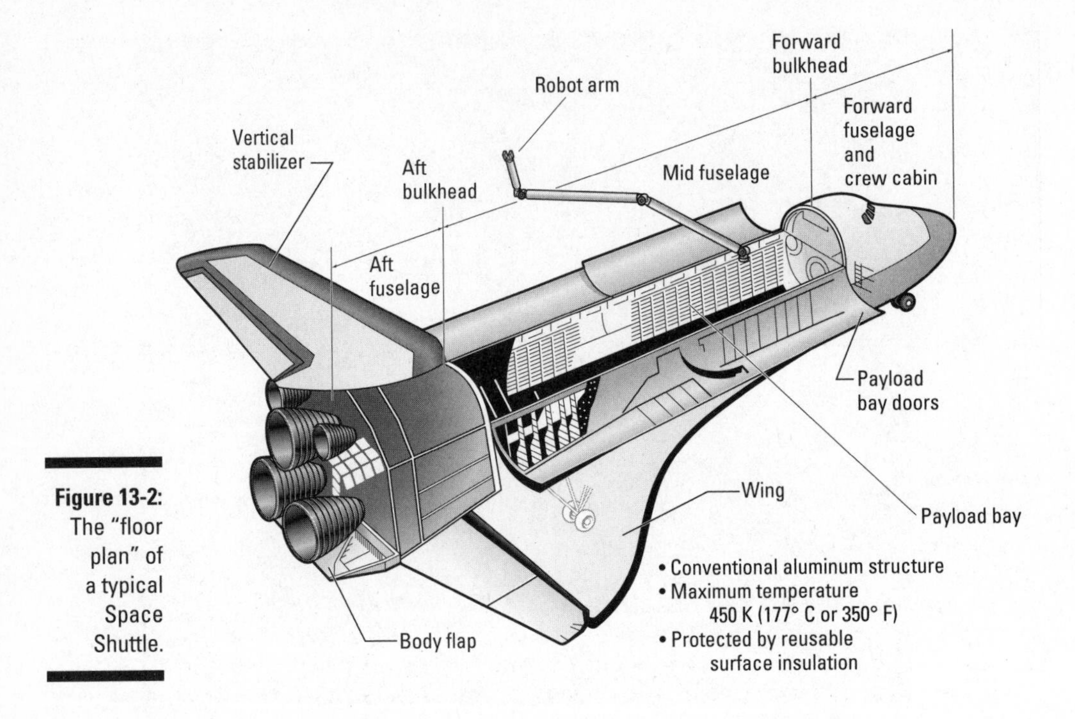


**Figure 13-2:** The "floor plan" of a typical Space Shuttle.

# Welcome Aboard: The Typical Course of a Space Shuttle Mission

Don't think you're astronaut material? Well, the next few sections are your chance to experience a Space Shuttle mission from launch all the way through landing. As a virtual passenger on the Space Shuttle, you're going to encounter three major phases of the mission (and configurations of your spacecraft):

- For launch, expect to be strapped into a seat with the Space Shuttle standing straight up on its engines, strapped to its boosters and fuel tank. After all conditions are safe for launch, the engines and solid rocket boosters ignite, leaving you squashed into your seat as you head into orbit.
- Once in orbit, you have time to perform scientific experiments, as well as practice your midair acrobatics and gaze out the window at home.
- Finally, it's time to strap yourself back into your seat for landing. Atmospheric reentry heats up the exterior of the orbiter, but the Space Shuttle's thermal tiles keep the interior (and you!) nice and cool. You only have one shot for a landing. Never fear though. Your expert pilot lands the spacecraft on the runway like a glider. Welcome back to Earth!

## Star Trekking

The first Space Shuttle orbiter was originally supposed to be called *Constitution,* but fans of the television show *Star Trek* staged a massive write-in campaign and convinced NASA officials to rename the orbiter *Enterprise.* Much of the cast of the original *Star Trek* series was on hand when *Enterprise* was dedicated. You can see the *Enterprise* in person at the new Udvar-Hazy Center in Virginia, an annex of the National Air and Space Museum (`www.nasm.si.edu/museum/udvarhazy`).

## Surveying the safety requirements for launch

Launching something that's as much of an investment of American time and resources as the Space Shuttle calls for some pretty intense safety precautions to protect both the orbiter and its crew. Weather is one of those issues that can cause some serious trouble for a Shuttle launch. That's why you can find NASA staff members carefully monitoring the weather nearby Florida's Kennedy Space Center (the official Space Shuttle launch site) on launch days.

In particular, they keep an eye out for storms with lightning. Apollo 12 was actually struck by lightning after launch on its way to the Moon, and although that mission was able to recover safely, mission controllers don't want to take that risk again! Consequently, launches don't take place when a potential lightning-causing cloud is within 10 nautical miles (18.5 kilometers) of the launch site. Weather must also be favorable at one of the abort landing sites located around the world, in case something goes wrong during launch.

Aside from positive weather conditions, in order for a Space Shuttle to be cleared for launch, it must successfully complete a range of preflight preparations and checks. All the onboard systems must be tested, and any critical problem that results halts the launch countdown.

Safely launching a Space Shuttle also requires being prepared for emergencies, particularly because history has shown that launch and landing are the two most dangerous parts of a spaceflight. Consequently, each Space Shuttle has five "abort modes" that allow it to safely terminate rather than complete its mission. (Want to know more about the Space Shuttle's abort modes? Check out the related sidebar nearby.)

## Aborting a Space Shuttle mission

A Space Shuttle doesn't have to complete its mission if it seems like something is wrong. Every Space Shuttle mission is designed so it can be terminated using one of the following five abort modes. The abort modes are progressed through sequentially with time after launch until the spacecraft successfully reaches orbit. Here's what each abort mode is and what it involves:

- **RSLS (Redundant Set Launch Sequencer):** This is the first abort mode. It can actually cancel the entire launch if flight computers detect an engine failure during the six-second period before the solid rocket boosters (SRBs) ignite. After the SRBs are ignited, however, they can't be turned off and the launch must proceed. Five RSLS aborts have taken place in the history of the Space Shuttle program.
- **RTLS (Return to Launch Site):** This second abort mode is available after the SRBs have burned out, two minutes after launch. The orbiter would turn around so that the firing main engines slow it down before descending to land back at Kennedy Space Center. This abort mode would involve landing about half an hour after takeoff, which is very risky. Perhaps that's why this abort mode has never been used.
- **TAL (Transoceanic Abort Landing):** This is the third abort mode. It would take place if the Space Shuttle took off but didn't have enough speed to reach orbit. The orbiter would land at a predetermined landing site across the Atlantic Ocean, about 25 minutes after takeoff. Several TAL sites in countries such as France, Spain, and Nigeria are always staffed and standing by during a Shuttle launch. Like the RTLS abort mode, TAL has never been used.
- **AOA (Abort Once Around):** This fourth abort mode would happen if the Shuttle has enough speed to orbit the Earth once and then land 90 minutes later. The circumstances for this mode are unlikely, and it too has never been used.
- **ATO (Abort to Orbit):** This fifth and final abort mode takes place when the orbiter can reach orbit, but at a lower level than originally intended. ATO was used in 1985 on mission STS-51-F, when one of the main engines shut down early (the mission was still successful despite the lower orbit).

# Launching the Space Shuttle

The Space Shuttle launch sequence proceeds as follows (note that negative numbers indicate time before launch and positive numbers indicate time after launch):

1. At T–9 minutes, an automatic launch hold takes place to make sure all systems check out; then the launch continues. The Ground Launch Sequencer, a computer program, is in charge of monitoring systems at this point.
2. At T–31 seconds, the Space Shuttle's onboard computers take over control of the launch.

3. At T–16 seconds, hundreds of thousands of gallons of seawater are pumped onto the launchpad to help protect both the orbiter and the neighboring buildings from the sound waves generated by the rocket thrusters.
4. At T–10 seconds, liquid oxygen and liquid hydrogen begin to flow into the main engines.
5. At T–6.6 seconds, the Space Shuttle's main engines ignite.
6. At T–3 seconds, all three engines must have full thrust. (If they don't, an abort mode is signaled.)
7. At T–0 seconds, the solid rocket boosters (SRBs) ignite (and can't be turned off). Explosive nuts release the Space Shuttle from the tower, and launch occurs.
8. At T+126 seconds, explosive bolts release the SRBs from the orbiter. The SRBs parachute into the ocean for recovery and reuse; the orbiter heads skyward by using its main engines.
9. At T+8 minutes, the main engines are shut off when their propellant is almost exhausted, and the external fuel tank is detached so it can burn up in the atmosphere.
10. At T+45 minutes, the Orbital Maneuvering System engines are fired, ensuring that the orbiter can reach a stable orbit.

## Living and working aboard the Shuttle

After reaching orbit, the Space Shuttle begins its mission (which usually lasts between 7 and 14 days). Although their most-common purpose these days is to bring supplies and *payloads* (cargo) to and from the International Space Station (described in Chapter 19), Space Shuttles are also capable of retrieving satellites from orbit, launching robotic spacecraft, and carrying advanced scientific instruments such as telescopes and instruments to map the Earth. Their large payload capacity was designed with space station construction in mind, but the sizeable payload bay has come in handy for a variety of other missions as well. (Check out the color section to see a Shuttle in orbit.)

### Van of the stars

Every Shuttle trip starts in a van — the Astrovan, that is. All Space Shuttle crews since 1984 have hitched a ride to the launchpad on the Astrovan. This NASA vehicle, essentially a modified motor home, transports the astronauts from their preflight housing out to the launchpad. The Astrovan has held up remarkably well despite its more than 20 years of NASA driving experience, mostly because it's only used for transporting astronauts to and from launches. It's specially designed to accommodate the astronauts in their launch suits.

The crew compartment is the main place where the astronauts live and work. It's separated into a flight deck, complete with navigational systems; a mid-deck with a galley, bunks, and exercise equipment; and a lower deck with a life-support system. The life-support system turns the Space Shuttle into an environment suitable for humans. It provides breathable air, drinkable water, food, electrical power, waste removal, temperature control, and a host of other essential functions. The Shuttle is also equipped with a communication system that allows the astronauts to talk with flight controllers back home.

Although much of the astronauts' time is taken up by required tasks, flight controllers make sure that they also get some scheduled relaxation time each day. Most astronauts find that looking out the windows at the Earth and taking pictures is one of their favorite ways to spend free time. Sunrises and sunsets are also beautiful to watch, and because they occur every 45 minutes in orbit, there are plenty of them! Astronauts also have time to read, exercise, and practice their acrobatics in microgravity.

If a particular Shuttle mission involves retrieving or deploying a satellite, or even deploying payloads into orbit, the astronauts use a tool called the *Remote Manipulator System* (RMS). This robotic arm is equipped with a camera and can be controlled from the flight deck; it can pick up and release payloads from the Shuttle's payload bay. Most current uses for the RMS involve the installation of modules and components on the International Space Station. The RMS also serves as an important safety device for inspecting the Shuttle's thermal tiles for damage after liftoff.

## Landing 101

The first part of coming home from a Space Shuttle mission involves reentering the Earth's atmosphere. During this journey, friction from the air heats up the outer surface of the Shuttle to more than 2,732 degrees Fahrenheit (1,500 degrees Celsius). The Shuttle's thermal tiles keep the orbiter and the astronauts safe and sound, though the fragility of this system was shown when Space Shuttle *Columbia* exploded during reentry in 2003 due to heat shield damage caused during launch (see Chapter 5 for details).

One major technological innovation of the Space Shuttle is its ability to return to Earth via the ground rather than the ocean. Unlike earlier American spacecraft that could land only by splashing down in the water, the Shuttle can leave Earth orbit and reenter the atmosphere on a carefully calculated trajectory that brings it to its landing site.

Computer programs control most aspects of the Space Shuttle's landing, though the pilot can land the spacecraft manually if necessary (for example, if a computer problem arises during reentry). However, the final approach and runway landing is usually done under manual control because the well-trained Space Shuttle pilots can land the Shuttle better than the computer can.

Regardless of who or what controls the landing process, the astronauts onboard the Space Shuttle have only one chance to land the spacecraft correctly — they can't circle around and try again because the orbiter doesn't have its own jet engines. And if you ever want a challenge, just try landing one of these puppies! In addition to having no propulsion source of its own during landing, the aerodynamics of the Shuttle are designed to help get rid of as much speed as possible before landing, unlike a conventional aircraft that's designed to stay aloft. That means the Shuttle basically returns to Earth in a barely controlled fall through the sky, requiring a skillful pilot to guide it to a safe landing. And although the pilot does have basic control over the orbiter's orientation, flying the spacecraft through the atmosphere is said to be similar to flying a brick!

As the Shuttle approaches the runway, it first deploys its landing gear. As the wheels touch down, the spacecraft then deploys a parachute to slow itself to a stop (see Figure 13-3).

**Figure 13-3:** The Space Shuttle uses a parachute to slow down as it lands.

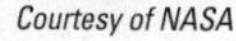

*Courtesy of NASA*

Although the Shuttle tries to land at Kennedy Space Center, weather conditions can force a landing at an alternate site. Out of the more than 20 such alternate Shuttle landing sites around the world, the two that have been used are Edwards Air Force Base in Southern California (49 landings) and White Sands Space Harbor in New Mexico (1 landing). If an alternate landing site becomes necessary, the Space Shuttle has to be attached to a Shuttle Carrier Aircraft and flown back home to Florida, because it has neither the means to take off under its own power nor engines that can let it fly in an atmosphere. NASA prefers to have the Shuttle land at Kennedy Space Center, because it costs time and money ($1.7 million to be exact) to transport it back from an alternate landing site. Seems funny to think that something as grand as the Space Shuttle has to hitch a ride back home, huh?

# The Early Days of the Space Shuttle Program

After the excitement of the Apollo Moon landings (Chapters 9 and 10) and the first American space station (Chapter 12), interest in space exploration waned in the U.S. Without the Cold War as a backdrop, NASA had to find its own justification for existence.

Enter the Space Shuttle. The futuristic design of this innovative spacecraft reinvigorated public interest in the space program, and NASA played up the idea of cheap, frequent, reusable access to space. Americans were justifiably proud of their new space vehicle, which was touted as able to make all sorts of scientific discoveries that could make life back on Earth better.

The following sections introduce you to the Space Shuttle fleet and highlight some of the most-notable achievements of the young Space Shuttle program. (Check out Chapter 14 for details on prime Shuttle missions of the late 1980s through the early 21st century.)

## The orbiter fleet

Once upon a time, the U.S. Space Shuttle program consisted of a fleet of five vehicles (all of which are listed in Table 13-1). The first two (*Columbia* and *Challenger*) are no longer operational due to devastating explosions (flip to Chapter 5 for the details). The remaining three are the Space Shuttles *Discovery, Atlantis,* and *Endeavour.* (Another vehicle, the *Enterprise,* was created for ground-landing test purposes and turned out to be too expensive to retrofit for spaceflight.)

**Table 13-1 The Space Shuttle Fleet**

| *Orbiter Name* | *Date of First Launch* | *Date of Last Launch* | *Number of Completed Missions* |
|---|---|---|---|
| Columbia | April 12, 1981 | January 16, 2003 | 28 |
| Challenger | April 4, 1983 | January 28, 1986 | 10 |
| Discovery | August 30, 1984 | TBD (2010?) | 35 (through 2008) |
| Atlantis | October 3, 1985 | TBD (2010?) | 29 (through 2008) |
| Endeavour | May 7, 1992 | TBD (2010?) | 22 (through 2008) |

*Columbia* and *Challenger* were the first two Space Shuttles to reach orbit (their notable achievements are described later in this chapter); *Discovery* was the third Shuttle to take flight. Lessons learned from the design of the first two orbiters allowed *Discovery* to be almost 7,000 pounds (3,175 kilograms) lighter than *Columbia.* On its maiden voyage in 1984, *Discovery* deployed a series of three communications satellites. Sixteen years later, *Discovery* was responsible for two rather notable missions:

- Deploying the Hubble Space Telescope
- Launching the solar explorer *Ulysses*

*Atlantis* took flight after *Discovery* and featured a number of improvements over its predecessors. For example, it was much quicker to build than *Discovery* because the thermal tiles were redesigned and their application became much faster as a result. *Atlantis's* big claim to fame is that it was the first Space Shuttle to successfully dock with Mir, the Russian space station (which we cover in detail later in this chapter).

Intended as a replacement for *Challenger, Endeavour* was the last Space Shuttle to join the fleet of orbiters. It pioneered a number of new features, including a drag chute to aid the Shuttle's glide landing, changes to the steering system, and an increased payload capacity. *Endeavour's* most-notable accomplishment to date is that it supported the longest *spacewalk* (the more-common term for an *extravehicular activity,* or EVA, performed in space) in history (eight hours) during its first spaceflight.

## The first flight: Columbia (1981)

The first Space Shuttle to take flight was *Columbia* during the 1981 STS-1 mission. STS-1 was essentially a test flight designed to ensure that all instruments, components, and life-support systems aboard the Space Shuttle were fully functional and ready for duty. During this no-pressure, decide-the-fate-of-the-entire-Space-Shuttle-program flight, Commander John Young and Pilot Robert Crippen proved that the Space Shuttle could be successfully orbited and returned to Earth. The mission was therefore deemed a success even though it uncovered a problem (specifically, damage to the heat-resistant tiling covering the spacecraft) that would have significant impact on future flights. Following the flight, changes were made to the way the launchpad was flooded with water at takeoff to absorb sound waves, resulting in far less tile damage on future flights.

Shuttle flights are usually numbered sequentially (though a numbering change employed for a few years in the mid-1980s used numbers and letters to designate the year, launch site, and flight number) and start with the STS designation (for *Space Transportation System,* the official name of the Space Shuttle program and also the official term for the combination of the orbiter, solid rocket boosters, and external fuel tank at launch). Missions sometimes fly out of sequential order due to changes in targets, training or equipment problems, or other issues.

### Fast facts about STS-1

Following are the basic facts about the STS-1 mission:

- **Launch date and site:** April 12, 1981; Kennedy Space Center, Florida
- **Spacecraft name and mass at launch:** Space Shuttle *Columbia;* 219,258 lb (99,453 kg)
- **Size of crew:** 2 astronauts
- **Number of Earth orbits:** 36
- **Mission duration:** 2 days, 6 hours, 20 minutes, 53 seconds

## Sally Ride, the first American woman in space (1983)

Among the Space Shuttle program's notable "firsts" was bringing America's first female astronaut into space. Sally Ride joined NASA's astronaut program in 1977 and spent years undergoing the various types of flight, survival, and navigation training required for astronauts. In 1983, she made history with her first spaceflight aboard the Space Shuttle *Challenger* (mission STS-7).

During their six days in space aboard *Challenger,* Ride and four other astronauts (Robert Crippen, Frederick Hauck, John Fabian, and Norman Thagard) became, at the time, the largest crew to travel in space. The astronauts deployed communications satellites while in orbit and performed a range of experiments in microgravity. Dr. Ride went on to fly on other Space Shuttle missions for a total of more than 340 hours in space. She was also one of the astronauts assigned to research the *Challenger* disaster of 1986.

## Notable early missions

Although each of the early Space Shuttle missions was significant, several from the 1980s stand out:

- **STS-5:** The fifth mission of Space Shuttle *Columbia* was the first fully operational flight of the Space Shuttle program (previous flights were designated test flights). Launching on November 11, 1982, the STS-5 mission sent four astronauts into space, the largest crew up to that point. The flight deployed two communications satellites, which were the first commercial satellites to be flown on the Space Shuttle.
- **STS-6:** Launched on April 4, 1983, with a crew of four, STS-6 was the first flight of Space Shuttle *Challenger.* Two astronauts performed the first spacewalk of the Shuttle program, testing out new spacesuits.

- **STS-41-D:** Launched on August 30, 1984, STS-41-D was the 12th flight of the Shuttle program and the first flight of *Discovery.* The crew of six astronauts included the first *payload specialist* (someone who isn't an official NASA astronaut; see Chapter 4 for more on this position), an employee of the McDonnell Douglas Corporation.
- **STS-51-J:** This 21st flight of the Shuttle program (and first flight of *Atlantis*) launched on October 3, 1985. The crew of five astronauts performed a classified mission for the U.S. Department of Defense.

Unfortunately, NASA soon ran into the problem that there was only so much that could be done in Earth orbit to capture the interest of the American public. In fact, by the mid-1980s, Space Shuttle launches had become routine as Americans were lulled into a false sense of invulnerability due to the lack of major disasters in the space program. The Teacher in Space program was designed to reinterest the public in the space program, and ironically the 1986 launch of *Challenger* drew a much larger viewing audience than previous flights. Unfortunately, the flight ended in tragedy just over a minute after it began (as we explain in Chapter 5).

# Mir, the Soviet/Russian Space Station (1986–2001)

Following the achievements of the Salyut space station program (see Chapter 12 for details), Soviet space research turned toward the creation of a more modular space station design that could accommodate more research labs, house more cosmonauts, and boast more-sophisticated capabilities. The space stations of the Salyut program were meant as one-shots, single modules to be visited by crews over the course of a few years, but not as permanent outposts in orbit. The many lessons learned and experiences gained from this series of stations made the Soviet Union the clear leader in long-term orbital missions.

Still smarting from the failure of their lunar exploration program, and faced with the futuristic Space Shuttle program from NASA, the Soviets saw the development of a long-term, modular space station as a clear way to meet both political and technical goals while boosting the space exploration profile of their country. The Mir program was developed, starting in 1976, to meet these goals. Ten years later, the station was launched into orbit. The Mir (which stands for "peace" in Russian) space station was a grand success that enhanced the reputation of the Soviet space program. Its existence also played a fundamental role in creating a new era of cooperation between the U.S. and the Soviet Union — cooperation that culminated with missions in which the Space Shuttle docked at Mir.

In the following sections, we describe the construction of Mir's modules, the collaboration of American astronauts and Russian cosmonauts, work and life on Mir, and the end of the Mir program.

## Building Mir, module by module

The basic design of the Mir space station focused on modules that were launched into orbit and attached together in space. The Core Module, which went up first in 1986, held the station's control center as well as living space for the cosmonauts. As more Mir modules were prepared and ready to go on the ground (see Table 13-2), other teams of cosmonauts launched and prepared the space station to accept its new arrivals.

**Table 13-2 Modules of the Mir Space Station**

| *Module Name* | *Launch Date* | *Launch Vehicle* | *Mass* | *Function* |
|---|---|---|---|---|
| Mir Core Module | February 19, 1986 | Proton-K | 44,313 lb (20,100 kg) | Control center, living space |
| Kvant-1 | March 31, 1987 | Proton-K | 22,000 lb (10,000 kg) | Scientific instruments, gyroscopes |
| Kvant-2 | November 26, 1989 | Proton-K | 43,300 lb (19,650 kg) | Scientific instruments, life-support systems, new airlock |
| Kristall | May 31, 1990 | Proton-K | 43,300 lb (19,650 kg) | High-resolution cameras, scientific experiments, docking port |
| Spektr | May 20, 1995 | Proton-K | 43,300 lb (19,650 kg) | Earth-observation equipment, solar arrays |
| Docking Module | November 12, 1995 | *Atlantis* (an American Space Shuttle) | 13,500 lb (6,100 kg) | Avenue for Space Shuttle to dock with Mir |
| Priroda | April 23, 1996 | Proton-K | 42,000 lb (19,000 kg) | Scientific experiments |

Because existing rockets had limits in terms of how big and how heavy their payloads could be, this modular approach to space station construction allowed for Mir to be much bigger than previous stations. Additionally, it was much cheaper to launch individual completed modules one at a time. Some of these modules could automatically dock with each other, whereas others were attached together by cosmonauts during spacewalks.

Because the launch of the station itself was something of a rush job due to political deadlines, a lot of work was left to be done in orbit to get Mir up and running. The first cosmonauts to reach Mir, Leonid Kizim and Vladimir Solovyov, got a ride on the Soyuz T-15 mission in March 1986, which remained with Mir for about 50 days before taking a trip to the older Salyut 7 space station, which was still in orbit. While inside Mir, Kizim and Solovyov helped set up the new space station and performed a number of experiments. Future missions brought more cosmonauts to Mir, where they were able to make repairs and do science as needed.

## Welcoming American astronauts

With the fall of the former Soviet Union in 1991 (and subsequent end of the Cold War), a new era of cooperation was suddenly possible between Americans and Russians. Joint space missions once again became a way to meet the two superpowers' political and scientific goals. As a way of discovering more about the business of constructing modular space stations, the U.S. became involved with the Mir project, sending both supplies and astronauts to the now-Russian space station.

### The Soviet shuttle that never took astronauts to space

With all their space station expertise, did the Soviets ever develop their own version of the Space Shuttle? The answer, perhaps surprisingly, is yes, although it never brought cosmonauts into space. The Buran program was the Soviet Union's answer to the American Space Shuttle. A reusable spacecraft that looked very similar to the NASA vehicle, the Buran had a few important differences:

- Although it launched like a rocket, the Buran was carried into orbit as payload on the large Energia rocket because it didn't have its own rocket engines.
- It could be flown completely automatically. The lack of a need for a human crew removed the necessity of a hair-raising first human flight of an untested spacecraft.

Five Burans were commissioned, but only one was ever deemed spaceworthy. The sole launch of the Buran program took place on November 15, 1988. The Buran spacecraft, with its life-support system and computer displays only partially completed, successfully launched into orbit, circled the Earth twice in just over three hours, and then made a completely automated landing on a runway at the Baikonur Cosmodrome. All seemed well until the Soviets realized the spacecraft had suffered damage to its thermal protection tiles during the flight — damage that would've been costly to fix.

Due to political circumstances in the soon-to-collapse Soviet Union, the project was initially put on hold before being canceled by the Russian government in 1993. No other Buran flights ever occurred.

In order for the Americans to contribute resources and astronauts to the Mir program, the Space Shuttle had to get involved. This exchange was conducted through the Shuttle-Mir program, which was also seen as the predecessor to the International Space Station (then in the planning stages). As part of the Shuttle-Mir program, a series of 11 Shuttle missions docked with Mir between 1994 and 1998. One of these missions launched an American astronaut into space in a Soyuz spacecraft and a Russian cosmonaut into space on the Space Shuttle. (Figure 13-4 shows the Space Shuttle *Atlantis* docked with Mir.)

**Figure 13-4:** The Russian space station Mir in orbit around the Earth and docked with the American Space Shuttle *Atlantis.*

*Courtesy of NASA*

## Fast facts about Mir

Following are the basic facts about the Mir space station:

- **Launch date and site:** February 19, 1986; Baikonur Cosmodrome, USSR
- **Spacecraft name and mass at launch:** Mir; 274,123 lb (124,340 kg)
- **Launch vehicle:** Proton rocket
- **Size of crew:** Launched unoccupied; visited by a total of 137 cosmonauts and astronauts
- **Number of Earth orbits:** About 88,000
- **Length of time in Earth orbit:** 5,519 days (4,592 of which were occupied)

The new era of cooperation in space began in 1994 with the STS-60 mission of Space Shuttle *Discovery,* which brought Russian cosmonaut Sergei Krikalev into space and included audio and video conversations with the Mir crew. In 1995, *Discovery* performed a rendezvous with Mir that included a close approach and a flight around the station as a dress rehearsal for an actual docking.

From 1996 through 1998, Mir was home to astronauts and cosmonauts working together to perform numerous tests and spacewalks. The astronauts also had a chance to test out the Russian spacesuits and equipment (and vice versa).

## Discovering how Mir residents spent their time

Daily life on Mir involved a combination of work, play, and exercise. Cosmonauts wore lightweight flight suits while they moved between the different modules, which connected together via nodes. As with any space station, the cosmonauts floated rather than walked to work; they also had to keep a close grip on portable equipment as they moved about during the day. (Check out the quarters on Mir in Figure 13-5.)

**Figure 13-5:** NASA astronaut Shannon Lucid exercises inside Mir.

Courtesy of NASA

Adjusting to weightlessness is one of the challenges faced by anyone who spends time in space. Mir residents were still able to brush their teeth, use the bathroom, shave, and perform most of the other basic functions that they did on Earth, but they had to do so rather differently (darn that lack of gravity). The station's zero-gravity space toilet, for example, used a hose with flexible tubing; each person aboard Mir was given his or her own funnel to use.

Astronauts and cosmonauts conducted significant experiments aboard Mir. Many were geared toward helping scientists understand more about how people and plants can live and function in the microgravity environment of space. Mir cosmonauts also performed astronomical observations and studied the Earth from orbit.

As the Mir space station aged, cosmonauts had to spend more and more time repairing it. In fact, on a typical mission with a mixed crew of three people, the two Russian cosmonauts devoted most of their time to space station maintenance while the American astronaut performed scientific experiments. The cosmonauts' quick-thinking abilities were put to the test on two occasions:

- On February 23, 1997, a small fire broke out in one of the modules due to a malfunctioning oxygen generator. The crew aboard Mir was able to quickly extinguish the flames, which did minor damage to nearby equipment.
- On June 25, 1997, a robotic supply ship crashed into one of Mir's solar panels while attempting to dock with the station. The collision resulted in a puncture to the hull of the struck module, which quickly lost pressure and had to be sealed off from the rest of the station to avoid a stationwide pressure loss. Partial power was eventually restored to Mir from the damaged module's solar panels thanks to a series of spacewalks, but the module couldn't be fixed and was left depressurized.

## Observing the end of an era

After almost ten years of continual habitation, the last crews left Mir in mid-1999. A brief, privately funded mission returned to the Russian space station in 2000 to try to salvage it as a commercial enterprise of some sort, but to no avail. With the Russian Federal Space Agency turning its support to developing the new International Space Station, Mir's long lifetime was officially over.

Flight controllers allowed Mir's orbit to decay as the drag of Earth's atmosphere gradually slowed it down. Robotic supply ships brought extra fuel to the station and then used their engines to bring it into an orbit that would intersect with Earth's atmosphere in a controlled way to ensure that most of the station burned up and the remnants fell harmlessly into the ocean.

The return of the 135-ton space station to Earth in March 2001 was visible with the naked eye. Observers reported seeing a bluish flare, followed closely by a sonic boom as the space station reentered the atmosphere. Mir completely broke apart upon reentry, as expected, and dropped into the Pacific Ocean off the coast of Australia.

# Chapter 14

# Prime Time for the Space Shuttle: From the 1980s into the 21st Century

### *In This Chapter*

- Shuttling major missions into space
- Conducting experiments in space

The Space Shuttle, as you find out in this chapter, became a showpiece of the American space program in the 1980s and continued its flights through the 1990s and into the early 21st century. During this time, the Space Shuttle launched interplanetary spacecraft such as *Galileo* and satellites such as the Hubble Space Telescope. The Shuttle astronauts also performed a wide variety of science experiments from space; examples include growing crystals in very low gravity and taking detailed topographic maps of the Earth.

The Space Shuttle program is slated for retirement in 2010. Its final few years of missions will be devoted almost entirely to helping complete the International Space Station (described in Chapter 19), with a single mission planned to repair and upgrade the Hubble Space Telescope one last time. The Space Shuttle is scheduled to be replaced by the Orion series of spacecraft, an Apollo-style, capsule-based design that we tell you about in Chapter 21.

## Launching Massive Missions from the Space Shuttle

In addition to its success as a new-and-improved vehicle for human space exploration, the Space Shuttle was also used as a launch platform for a variety of massive robotic space missions. Some of the most important missions launched from the Space Shuttle include the *Galileo* spacecraft and the Hubble Space Telescope.

## Atlantis gives Galileo a lift (1989)

Although the Space Shuttle had many historic launches, one notable mission was STS-34, otherwise known as the mission that sent the robotic spacecraft *Galileo* into orbit. This probe, launched in October 1989 aboard Space Shuttle *Atlantis,* was designed to photograph and study Jupiter and its moons (a combination called the *Jovian system*).

Here's a fun bit of trivia for you: *Galileo* was actually ready to go before *Atlantis* was! Because of the Space Shuttle *Challenger* explosion in 1986 (see Chapter 5), the *Galileo* spacecraft sat — fully assembled, no less — in a warehouse for more than two years before the Space Shuttle program finally returned to flight.

*Galileo* was designed to fit inside the Space Shuttle's payload bay, making it the main payload for mission STS-34. But in order to fit inside the payload bay, the probe's main antenna had to be folded like an umbrella. Although *Galileo* was successfully launched from the Space Shuttle, when engineers went to open the antenna, they found that one of the ribs had become stuck during the probe's long storage period. Because the main antenna was unable to open fully, the amount of data the *Galileo* spacecraft could return to Earth was severely limited (although the mission was still considered a success, as you can read about in Chapter 17).

## Discovery launches the Hubble Space Telescope (1990)

One of the most-amazing objects launched into space by the Space Shuttle was the Hubble Space Telescope. A giant optical telescope sized to fill the Shuttle's payload bay, the Hubble is currently the largest space telescope of its kind. It was designed via a massive international collaboration between the European Space Agency (ESA) and NASA.

### Fast facts about STS-34

Following are the basics facts about the STS-34 mission:

- **Launch date and site:** October 18, 1989; Kennedy Space Center, Florida
- **Spacecraft name and mass at launch:** Space Shuttle *Atlantis;* 257,568 lb (116,831 kg)
- **Size of crew:** 5 astronauts
- **Number of Earth orbits:** 79
- **Mission duration:** 4 days, 23 hours, 39 minutes, 20 seconds

## Fast facts about STS-31

Following are the basics facts about the STS-31 mission:

- **Launch date and site:** April 24, 1990; Kennedy Space Center, Florida
- **Spacecraft name and mass at launch:** Space Shuttle *Discovery;* 259,233 lb (117,586 kg)
- **Size of crew:** 5 astronauts
- **Number of Earth orbits:** 80
- **Mission duration:** 5 days, 1 hour, 16 minutes, 6 seconds

Like *Galileo* (which we describe in the preceding section), Hubble had to wait its turn for launch in the aftermath of the 1986 *Challenger* disaster. In April 1990, mission STS-31 finally received the green light to fly (changes in the launch schedule caused STS-31 to end up flying later than STS-34, which launched *Galileo*), and Space Shuttle *Discovery* deployed Hubble.

To make sure it was completely free of the interfering effects of Earth's atmosphere, Hubble needed to be placed in an orbit high above Earth. To achieve this feat, *Discovery* had to reach a higher altitude than any Space Shuttle had reached before. As a result, astronauts were able to take both still photographs and IMAX movies of the Earth from this high orbit.

Hubble was designed to be upgraded and serviced by future Space Shuttle missions, and this design capability proved fortunate when the telescope's primary mirror was discovered to be misshapen, resulting in blurry images. Astronauts were able to restore Hubble to crystal-clear imaging capability during a marathon servicing mission, and they've visited the telescope for subsequent upgrades as well (see Chapter 18 for more info).

### Other notable launches

From the late 1980s through the 1990s, the Space Shuttle had a number of other important missions to launch spacecraft, telescopes, and space station supplies. Missions flown during this time period include the following:

- **STS-30:** Space Shuttle *Atlantis* launched the *Magellan* spacecraft on May 4, 1989, so it could begin its voyage to Venus (which we describe in Chapter 16). This was the first launch of an interplanetary space probe by the Space Shuttle.

- **STS-41:** On October 6, 1990, Space Shuttle *Discovery* launched the *Ulysses* spacecraft, which used a flyby of Jupiter to skew its orbit out of the flat plane of the solar system so it could enter an orbit that takes it over the poles of the Sun. *Ulysses* was jointly built by NASA and ESA. It's still studying the Sun's magnetic field.

  You can visit either of these Web sites to find out more about the amazing discoveries made by the *Ulysses* spacecraft:

  - `ulysses.jpl.nasa.gov`
  - `ulysses-ops.jpl.esa.int`

- **STS-37:** Space Shuttle *Atlantis* launched the Compton Gamma Ray Observatory on April 5, 1991. This telescope was designed to observe strange high-energy outbursts from distant stars and galaxies (as explained in Chapter 18). It remained in low Earth orbit until 2000.

- **STS-49:** On May 7, 1992, Space Shuttle *Endeavour* took its first flight with a powerhouse crew of seven astronauts. Even though the Space Shuttle wasn't really intended to retrieve errant satellites, spacewalking astronauts succeeded in fetching the *Intelsat VI* from low Earth orbit, attaching it to a new motor, and relaunching it into the correct orbit. In the process, this mission supported the world's first three-person simultaneous *spacewalk* (an *extravehicular activity* [EVA] done in space), as well as the longest spacewalk at the time (8 hours and 29 minutes).

- Between 1994 and 1998, 11 separate Space Shuttle missions docked with the Mir space station to bring crews and supplies. (Chapter 13 has full details on Mir.)

- Beginning in 1998, the Space Shuttle's main purpose became the construction of the International Space Station (ISS). This multinational station couldn't have been built without the Shuttle's capability of lifting large *payloads* (cargo, such as the various modules that make up the ISS) into orbit. (Flip to Chapter 19 for the scoop on the ISS.)

- **STS-93:** On July 23, 1999, Space Shuttle *Columbia* launched the Chandra X-Ray Observatory, a telescope that orbits the Earth and makes images of the sky using high-energy X-rays, which are absorbed by the Earth's atmosphere and can't be observed from the ground. This telescope is still successfully collecting data (see Chapter 18).

The Space Shuttle has also been used to launch a series of tracking and *data-relay satellites,* which are communication satellites that NASA uses to talk to the Shuttle and the ISS. Five of these essential satellites were successfully launched on Space Shuttle missions between 1988 and 1995.

Some Shuttle missions have been classified, such as those used for launching U.S. Department of Defense satellites. The Shuttle has also launched satellites that were part of the Defense Satellite Communications System.

# Doing Science aboard the Space Shuttle

The Space Shuttle wasn't just a tool for ferrying around astronauts, satellites, space probes, and supplies. It was also a unique opportunity for doing science in space. Toward that end, NASA and the European Space Agency (ESA), in the 1980s, committed to developing a mobile laboratory, Spacelab, that could function inside the Space Shuttle.

Spacelab provided a platform for collaboration between the United States and Europe, considering the first entire laboratory module was provided by ESA, as were later components. Astronauts from several different countries eventually conducted research and experiments within Spacelab.

Other science payloads aboard the Space Shuttle involved astronomical and remote sensing equipment that could be used by astronauts to study the stars, the growth of materials in space, or the Earth.

## Spacelab (1983–2000)

The basic organization of Spacelab (the Space Shuttle's laboratory module — not to be confused with the space station Skylab, which we cover in Chapter 12) revolved around differing configurations of modules and pallets in the Space Shuttle's payload bay. What's the difference?

- *Modules* were pressurized to allow the astronauts to do their work.
- *Pallets* contained the large, remotely operated equipment, such as large telescopes, that had to have a direct output into space. Pallets were moveable, nonpressurized segments, so they couldn't be used for human work without a spacewalk.

Spacelab modules and pallets could be customized to accommodate the research planned for a particular mission (see Figure 14-1). Each overall module consisted of a core module, an experiments module, and between one and three pallets, depending on the mission and its configuration. Access tunnels and adapters were also included for astronauts to enter and leave Spacelab from the crew cabin on the Space Shuttle's flight deck. Much like the reusable Space Shuttle orbiter, Spacelab could be brought back to Earth after its mission was completed and reconfigured to support research objectives on a future mission.

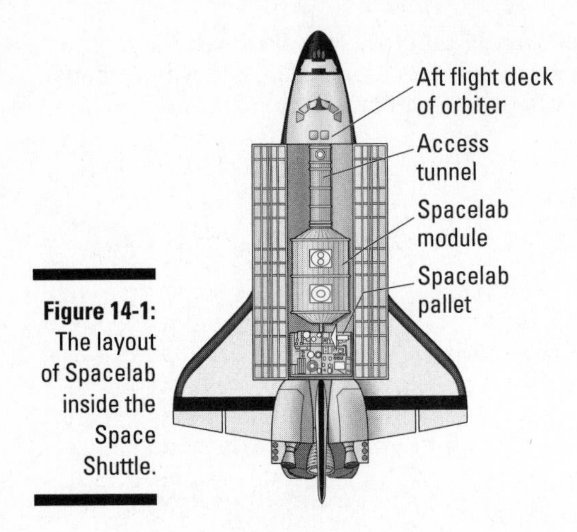


**Figure 14-1:** The layout of Spacelab inside the Space Shuttle.

The first Spacelab, Spacelab 1, was launched aboard Space Shuttle *Columbia's* STS-9 mission in November 1983. All in all, 25 Space Shuttle missions launched with Spacelab components. The final flight with a Spacelab element was STS-99, aboard Space Shuttle *Endeavour* in February 2000.

Spacelab was a proving ground, in a sense, for the Space Shuttle's scientific capabilities. Numerous experiments were conducted within Spacelab in areas of biology, crystallography, and, of course, astronomy. In addition, Spacelab proved that real science could be done in *microgravity* (an area with a very small gravitational force), knowledge that proved immensely useful for future work and research.

Spacelab also broke new ground in that it was a hands-on experience for the astronauts. Previously, most science in space was conducted remotely through ground-operated instruments. But thanks to Spacelab, *payload specialists,* who were experts in their scientific fields, were specifically trained to become scientist-astronauts. Spacelab was thus a place where scientists could conduct their own experiments and study the resulting data, in addition to fixing broken equipment and otherwise supporting ongoing research.

## U.S. Microgravity Payloads (1992–1997)

Space Shuttle missions of the 1990s offered the opportunity to test the effects of microgravity. Astronauts in orbit experience weightlessness because they're in free-fall as they move around the Earth, but some slight gravitational effects, dubbed microgravity, still exist due to the gravitational attraction of objects to each other and to the Space Shuttle itself. These effects can't be tested on Earth, so they must be tested in space.

Science-heavy payloads known as the U.S. Microgravity Payloads were deployed on specific Shuttle flights from 1992 through 1997. These payloads consisted of a series of experiments that studied things like material properties, combustion, and the physical behavior of fluids, all under microgravity conditions. These experiments have had impacts on both fundamental science and commercial applications.

One interesting experiment type studied the growth of crystalline structures in microgravity. It found that without the pull of gravity, these structures can develop intricate formations that help scientists learn about the theoretical basis of crystal growth under ideal conditions. Such crystals also have important commercial possibilities. For example, large, uniform, pure-protein crystals can be grown in space and studied by pharmaceutical companies back on Earth to reveal fine structural details. This information could help scientists understand how these proteins interact with the human immune system so they can use this information to design more-effective medicines to save lives.

## The Wake Shield Facility (1994–1996)

The Wake Shield Facility (WSF), a large disk that measured 12 feet (3.7 meters) in diameter, was a stand-alone scientific research payload designed to study materials grown in the conditions of pure vacuum. The Space Shuttle launched it into space three times during missions in the mid-1990s. The Shuttle released the WSF into space independently before retrieving it a few days later. With the help of the Shuttle's robotic arm, the WSF then flew behind the spacecraft to take advantage of the vacuum that exists in the Shuttle's wake.

The Space Shuttle actually deflects most of the remnants of Earth's atmosphere that exist at the 186-mile (300-kilometer) mark above the planet. The design of the WSF served to deflect any remaining particles, creating a near-perfect vacuum environment. The WSF used this vacuum to grow thin films of material. Materials grown on the WSF included crystalline semiconductor thin films. Such materials science research has helped the semiconductor industry develop better computer chips and improve other devices, such as photocells.

## The Shuttle Radar Topography Mission (2000)

The Shuttle Radar Topography Mission (SRTM) had a lofty goal: to produce three-dimensional maps of the Earth at a very high resolution. The project consisted of a science payload launched on mission STS-99, aboard Space Shuttle *Endeavour* in 2000, and it used radar mapping to gather topographic data over 80 percent of the Earth's surface.

To accomplish this goal, two radar antennae were located in different parts of the Space Shuttle: one was in the payload bay, and another extended out from the Shuttle on a long mast (see Figure 14-2). Radar signals went out from the two antennae, bounced off the Earth's surface, and then returned to space where they were detected again by the antennae. By carefully correlating the differences in how these signals returned to the two separated antennae, computer algorithms could use this data to determine the height of the surface under the radar, resulting in maps that show not only shapes of surface features but also their altitudes. This data could later be pieced together and composited into global maps.

**Figure 14-2:** The SRTM extended a 200-meter mast from the Space Shuttle to produce stunning 3-D maps of the Earth's surface.

*Courtesy of NASA*

The SRTM mission was important because it created the best high-resolution topographic mapping of the Earth that had ever been done (see the color section for an example). This information gave scientists critical data on how high mountains, continents, and other features of the Earth really were, as opposed to their previous best ground-based estimates. Knowing the absolute heights of various geologic features is crucial to understanding the geophysics related to their formation and subsurface structure.

# Chapter 15

# Revenge of the Red Planet: Successes and Failures in Mars Exploration

### *In This Chapter*

- Failing to succeed: The Phobos spacecraft and *Mars Observer*
- Catching a break with *Mars Pathfinder* and *Mars Global Surveyor*
- Encountering problems with *Mars 96, Mars Climate Orbiter,* and *Mars Polar Lander*

Robotic exploration of Mars took a long hiatus following the failure of NASA's 1976 Viking missions to detect life, although not for lack of trying. The Soviet Union's Phobos spacecraft failed due to computer glitches in the late 1980s, and the United States' *Mars Observer* spacecraft was lost in the early 1990s, further depressing the status of NASA's planetary exploration program. Then at long last came the Mars Pathfinder mission (1996–1997), which used the emerging Internet to capture the imagination of Americans. This successful NASA mission was closely followed by the *Mars Global Surveyor* orbiter. Despite the loss of two other NASA missions and one Russian mission, the 1990s, as you find out in this chapter, were marked by a string of successful missions that revolutionized society's view of Mars.

## Stumbles on the Road to Mars

By many accounts, early exploration of Mars courtesy of the United States' Viking missions was, in some ways, almost too successful. The orbiters produced amazing maps of the surface, and the landers made soil measurements and watched the seasons change. Additionally, all components of Viking were incredibly durable, lasting for years beyond their expected lifetime. However, the Viking missions (which you can read about in Chapter 11) had failed where it really counted: finding life on Mars.

### The Mars "curse"

Space exploration is a difficult business, and not all missions are successful. Mars missions in particular have had a bad track record: Between 1960 and 2008, 42 missions were launched to Mars, but only 18 have been completely successful, with 4 more partial successes (missions that either landed or reached orbit successfully but stopped transmitting data after sending back only a few pictures or measurements). This success rate, of a little more than 50 percent, has led to humorous speculation about a "Galactic Ghoul" that survives on a diet of Mars space probes.

A far more likely explanation is that Mars is a fascinating yet technically challenging target. On one hand, the dramatic scientific interest in exploration of the Red Planet has led to the launch of many missions. On the other hand, the difficulty of landing in the thin Martian atmosphere, plus the technical complexity of many of these spacecraft, has led to many failures. As scientists discover more about Mars, and about building interplanetary spacecraft, Mars missions are becoming more reliable, and the success rate is slowly creeping up.

Despite the Viking landers' suites of instruments specifically designed to look for signs of life, nothing was sent back to Earth to indicate life exists on Mars. If the idea of potential life on Mars had excited the public worldwide, the Viking missions' negative results definitely soured public support for future Mars exploration. Building scientific and political momentum for a new crop of Mars missions took time, and those plans suffered further setbacks from a series of failed Soviet efforts in the late 1980s and disappointing American attempts in the early 1990s.

## Computer failures for Phobos 1 and 2 (1988–1989)

Renewed exploration of the Red Planet didn't start out well. In fact, the concept of Murphy's Law comes to mind. . . . The first post-Viking Mars missions were a pair of ambitious spacecraft developed by the Soviet Union. As part of the Phobos program, the Soviets sent a pair of robotic spacecraft, *Phobos 1* and *Phobos 2,* to Mars to study both the planet itself and its two small moons, Phobos and Deimos. The missions launched a few days apart in July 1988.

Both Phobos spacecraft featured a new design that consisted of an instrumentation section surrounded by electronics, sphere-shaped tanks that contained fuel for controlling the spacecraft's altitude and orbit, and winglike solar panels for power.

*Phobos 1* seemed to go as planned until it dropped communication less than a month after launch. The Soviets eventually discovered that a software glitch had turned off the spacecraft's attitude thrusters, which were responsible for controlling the orientation of the spacecraft so that it could use its solar panels to keep the batteries charged. Because *Phobos 1* never made it into orbit around Mars, no useful mission data was collected.

*Phobos 2* successfully orbited Mars and managed to return about 40 images before losing contact with Earth in March 1989, just before it was supposed to deploy a small lander onto the surface of Phobos. This time a computer failure was to blame. Future Soviet probes to Mars and other planets were all put on hold thanks to the 1991 breakup of the Soviet Union. The Russian Federal Space Agency took over the Soviet Union's space work and began putting all of its dwindling resources into keeping Russia's cosmonaut program and Mir space station running (you can read about Mir in Chapter 13).

## Mars Observer goes missing in action (1992)

NASA's *Mars Observer* was supposed to study the surface material of Mars in depth, in addition to determining the nature of Mars's gravitational and magnetic fields and obtaining details on Martian weather and atmosphere. It was also meant to be the first of a series of missions, much like the Mariner and Pioneer mission series of two decades earlier (see Chapter 11). The Observer missions were intended to study the inner solar system, but Mars Observer was the only one of these missions to launch. Needless to say, it wasn't a crowning achievement for NASA (perhaps in part due to the fast-track, low-cost nature of the project).

Launch presented the first bad omen. A routine inspection of the spacecraft (which was already tucked inside its launch vehicle) prior to launch showed that its surface had been contaminated with debris, including metal filings and paint chips, due to ventilation with an impure source of nitrogen gas. The spacecraft had to be taken off the rocket booster for cleaning and sterilization. Fortunately, the delay was relatively minor, and *Mars Observer* was launched into space a day later on September 25, 1992.

At first all seemed normal, and *Mars Observer* successfully traveled through interplanetary space toward Mars. However, the spacecraft lost contact with Earth a few days before it was scheduled to arrive in Martian orbit. All future attempts to contact it failed, and a panel was ultimately convened to determine the cause of the failure. Scientists later decided that the most likely problem was a rupture of a contained propellant line after the system was pressurized just before the planned rocket burn to enter orbit. The mission was a failure in terms of gathering data, but NASA engineers were able to upgrade future missions to account for what they think derailed *Mars Observer.*

# Success at Last! Mars Pathfinder and Mars Global Surveyor

The failure of Mars Observer, which was a big flagship mission, left NASA's planetary exploration program in an uproar. Scientists and engineers had lost large chunks of time and effort, and the agency itself had lost funding from the federal government. NASA was therefore ready for a new philosophy, summarized by the mantra "faster, better, cheaper," and the new Discovery Program (which started with the *NEAR* spacecraft, as you find out in Chapter 16) was the emblem of this approach.

Instead of risking it all by sending a single, huge mission into space, Discovery missions were designed to support innovations and try out new techniques. Mars Pathfinder, the second Discovery mission, was a serious shake-up of the Mars Exploration Program at NASA; *Mars Global Surveyor* (though not actually a Discovery mission) became the first orbiting post-Viking success story with its detailed views of the Martian surface.

## Mars Pathfinder: Innovative design meets the little rover that could (1996–1998)

As part of NASA's Discovery Program, the Mars Pathfinder mission was seen mainly as a testbed for trying out technologies, innovations, and equipment intended for later missions. It was also responsible for conducting experiments designed to reveal more about the structure, topography, and composition of the Martian surface.

The mission design of Mars Pathfinder was ambitious and risky, including many technologies never before attempted by NASA. Most notably, it included not only a lander but also a small rover, the iconic *Sojourner.* The *Pathfinder* spacecraft itself also featured an innovative landing technique. Instead of using the standard retrorockets that had aided the landing of the Viking spacecraft two decades before, *Pathfinder* slowed its descent with parachutes. It then inflated giant airbags so it could bounce along the Martian surface before coming to a rest. If any Martians had been watching, they would've thought a giant beach ball had come to visit!

On July 4, 1997, seven months after its launch, the solar-powered *Pathfinder* bounced to a safe landing in the Red Planet's Ares Vallis area. After the *Sojourner* rover rolled out of the lander, it got right to work by driving over to interesting rocks to examine them. Valuable data was sent back over a three-month period, about two months longer than the mission was expected to last.

The six-wheeled *Sojourner* rover, a veritable lightweight at only 23 pounds (10.4 kilograms), became the first successful rover to explore Mars. *Sojourner* contained three separate cameras, a laser tool for avoiding hazardous rocks and craters, and many other instruments designed to take measurements of Martian soil and surface properties.

The *Pathfinder* lander wasn't sitting idly by while *Sojourner* worked. It was busy taking stereo images of the landing site, monitoring *Sojourner*'s progress, and studying the weather. Figure 15-1 shows a view of the *Pathfinder* lander and the *Sojourner* rover on Mars, with a map of where the rover traveled during its mission.

Mars Pathfinder, and in particular the *Sojourner* rover, marked one of the first NASA missions to fully take advantage of the emerging Internet. Millions of people from around the world followed the mission via the Web site, logging in daily to see the latest pictures. The novelty of watching a rover drive around the surface of another planet captured the imagination of the general public, and *Sojourner* had a huge impact on society's renewed interest in Mars exploration.

Interested in checking out archived *Sojourner* and *Pathfinder* images for yourself? Go to `mars.jpl.nasa.gov/MPF/index1.html`.

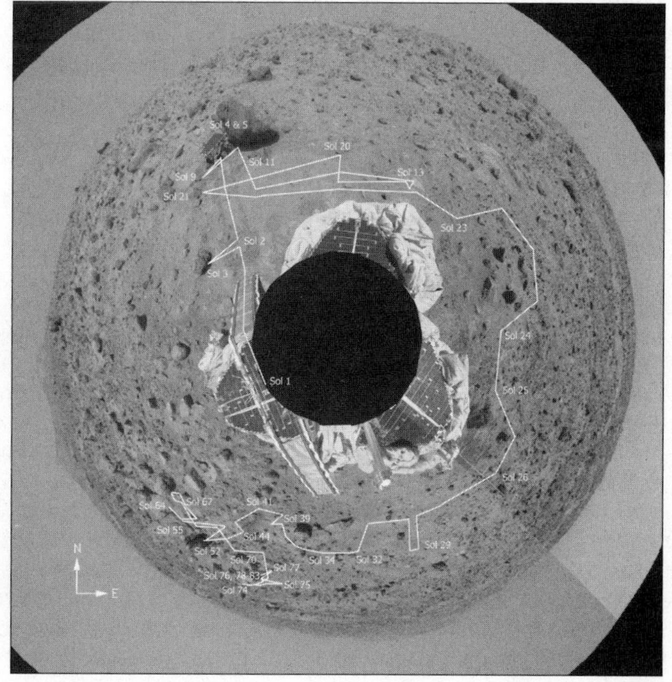


**Figure 15-1:** A view of *Mars Pathfinder* and *Sojourner,* with a map of *Sojourner's* travels.

*Courtesy of NASA/JPL*

### Fast facts about Mars Pathfinder

Following are the basic facts about the Mars Pathfinder mission:

- **Launch date and site:** December 4, 1996; Cape Canaveral Air Force Station, Florida
- **Spacecraft names and masses at launch:** Mars *Pathfinder* Lander, 582 lb (264 kg); *Sojourner* Rover, 23 lb (10.5 kg)
- **Launch vehicle:** Delta II 7925
- **Number of Mars orbits:** 0 (landing mission)
- **Mars landing date:** July 4, 1997
- **Length of time on Mars surface:** 3 months

## Mars Global Surveyor: Pictures and topography from orbit (1996–2006)

The Mars Global Surveyor (MGS) mission, a separate mission from the Discovery Program, actually marked the beginning of the United States' return to Mars because it launched a month before the Mars Pathfinder mission. However, it arrived in orbit around the Red Planet later than *Pathfinder.* To date, the Pathfinder mission still gets all the attention (largely because of *Sojourner*), making MGS the first runner-up to Pathfinder's Miss Universe.

The MGS mission had several major scientific goals:

- Study and map the surface characteristics of Mars
- Discover the true shape of the planet, plus details on its topography
- Find out about Mars's magnetic field
- Record weather patterns and atmospheric details

Equipment and instrumentation was packed into the spacecraft in order to accomplish these goals. The Mars Orbiter Camera took images (see one for yourself in Figure 15-2), the Mars Orbiter Laser Altimeter took topographic measurements, and a host of other instruments measured magnetic fields and thermal emissions.

Launch proceeded on November 7, 1996, and the orbiter reached Mars orbit on September 11, 1997 (around the time that the Mars Pathfinder mission was reaching its conclusion). In order to get closer to the surface, the spacecraft used a technique called *aerobraking* to lower its altitude; this is a fuel-saving maneuver that uses atmospheric drag rather than the spacecraft's engines to slow down and change the shape of the spacecraft's orbit.

## Fast facts about Mars Global Surveyor

Following are the basic facts about the Mars Global Surveyor mission:

- **Launch date and site:** November 7, 1996; Cape Canaveral Air Force Station, Florida
- **Spacecraft name and mass at launch:** *Mars Global Surveyor;* 2,337 lb (1,060 kg)
- **Launch vehicle:** Delta II 7925
- **Date Mars orbit achieved:** September 11, 1997
- **Number of Mars orbits:** About 7,300
- **Length of time in Mars orbit:** 10 years

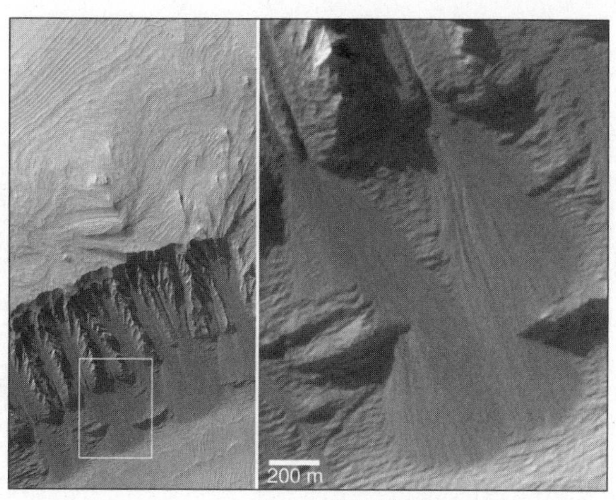

**Figure 15-2:** A high-resolution view of Mars from the *Mars Global Surveyor.*

Courtesy of NASA/JPL/Malin Space Science Systems

The MGS mission made significant scientific discoveries, including observations of surface features that changed over the course of the mission and could, perhaps, be due to liquid water near the surface of Mars. Other significant contributions of the MGS mission include its study of key facets of the Martian climate and the info it gathered that could one day lead to human Mars exploration.

Despite being scheduled for completion in January 2001, *MGS* continued returning data to Earth for almost ten years until communications were lost in late 2006. During the extended portion of the mission, the spacecraft was able to take images of potential landing sites for future Mars missions and continue recording weather and atmospheric data.

# Additional Flops in Mars Exploration in the Late 1990s

Mars exploration was like one giant roller coaster ride. After reaching a high point with the success of NASA's 1996 missions (described earlier in this chapter), the success rate of Mars exploration plummeted, starting with a failed Russian launch in 1996. Then in 1998 and 1999, that success rate dipped even lower with the failure of two NASA missions. The following sections delve into what went wrong in these three cases, as well as what each mission would've accomplished had it succeeded.

## Russia's Mars 96 fails to leave Earth's orbit (1996)

Following the breakup of the Soviet Union in 1991, the Russian space program was in chaos. Although an infusion of funds from NASA was able to help sustain Russian human spaceflight to the space station Mir, and later the International Space Station, Russia's robotic space program wasn't so fortunate. Many of these missions were canceled or delayed for years.

In 1996, a Russian Mars mission, called Mars 96, was resurrected. Its goals were to explore the surface of Mars and create maps of the Red Planet's structure, topography, and composition. Mars 96 also was charged with studying Mars's magnetic field and examining gamma rays and other phenomena on its way to the planet.

The *Mars 96* spacecraft was similar in design to the earlier Soviet Phobos spacecraft, and it was supposed to have made some improvements based on the Phobos failures that we describe earlier in this chapter. Stuffed to the gills with equipment and instruments, *Mars 96* was actually one of the most massive space probes in the history of space exploration, weighing around 13,624 pounds (6,180 kilograms).

The spacecraft carried two landers, which in turn carried a series of instruments to study the surface of Mars; *Mars 96* also carried two penetrators designed to hit the Martian surface at a high speed to drive the pointed probe into the surface. Both penetrators were equipped with cameras, sensors, and other measuring equipment for analyzing the soil and material underneath the Martian surface. The idea was that after the spacecraft arrived at Mars, it'd deploy the landers. Then the penetrators would land and burrow into the surface separately.

The Russians had high hopes that *Mars 96* would be a triumphant success for a space program that had a history of Mars failures. Unfortunately, the Russian space program continued to have bad luck at Mars.

Although liftoff initially succeeded, the spacecraft's launch vehicle failed in its final stage, leaving *Mars 96* unable to leave Earth orbit. The spacecraft fell back to Earth about five hours later. Initially, Russian scientists thought that the spacecraft, with its radioactive plutonium fuel, had burned up safely in the atmosphere, with any remnants falling into the Pacific Ocean. However, reanalysis of the trajectory, as well as reports from witnesses on the ground, suggest that *Mars 96* may have crashed in the Andes Mountains somewhere in Chile or Bolivia. It has yet to be found.

The failure of *Mars 96* was a severe blow to the Russian robotic space exploration program. No other missions have been launched since Mars 96, although Russia is currently working on a new Mars mission, Phobos-Grunt, that could launch as soon as late 2009 and would return a sample from the Martian moon Phobos.

## Mars Climate Orbiter and Mars Polar Lander embarrass NASA (1998–1999)

The 1996 Mars Pathfinder and Mars Global Surveyor missions set the stage for what was to be an ambitious program of Mars exploration by NASA. However, that program was derailed almost before it began thanks to two mission failures. The Mars Climate Orbiter was part of a two-spacecraft program with the Mars Polar Lander. These two Mars missions were intended to provide detailed studies of the weather, atmosphere, water, and general climate on Mars, but neither fulfilled its goals.

The loss of *Mars Climate Orbiter* and *Mars Polar Lander* resulted in severe setbacks for the just-rebounding NASA Mars exploration program. Both *Mars Climate Orbiter* and *Mars Polar Lander* were lost due to minor, yet fatal, errors that would've been found with sufficient time and staff to perform quality checks and troubleshooting. Future Mars missions, such as Mars Odyssey and Phoenix Mars Lander (see Chapter 20), successfully learned from the mistakes of these two failed missions, but the tradeoff was that these later missions were far more costly.

### Mars Climate Orbiter makes an expensive navigational error (1998)

The *Mars Climate Orbiter* was designed with equipment for monitoring weather, noting wind strength and other atmospheric conditions, recording the amount of water vapor in the atmosphere, and generally giving scientists an understanding of how climate works on Mars. Stabilized around three axes, the orbiter contained cameras and other recording equipment, antennae for data relay, solar panels for power, and a host of other instruments.

The spacecraft made it to Mars on schedule but lost contact while entering Martian orbit. A study of the failure revealed that the spacecraft mistakenly entered orbit far closer to Mars than was safe. Consequently, forces on *Mars Climate Orbiter* from friction and heating in the Martian atmosphere resulted in the spacecraft's destruction.

The reason for the *Mars Climate Orbiter's* failure was ultimately determined to be one that plagues anyone who has ever taken math or physics: units. The spacecraft's software used pounds as the unit of force for the thrusters, but the spacecraft units were newtons, which means the flight instructions were off by a scientifically substantial amount (specifically, a factor of 4.45). Although an error in units may cost you a good grade in a science class, in this case, it caused a $300 million mistake that didn't fulfill any of its mission goals.

### Mars Polar Lander loses touch (1999)

The Mars Polar Lander was a 1999 mission scheduled to study the climate, weather, and atmosphere of Mars. It had similar goals to the Mars Climate Observer mission (see the preceding section), but the spacecraft itself consisted of a lander that would've landed near the South Pole of Mars. This lander had a hexagonal base; robotic arms for taking samples; and an array of imaging devices, sensors, and antennae.

*Mars Polar Lander* was sent to Mars alongside two small probes designed to burrow into the Martian surface. These probes were quite small, around 5 pounds (2.3 kilograms) each and surrounded by an *aeroshell* (a protective exterior shield) that would've allowed them to impact the surface, separate from the shells, and penetrate the soil.

The spacecraft and probes reached Mars orbit successfully following their launch on January 3, 1999, but they lost contact with Earth before landing. The failure of *Mars Polar Lander* was ultimately blamed on a software design error that caused the descent rockets to cut off too far above the surface of Mars, resulting in a crash-landing. The probes were also never heard from again, and the cause of their failure is still unknown (though likely unrelated to the loss of *Mars Polar Lander*).

# Chapter 16

# Exploring the Inner Solar System

### In This Chapter

- Seeing Venus via *Magellan* and *Venus Express*
- Studying the Moon and the Sun
- Investigating asteroids and comets

Mars isn't the only planet worth visiting, and in this chapter, you discover that exploration of the rest of the inner solar system continued from the late 1980s through the 1990s and into the early 21st century thanks to a series of missions operated by NASA and the European Space Agency (ESA). Some missions performed radar mapping of Venus, visited the Sun, and took detailed maps of the Moon's geology and surface composition. Other missions touched down on an asteroid and retrieved samples of comet materials. Only one thing was certain: The inner solar system held a wealth of possibilities for new discoveries and achievements.

## Magellan Peeks through the Veil of Venus (1989–1994)

Little was known about the surface of Venus throughout most of the 20th century because previous exploration of the planet in the 1970s had been foiled by thick clouds that prevented astronomers from seeing through to the surface. These clouds parted, but only briefly, for the *Venera 15* and *16* spacecraft, which performed radar mapping from orbit (see Chapter 11 for details). These preliminary maps were intriguing, but a full understanding of the nature of the surface of Venus required a global, high-resolution map.

The United States' Magellan mission, launched in 1989 by the Space Shuttle *Atlantis,* used a sophisticated radar system that was specially designed to be able to see through the clouds of Venus. This system allowed the spacecraft to map the surface with unparalleled resolution (approximately ten times that of the Venera missions), revealing interesting details about the planet. The next couple sections clue you in to how this new mapping system worked and what details it uncovered.

## Mapping with radar

*Magellan*'s radar mapping used a 12-foot (3.7-meter) diameter antenna to bounce radar signals off the surface of Venus (see Figure 16-1). Called *Synthetic Aperture Radar* (SAR), this method sends radar signals out at an angle and captures the return signals. Instead of simply receiving a radar signal from one point on the planet's surface, the SAR technique builds up an image of the surface using the motion of the spacecraft itself. *Magellan* was in orbit over the poles of Venus, which allowed it to map more than 98 percent of the planet's surface.

Radar images are different from images taken with a visible-light camera. Rather than the darks and brights in a photo that correspond to surface regions whose colors may be dark and bright, radar images show smooth areas as dark and rough areas as bright. Smoothness or roughness relates to the physical characteristics of the planet's surface as compared to the 5-inch (13-centimeter) wavelength of the radar system. Areas of the surface covered with rocks smaller than 5 inches across are considered "smooth" and appear dark, whereas areas with rocks larger than 5 inches across are considered "rough" and appear bright.

**Figure 16-1:** The *Magellan* spacecraft with its radar antenna, and attached booster rocket, in the Space Shuttle's payload bay before launch.

Courtesy of NASA

## Fast facts about Magellan

Following are the basic facts about the Magellan mission:

- **Launch date and site:** May 14, 1989; Kennedy Space Center, Florida
- **Spacecraft name and mass at launch:** *Magellan;* 7,612 lb (3,460 kg)
- **Launch vehicle:** Space Shuttle *Endeavour* and Inertial Upper Stage booster rocket
- **Number of Venus orbits:** 4,225
- **Length of time in orbit:** 4 years, 2 months

*Magellan*'s radar images have been combined with altimetry data to produce three-dimensional stereo views of Vensus's surface, made by combining slightly different views of the surface taken on different orbits. The SAR images provided the side view, whereas the altimetry data provided the view you'd get if you looked straight down at the planet. Together, they created a three-dimensional picture of the Venusian surface.

## Uncovering details of Venus's surface

The Magellan images revealed a Venusian surface that's surprisingly Earth-like in many ways but lacking common geologic features found on the Earth. For example, most of Venus's surface features are volcanic in nature, including huge lava plains, large shield volcanoes, and smaller lava domes. Lava channels that can be up to 3,700 miles (6,000 kilometers) long suggest that the lava was very fluid and likely erupted at a high rate. However, because Venus is a very hot planet, with a surface temperature as high as 850 degrees Fahrenheit (454 degrees Celsius), water isn't stable at its surface, meaning the familiar geologic features carved by water on Earth (and Mars) are missing on Venus. In addition, *Magellan* found no evidence of plate tectonics on Venus, meaning that volcanic activity on Venus operates very differently from Earth. (If you're curious to check out an image taken by *Magellan,* flip to the color section.)

One surprise found on Venus was that the surface appeared to be uniformly younger than expected, as shown by the small number of impact craters, with an average surface age of about 500 million years. In geologic time, this is quite young. Scientists are still debating whether a single cataclysmic resurfacing event, such as a global volcanic eruption, could've occurred 500 million years ago, or whether the apparent surface age comes from ongoing geologic activity that slowly covers in parts of the planet's surface.

After *Magellan* accomplished its mapping goals, the spacecraft's systems began to fail, and it was given one final task: to test a new technique called *aerobraking* (using a planet's atmosphere to change the orbit of a spacecraft). Spacecraft controllers flew the spacecraft deep into Venus's atmosphere until it burned up. The technique destroyed *Magellan* but helped engineers figure out how to use aerobraking to change the orbits of other spacecraft. Consequently, this technique has since been used successfully by *Mars Global Surveyor* (Chapter 15) and *Mars Odyssey* (Chapter 20).

# Dedicated Spacecraft Visit the Moon and the Sun

The prominence of the Moon and the Sun in the sky may make them seem familiar, but both celestial bodies warranted dedicated missions in the 1990s for the purpose of revealing their secrets:

- After the heyday of American lunar exploration during the Project Apollo years (see Chapters 9 and 10), little further study of the Moon was done until the Clementine mission of the 1990s. This mission produced global digital maps of the Moon and was followed by the Lunar Prospector mission, which studied the Moon's composition from orbit.
- Although the Sun is bright enough to be studied from Earth-based observatories, some aspects of it are still best studied by a dedicated spacecraft. Enter the *Solar and Heliospheric Observatory (SOHO)*.

## Investigating the Moon

To commemorate the 20th anniversary of the Apollo 11 Moon landing in 1989, President George H. W. Bush announced a new space exploration initiative that would've sent humans back to the Moon and on to Mars. The program never really got off the ground though because of the huge $500 billion price tag, but it did spur some renewed interest in the Moon as a target for observation.

### Clementine (1994)

The Clementine mission, a joint effort of NASA and the Ballistic Missile Defense Organization, took American space exploration back to the Moon when it launched on January 25, 1994. Although the primary purpose of Clementine was to test how spacecraft and sensors behave when in space for long periods of time, NASA was also able to sneak in some science from a variety of different instruments.

## Fast facts about Clementine

Following are the basic facts about the Clementine mission:

- **Launch date and site:** January 25, 1994; Vandenberg Air Force Base, California
- **Spacecraft name and mass at launch:** *Clementine;* 500 lb (227 kg)
- **Launch vehicle:** Titan 23G
- **Number of lunar orbits:** 360
- **Length of time in lunar orbit:** 5 months

The eight-sided *Clementine* spacecraft, which measured 5.9 feet by 3.6 feet (about 1.8 meters by 1.1 meters), had solar panels to provide it with power and a dish antenna so it could relay data back to Earth. It carried seven experiments onboard, including a charged particle telescope and a laser altimeter.

*Clementine* actively observed the Moon using ultraviolet and infrared imaging, laser altimetry, and other measurements. With the combined results of this various data, scientists pieced together a multispectral picture of the Moon's surface, including a more in-depth study of lunar surface material and a modern digital global map of the Moon's surface. These maps help scientists better understand the geologic history of the Moon and have been used as a basis for future lunar exploration.

### *Lunar Prospector (1998–1999)*

Part of the NASA Discovery series of missions (which we introduce in Chapter 15), the Lunar Prospector mission was charged with mapping the composition of the lunar surface. The small, cylindrical spacecraft launched on January 7, 1998, and entered lunar orbit four days later. During its voyage into lunar orbit, the spacecraft deployed three 8.2 foot (2.5 meter) arms that contained instruments and equipment for measuring the Moon's gravity, atmosphere, temperature, and other quantities necessary for creating a surface compositional map of the Moon.

*Lunar Prospector* sent home data concerning the Moon's atmosphere and crust, as well as the potential for ice on the satellite's surface. A major goal of the mission, which fit into the *Prospector* name, was to explore the Moon from the perspective of resources that might be available to help support either future human exploration or commercial enterprises. If water were found on the Moon, for instance, it would be a very important resource for future human bases.

## Fast facts about Lunar Prospector

Following are the basic facts about the Lunar Prospector mission:

- **Launch date and site:** January 7, 1998; Cape Canaveral Air Force Station, Florida
- **Spacecraft name and mass at launch:** *Lunar Prospector;* 348 lb (158 kg)
- **Launch vehicle:** Athena II
- **Number of lunar orbits:** 7,060
- **Length of time in lunar orbit:** 570 days

Some space probes burn up in the Earth's atmosphere when they return; others are left to continue orbiting until their rechargeable power supplies are completely drained. In the case of *Lunar Prospector,* NASA officials decided to crash the probe into the Moon's surface, with the hope that the resulting impact plume would contain evidence that water existed within permanently shaded craters near the lunar poles. *Lunar Prospector* did crash in the appointed spot, a crater located at the Moon's South Pole, but no water was detected.

# Sending SOHO to spy on the Sun (1995–present)

The Sun's cycles of internal activity can affect life on Earth. For example, solar flares can knock out communications satellites or risk the lives of astronauts in orbit, making a detailed study of the Sun a high-priority mission for both the U.S. and Europe. Consequently, both nations set their sights on the Sun in 1995 with the launch of the Solar and Heliospheric Observatory (SOHO), a collaborative project between NASA and the European Space Agency (ESA). SOHO's primary mission was to study and discover more about the Sun from close range. Specifically, its mission goals were to understand the core structure of the Sun, study the solar corona, and demystify the details behind solar winds.

The observatory consists of two separate modules, a Service Module and a Payload Module:

- The Service Module powers the spacecraft, controls its temperature, and sends data back to Earth.
- The Payload Module houses all the spacecraft's instruments. Instrumentation on *SOHO* came from both European and American developers. Twelve main instruments are designed to work together to relay info back to Earth, and scientists worldwide use the data gleaned

from these instruments. The major instruments include ultraviolet imagers and solar wind and energetic particle detectors.

*SOHO* orbits between the Sun and the Earth, around a stable point known at the First Lagrangian Point, or L1. At L1, the gravity from the Sun and the Earth balances out, resulting in a particularly stable orbital configuration. *SOHO* orbits the L1 point so it can always communicate with Earth.

Originally intended to operate for about two years, *SOHO* is still operating as of this writing and has returned valuable data regarding solar winds, solar structures, coronal waves, and new phenomena such as solar tornadoes (see Figure 16-2). *SOHO* has also spotted hundreds of comets and played a large role in helping meteorologists give more-accurate weather predictions.

To see images from *SOHO* and read updates on its findings, head to `sohowww.nascom.nasa.gov`.

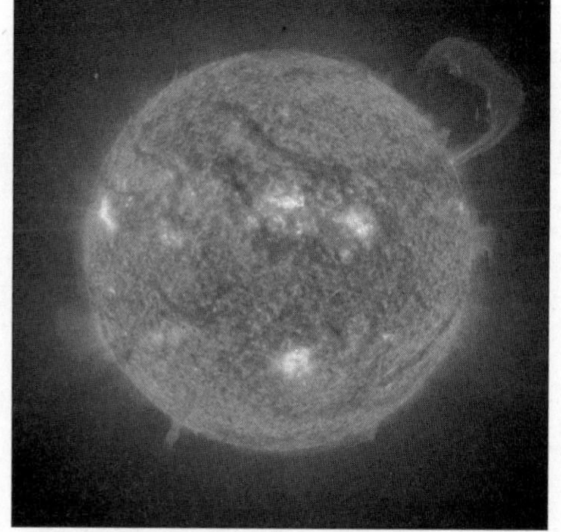

**Figure 16-2:** A huge solar prominence erupts from the Sun in this ultraviolet image taken by *SOHO.*

Courtesy of ESA/NASA/SOHO

## Fast facts about SOHO

Following are the basic facts about the SOHO (Solar and Heliospheric Observatory) mission:

- **Launch date and site:** December 2, 1995; Cape Canaveral Air Force Station, Florida
- **Spacecraft name and mass at launch:** *SOHO;* 4,100 lb (1,850 kg)
- **Launch vehicle:** Atlas II-AS
- **Number of solar orbits:** 13, as of this writing
- **Mission duration:** 13 years to date

# Venus Express Studies the Atmosphere (2005–Present)

In 2005, the European Space Agency (ESA) brought exploration of Venus back into the spotlight with Venus Express, the agency's first foray into studying Venus. The origins of this mission go back to 2001, when there was an interest in recycling the Mars Express mission design (see Chapter 20) for a trip to Venus. Because the conditions of Venus are quite different from those of Mars, though, many changes had to be made. For instance: Venus is about half the distance from the Sun as Mars is, so any spacecraft exploring Venus has to be able to withstand more heat. Of course, that closer proximity to the Sun also means the probe's solar panels can gather and generate more power.

With these and other changes in mind, the *Venus Express* was prepared and launched in November 2005, reaching Venus about five months later. As of late 2008, *Venus Express* has made a number of interesting discoveries, including observations of frequent lightning, a vortex in the atmosphere over the South Pole of the planet, and evidence that liquid water could've existed on Venus long ago.

*Venus Express* has completed its 500-day mission and is on an extended mission until at least late 2009. It's expected to take measurements and images of the Venusian atmosphere, particularly clouds, as well as perform global mapping of the planet's surface through temperature studies. It'll be the first such mission to globally map the atmospheric composition of Venus and make a comprehensive study of Venusian temperatures.

Check out these great Web sites to see the latest pictures and results from *Venus Express:*

- www.esa.int/SPECIALS/Venus_Express
- www.venus.wisc.edu

## Fast facts about Venus Express

Following are the basic facts about the Venus Express mission:

- **Launch date and site:** November 9, 2005; Baikonur Cosmodrome, Kazakhstan
- **Spacecraft name and mass at launch:** *Venus Express;* 2,800 lb (1,270 kg)
- **Launch vehicle:** Soyuz-Fregat rocket
- **Number of Venus orbits:** 1 orbit every 24 hours since reaching final Venus orbit on May 7, 2006
- **Length of time in Venus orbit:** More than 3 years to date

# Small Missions of Large Importance

Supplementing the planets and satellites of the inner solar system, smaller bodies (think comets and asteroids) provide leftover remnants of the early days of our solar system. Despite difficulties in targeting them, a number of NASA missions in the late 20th and early 21st centuries investigated these small objects. The next few sections detail the most-exciting of these missions.

## NEAR orbits an asteroid (1996–2001)

*NEAR (Near-Earth Asteroid Rendezvous)* made history as the first spacecraft to orbit an asteroid. Launched in February 1996, *NEAR* was destined to rendezvous with the near-Earth asteroid 433 Eros to help provide more information about the composition, magnetic field, mass, and other attributes of the asteroid.

Although small by planetary standards, the asteroid 433 Eros is certainly nothing to sneeze at. The largest dimensions of the potato-shaped asteroid are 8 miles and 20 miles (13 kilometers and 33 kilometers); for comparison you could fit more than 200 football fields inside it, making it one of the largest near-Earth asteroids. *NEAR* sent back high-resolution images (you can see one in Figure 16-3) and other measurements, and it was a successful first launch of NASA's Discovery Program of low-cost spacecraft (flip to Chapter 15 for more on the Discovery Program).

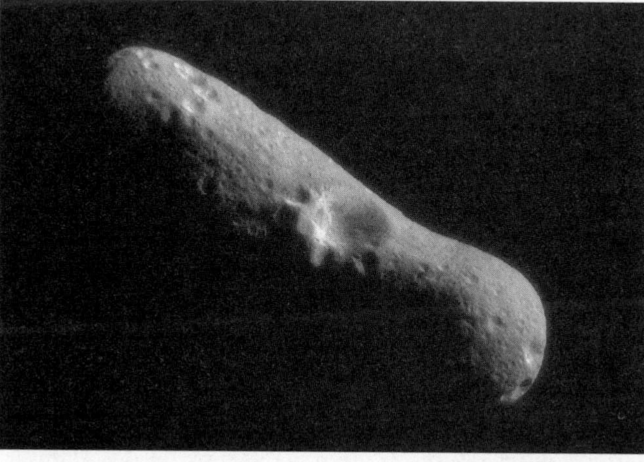

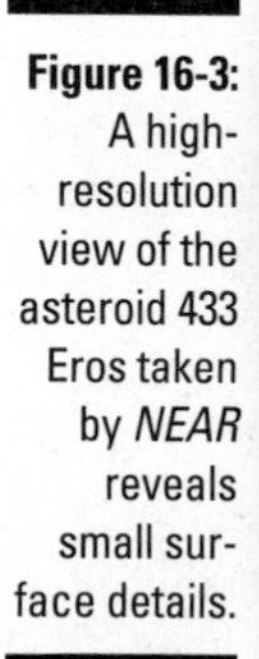

**Figure 16-3:** A high-resolution view of the asteroid 433 Eros taken by *NEAR* reveals small surface details.

Courtesy of NASA/JPL/JHUAPL

## Fast facts about NEAR

Following are the basic facts about the NEAR (Near-Earth Asteroid Rendezvous) mission:

- **Launch date and site:** February 17, 1996; Cape Canaveral Air Station, Florida
- **Spacecraft name and mass at launch:** *NEAR;* 1,074 lb (487 kg)
- **Launch vehicle:** Delta II
- **Number of 433 Eros orbits:** 230
- **Length of time in 433 Eros orbit:** 1 year

Even though *NEAR* was never built as a lander, in February 2001, at the end of its mission, the probe was commanded to go into a lower and lower orbit until it eventually touched 433 Eros's surface. The asteroid's extremely low gravity allowed the probe to survive the landing, creating a first in the grand scheme of space probe research: Something that wasn't a lander landed!

See more images (and movies!) made by *NEAR* at its official Web site: `near.jhuapl.edu`.

# *Stardust brings home precious particles (1999–present)*

The Stardust mission was the first sample return of particles from a comet. The spacecraft launched in February 1999 and used a gravity-assisted trajectory to reach Comet Wild 2 in January 2004. As the spacecraft performed a close flyby of the comet, it collected samples from the comet's *coma* (essentially the ice and dust particles that make up the comet's "tail"). The comet particles were captured in a very low-density collector called an *aerogel,* which preserved their delicate structure. The capsule containing the sample returned to Earth in 2006, landing with a parachute in the Utah desert. The main *Stardust* spacecraft continued past Earth and is en route to fly past Comet 9P/Tempel 1 in 2011.

The Stardust mission is historically significant because it marks the first time scientists had an opportunity to study an actual sample that came from primitive comet material. Because comets are thought to be primordial remnants from when the solar system was young, researchers hope to use this information to study the origin and evolution of the early solar system. Preliminary analysis of the sample found a number of interesting compounds, including nitrogen and organic compounds.

### Fast facts about Stardust

Following are the basic facts about the Stardust mission:

- **Launch date and site:** February 7, 1999; Cape Canaveral Air Force Station, Florida
- **Spacecraft name and mass at launch:** *Stardust;* 661 lb (300 kg)
- **Launch vehicle:** Delta II
- **Number of Earth orbits:** 2, as of this writing
- **Mission duration:** 7 years to date

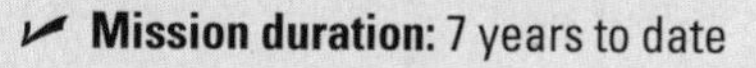

Want to keep up with *Stardust?* Put the following in your Web browser's Favorites list:

- stardust.jpl.nasa.gov
- www.nasa.gov/mission_pages/stardust/main

## Deep Impact crashes into a comet (2005–present)

Deep Impact wasn't just the name of a NASA mission launched in January 1995 — it was what scientists wanted their spacecraft to leave on Comet 9P/Tempel 1 so they could study the comet's insides. The plan was to send part of the spacecraft on a collision course with the comet and use the impact to eject material from inside the comet into outer space, where it could be studied.

The spacecraft used during the Deep Impact mission was designed in two sections:

- A "Smart Impactor" was for crashing into the comet. It was armed with a camera so it could take pictures of the comet's nucleus right up to the point when the impactor itself was destroyed.
- A flyby spacecraft was for photographing the crash and continuing on with — you guessed it — a flyby. The flyby spacecraft was solar-powered and shielded from flying comet debris (as well as flying spacecraft debris!) and boasted different types of cameras and other measuring devices.

## Fast facts about Deep Impact

Following are the basic facts about the Deep Impact mission:

- **Launch date and site:** January 12, 2005; Cape Canaveral Air Force Station, Florida
- **Spacecraft name and mass at launch:** *Deep Impact;* 1,430 lb (650 kg)
- **Launch vehicle:** Delta II
- **Number of Earth orbits:** 0
- **Mission duration:** 4 years to date

Impact occurred on July 4, 2005. The impactor had detached successfully from the flyby spacecraft five days earlier and positioned itself so that the comet would fly into it. The impact made a large crater and ejected a large plume of material, as NASA scientists had hoped it would. However, images and compositional measurements taken by the flyby spacecraft showed more dust, and less ice, than scientists had anticipated. Due to the composition and chemistry, scientists were able to better pinpoint where the comet may have originated — the far reaches of the Kuiper Belt (see Chapter 2 for more on this part of the solar system).

*Deep Impact* is currently heading to a flyby of Comet Hartley-2 in 2010. The *Stardust* spacecraft (described in the preceding section) will fly past Comet 9P/Tempel 1 in 2011 to study any changes that have occurred since *Deep Impact*'s visit.

For all the *Deep Impact* info you desire (and then some!), check out either `deepimpact.umd.edu` or `www.nasa.gov/mission_pages/deepimpact/main`.

# Chapter 17

# Voyages to the Outer Solar System

### In This Chapter

- Trekking to Jupiter with *Galileo*
- Seeing the Saturn system through the eyes of *Cassini* and *Huygens*

Travel times to the outer solar system are long and conditions are harsh, meaning that next-generation orbiter missions, following the flybys of the Pioneer and Voyager missions (which we tell you about in Chapter 11), needed to be large. Two such flagship missions, Galileo and Cassini, visited Jupiter and Saturn, respectively, with impressive results, as you discover in this chapter:

- NASA's *Galileo* spacecraft arrived in orbit around Jupiter in 1995. *Galileo's* entry probe descended through the clouds of Jupiter to take measurements, while its orbiter made multiple flybys of Jupiter and its moons over the next eight years.
- NASA's *Cassini* spacecraft reached Saturn orbit in 2004, and the accompanying *Huygens* probe (built by the European Space Agency, or ESA) landed on the surface of Titan in early 2005. *Cassini's* mission continues until at least 2010.

## Galileo Journeys to Jupiter (1989–2003)

Exploration of the outer solar system began in earnest with the United States' Galileo mission in the late 1980s. Named after the Renaissance astronomer Galileo Galilei, who discovered Jupiter's four large moons, the Galileo mission to Jupiter featured a robotic spacecraft designed to study the *Jovian system* (Jupiter and its moons) in detail. The *Galileo* spacecraft, which was designed as a follow-up to the Voyager flybys of the Jovian system in the late 1970s, was launched into orbit by Space Shuttle *Atlantis* on October 18, 1989 (see Chapter 14 for details). It was the first spacecraft to orbit a planet in the outer solar system.

*Galileo* consisted of an orbiter and an atmospheric probe, together standing about 23 feet (7 meters) tall (see Figure 17-1). One section contained instruments and equipment; another section was devoted to a powerful high-gain antenna for transmitting data. Because *Galileo* would be too far from the Sun, solar panels weren't an option for keeping the spacecraft powered, so two radioisotope thermoelectric generators (RTGs) with radioactive plutonium were used instead (we explain RTGs in more detail in Chapter 3). Instruments onboard *Galileo* included many devices geared toward studying surface geology, composition, radiation, particles, magnetic fields, and other aspects of Jupiter and its moons.

**Figure 17-1:** The *Galileo* spacecraft being prepared for launch.

Courtesy of NASA

Although hampered in data return by its crippled high-gain antenna, the *Galileo* spacecraft was otherwise a triumph of engineering in that it returned images and other measurements of Jupiter for around eight years. This duration surpassed all estimates of the amount of radiation damage the spacecraft could endure and still operate.

The following sections highlight *Galileo*'s story, from the mission's exciting yet problematic launch to its ending sacrifice.

## Kicking things off with priceless views and a broken antenna

For the *Galileo* spacecraft, the journey to Jupiter turned out to be almost as exciting as the destination. With a looping trajectory that took the spacecraft around Earth twice and Venus once, NASA mission planners were able to design a flight path that used the "gravitational slingshot" method (described in Chapter 3) to hurl *Galileo* toward Jupiter. Along the way, *Galileo* took spectacular pictures of the Earth and its moon together; it also performed the first close-up flybys of two asteroids, Gaspra and Ida. *Galileo* even had prime seating when Comet Shoemaker-Levy 9 collided with Jupiter in July 1994 and returned images back to Earth before Jupiter rotated enough to be seen by Earth-based telescopes. The spacecraft finally arrived at Jupiter in 1995.

Despite this exciting start, the mission almost self-destructed. The high-gain antenna, which sat like an umbrella at *Galileo*'s peak, had been folded and tucked behind a sunshade for most of the journey to Jupiter. When the time came to open the antenna, it only opened partway because of friction in the antenna ribs. No amount of maneuvering by scientists on Earth could open the antenna completely, so the mission had to rely on the less-powerful, slower-transmitting low-gain antenna instead. Although the Galileo mission still performed useful science, its total data return was severely limited by the loss of the high-gain antenna.

Scientists determined that the antenna rib likely got stuck during the years that *Galileo* sat in storage, awaiting its turn to be launched whenever the Space Shuttle returned to flight following the *Challenger* disaster in 1986 (see Chapter 5). Fortunately, engineers were able to come up with innovative image-compression algorithms that could be performed on the spacecraft's rudimentary computer, which helped increase the total data download and allowed the weaker antenna to still meet most of the mission objectives.

## Probing the atmosphere for a short while

The probe component of *Galileo,* designed to study Jupiter's atmosphere, was jettisoned from the main *Galileo* spacecraft in July 1995, several months before the latter reached Jupiter. When the probe reached Jupiter's atmosphere in December, it slowed down as it flew through, collecting data along the way via its onboard instruments. These instruments included mass spectrometers, cloud-detection equipment, sensors for measuring temperature and pressure, and particle detectors for studying Jupiter's radiation belts.

The *Galileo* probe sent data back to the orbiter (for later relay to Earth) for nearly an hour before it stopped communicating. As expected, the probe most likely melted and, eventually, disintegrated as it moved deeper into Jupiter's atmosphere. Scientists are still puzzling over the probe's findings, which included the fact that Jupiter's atmosphere possesses far less water than originally predicted. The probe also found higher-than-expected concentrations of certain chemical elements called noble gases (think argon, krypton, and xenon, among others). The presence of these elements, which don't react with most substances, suggests that Jupiter may have formed in a colder part of the solar system than originally thought.

## Studying the satellites

The Galileo mission answered a number of questions about Jupiter's *satellites* (moons). Its four large satellites — Io, Europa, Ganymede, and Callisto — are particularly interesting because of their varied appearances:

- The Voyager missions of the late 1970s discovered volcanoes on Io, but *Galileo* one-upped them by confirming that Io is the most volcanically active body in the solar system. (In fact, it's about 100 times more volcanically active than Earth is today!) Io's tidal heating results in material being ejected in volcanic plumes and lava flows, which cover the moon's surface. (*Tidal heating* is the way that the interior of a planet or satellite is heated due to the squishing and pulling that takes place due to varying gravitational forces.)
- Another major discovery by *Galileo* was evidence for a liquid water ocean under the icy surface of Europa. Images from the Galileo mission showed a cracked surface with iceberglike features; gravitational measurements showed that the surface is made up of a layer with the density of water and/or ice and is about 62 miles (100 kilometers) thick. Results from *Galileo*'s magnetometer gave the most decisive evidence for an ocean on Europa, with the discovery of an induced magnetic field consistent with a large subsurface volume of salty water, similar to terrestrial seawater. Although not yet definitively confirmed, *Galileo* results very strongly suggest that Europa's subsurface contains an ocean with more water than all of Earth's oceans combined. And where there's water, there's the possibility for life. (Check out an image of Europa taken by *Galileo* in the color section.)
- *Galileo* also took images of Ganymede, the only moon in the solar system with its own magnetic field, and Callisto. Callisto is an old cratered body, whereas Ganymede has undergone some cracking and faulting in the past due to the slight degree of tidal heating that it receives. Interestingly, magnetic field measurements suggest that both Ganymede and Callisto could also have subsurface liquid oceans, though these bodies of water would be much farther beneath the surface than the ocean on Europa.

Io, Europa, and Ganymede are in an orbital resonance with Jupiter. An *orbital resonance* is the state that occurs when orbiting planets, satellites, or other celestial bodies exert a specific gravitational force upon the other bodies in the circuit. So for every time Ganymede goes around Jupiter once, Europa goes around the planet twice, and Io goes around it four times. This resonance keeps the orbits noncircular and helps drive tidal heating, because the surfaces of the satellites are squeezed and flexed. Tidal heating is most prevalent on Io because it's closest to Jupiter, but tidal heating is significant on Europa as well. The degree of heating is reflected in the moons' surface geologies.

## Making a heroic sacrifice at mission's end

After two years of study, *Galileo* completed its primary mission in 1997, but the length of its overall mission was extended multiple times to add additional flybys of the satellites and studies of Jupiter's atmosphere and rings. Data from *Galileo* was used to study the composition of Jupiter's atmosphere, including clouds made from ammonia ice, and measure wind speeds around the Great Red Spot (a giant storm in Jupiter's atmosphere) and other features. Images also revealed the fine structure of Jupiter's faint ring system. *Galileo* spent a total of eight years in orbit around Jupiter and managed to complete almost all of its mission objectives despite its nonfunctional high-gain antenna.

At the end of *Galileo*'s mission, NASA mission planners faced a painful choice about the fate of the aging spacecraft, which was running out of propellant for the thrusters that refined its orbit and controlled its position. In addition, the computer chips and other components that controlled the spacecraft were far over their design tolerances for radiation and would soon cease operating. Such spacecraft are normally left to have their orbits decay naturally, but mission designers faced an additional complication when operating in the Jovian system: The discovery of a likely subsurface ocean on Europa (see the previous section) meant that the prospects for life there were high enough that planetary protection constraints came into play.

*Galileo,* as an orbiter, was never sterilized after being built on Earth. In addition to the possible microbes that could still be living inside the spacecraft, another danger came from the nuclear power source that fueled it. Rather than run the risk of an out-of-control spacecraft crashing into Europa's surface in the future (and possibly contaminating a subsurface biosphere), NASA mission planners made the difficult choice to send the still-functioning *Galileo* deep into Jupiter's atmosphere, where it could burn up without causing any potential damage to Europa. In September 2003, *Galileo* sent its final signals back to Earth before entering Jupiter's atmosphere.

### Fast facts about Galileo

Following are the basic facts about the Galileo mission:

- **Launch date and site:** October 18, 1989; Kennedy Space Center, Florida
- **Spacecraft name and mass at launch:** *Galileo;* 5,650 lb (2,560 kg)
- **Launch vehicle:** Space Shuttle *Atlantis*
- **Number of Jupiter orbits:** 34
- **Length of time in Jupiter orbit:** 8 years

# Cassini-Huygens Views the Saturn System (1997–Present)

Cassini-Huygens, a flagship-style NASA mission to the Saturn system, was nearly 20 years in the making. As far back as 1982, scientists were discussing ways to tag-team a mission to Saturn with one to Titan, one of Saturn's moons. The final product of these brainstorms, Cassini-Huygens, was a joint venture between NASA, the European Space Agency (ESA), and the Italian Space Agency (ASI). The mission consisted of two basic parts: The *Cassini* orbiter, named after a famous Italian-French astronomer, was built by NASA; the *Huygens* space probe, named after a famous Dutch astronomer, was built by ESA. ASI built the high-gain antenna for *Cassini,* as well as the radar system used to see through Titan's clouds.

The Cassini-Huygens mission's goals focused on Saturn and its large moon Titan. The *Cassini* orbiter was charged with studying Saturn's rings, clouds, and other atmospheric conditions; investigating its satellites; and performing a wide-ranging investigation of Titan's atmosphere and surface. The *Huygens* probe was responsible for returning surface and atmospheric data during its trip through Titan's atmosphere and for about an hour and a half after landing. The mission launched on October 15, 1997, complete with a hefty science *payload* (or cargo, specifically the scientific experiments and equipment required for the mission). In the following sections, we offer insight into the massive inventory of instrumentation aboard *Cassini* and delve into the details of what the mission has found.

The *Cassini* orbiter was recently approved for an extended mission, which will continue until at least mid-2010. Check out these Web sites for the latest pictures and discoveries from *Cassini:*

- `saturn.jpl.nasa.gov`
- `www.esa.int/SPECIALS/Cassini-Huygens`

## *Cassini's load of complex instruments*

The *Cassini* orbiter (see Figure 17-2) is the size of a school bus, measuring 22 feet (6.8 meters) high by 13 feet (4 meters) wide. As one of the most-complex spacecraft ever built, the orbiter contains a number of complex scientific instruments, including the following:

- The Radio Detection and Ranging (RADAR) instrument for bouncing signals off Titan's surface to see through the clouds
- The Imaging Science Subsystem (ISS) for taking pictures of Saturn and its moons under a range of visible light conditions
- The Visible and Infrared Mapping Spectrometer (VIMS) for gauging the composition of surfaces, atmospheres, and even planetary rings
- A cosmic dust analyzer for studying the small particles orbiting Saturn
- A composite infrared spectrometer for measuring temperatures
- An ion and neutral mass spectrometer for studying charged and neutral particles
- A radio and plasma wave experiment for detecting radio waves from Saturn
- A UV imaging spectrograph for taking images and spectra at UV wavelengths
- A magnetometer for studying Saturn's magnetic field

Because *Cassini* was intended to operate far from the Sun, NASA engineers designed it to run off of three radioisotope thermoelectric generators (RTGs; see Chapter 3 for the full scoop on RTGs). These RTGs were designed to be long-lasting, and they still provide sufficient power more than ten years after *Cassini*'s launch.

More than 70 pounds of plutonium were rocketed into space with *Cassini,* causing no small amount of protesting from environmentalists and others concerned about the dangers of sending this amount of radioactive material into space. The only real risk was from the spacecraft failing and crashing immediately after launch, like *Mars 96* did (see Chapter 15), or during its close Earth flyby en route to Saturn. The RTGs were actually designed to be strong enough to withstand some sort of catastrophic landing back on Earth, making the chances of sufficient plutonium dispersal to be hazardous extremely low. In any event, *Cassini*'s launch and Earth flyby took place uneventfully.

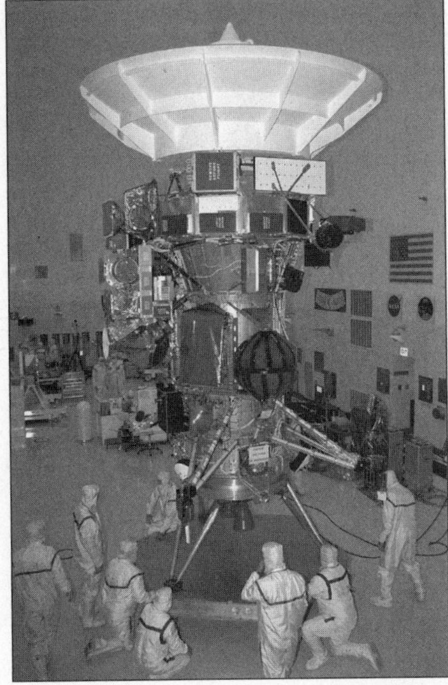

**Figure 17-2:** The *Cassini* orbiter before launch.

Courtesy of NASA/JPL-Caltech

## Cassini takes images of Saturn and its satellites (2004–present)

Upon arriving at Saturn on July 1, 2004, *Cassini* flew through the plane of Saturn's rings and had its only close approach to the dark outer moon Phoebe. *Cassini* has made almost 100 orbits of Saturn as of this writing, with each orbit targeted with a close flyby of one or more satellites. The spacecraft's discoveries are many and varied. For example, *Cassini* took detailed images of Phoebe, including some that led scientists to believe that water ice exists in the moon's subsurface layer. *Cassini* images also

- Showed a huge ridge encircling the middle of the moon Iapetus like a belt.
- Found hurricanes at Saturn's North and South Poles; winds in these stable storms circulate like a whirlpool at speeds of about 325 miles (530 kilometers) per hour — twice as fast as the fastest winds in similar features on Earth.
- Discovered four new small moons, including one, Daphnis, that orbits inside Saturn's rings.

- Spotted features in Saturn's rings called spokes, as well as incomplete "ring arcs" that extend before and after some small moons in their orbits but don't form a complete ring.

However, the true stars of the Saturn system turned out to be the moons Enceladus and Titan; we describe *Cassini*'s discoveries on these moons in more detail in the following sections.

### Geysers on Enceladus (2005)

Tiny Enceladus, which is the sixth-largest of Saturn's moons, was officially discovered in 1789, but little was known about it until NASA's Voyager missions in the late 1970s and early 1980s, which took low-resolution pictures that showed a bright surface. In 2005, *Cassini* flew past Enceladus and spotted startling signs of activity for such a small moon: Huge geysers of material were observed being ejected from Enceladus's surface (see Figure 17-3). The geyser sites are located at the moon's South Pole, and the plumes contain primarily water ice (in addition to other materials). The geysers are associated with areas of extremely high temperatures, consistent with pockets of warm, liquid water under the surface that could act as reservoirs.

**Figure 17-3:** *Cassini* observed active geysers of material jetting off Enceladus.

Courtesy of NASA/JPL/Space Science Institute

The discovery of geysers on Enceladus means that this small satellite joins only Jupiter's moon Io and Neptune's moon Triton as extraterrestrial sites of ongoing geologic activity. Some material ejected from the geysers falls back onto the surface of Enceladus, and the rest of it goes into orbit. Enceladus is covered with fresh, bright frost from the deposited geyser material, explaining the very bright surface first seen by *Voyager.* Enceladus orbits Saturn embedded in a tenuous ring called the E ring, and scientists now believe that the geyser material that escapes Enceladus is actually the source of this ring. The mechanics of what drives the geysers, or provides the heat source for them, are still the subject of vigorous scientific debate.

### Views of Titan's surface (2005)

Instruments onboard *Cassini* were specifically designed to see through the moon Titan's thick, hazy atmosphere in order to provide scientists with a clear view of the surface below. The innovative RADAR instrument has proved to be the real star of Titan surface studies. Similar to the radar mapper on the *Magellan* spacecraft that observed Venus (see Chapter 16), *Cassini*'s RADAR instrument sends radio signals out that bounce off Titan's surface and are collected by the spacecraft. The motion of the spacecraft helps build up an image, and a long, narrow image "noodle" is collected on each close radar flyby.

Data and images gathered by *Cassini* in 2005 have shown Titan to possess a surprisingly familiar surface. Although Titan's surface temperatures are cold enough that exotic substances such as ethane and methane are liquid, these materials flow over the surface and form lakes and valleys just like liquid water does on Earth. Much of Titan's surface currently appears dry, but lakes of methane have been found near the moon's North Pole. These are the first lakes found beyond Earth, with implications for hydrocarbons and organic chemistry creating possibilities for life on Titan.

## The Huygens probe lands on Titan (2005)

To complement the orbital observations of Titan from the *Cassini* orbiter (described in the previous section), the *Huygens* space probe, which hitched a ride to Saturn aboard *Cassini,* was designed to take detailed measurements of Titan's atmosphere and surface from the moon itself. Because the probe was too far from Earth to accept data from or send data to NASA directly, its findings were relayed through *Cassini.* In the following sections, we describe *Huygens*'s descent through Titan's atmosphere and its landing on the moon's surface.

### Parachuting and photographing through the thick atmosphere

On December 25, 2004, the *Huygens* probe was set free from *Cassini* courtesy of a spring mechanism. It then traveled independently, entering Titan's atmosphere just over two weeks after its launch. As the probe slowly parachuted to Titan's surface, it took images and analyzed Titan's winds and the chemical composition of its atmosphere.

Images sent back from *Huygens* during its descent showed what appeared to be drainage channels: Dark swatches crossed larger, light swatches as they led into very large (sealike) dark swatches. These channels are much smaller than those that can be seen from the orbiting *Cassini* spacecraft, supporting the theory that liquid has traveled over much of Titan's surface at various times. In all, *Huygens* sent about 700 images to *Cassini* for relay back to Earth. Unfortunately, a software glitch resulted in the loss of about half of those images.

### *Landing on a world of ice*

*Huygens* was primarily designed to study Titan's atmosphere, but NASA engineers (those ol' Boy Scouts) wanted to be prepared in case the probe survived its landing. Of course, they didn't know whether *Huygens* was headed for dry land or a global ocean, so they made sure it could survive a landing on either a wet or dry surface — if it had landed in liquid, it would even have floated!

As it turned out, *Huygens* landed on a dry riverbed, surrounded by rounded ice pebbles (see Figure 17-4). No liquid was detected, though a slight increase in methane vapor was detected shortly after the probe's landing.

*Huygens*'s battery was intended to last for three hours at most, and much of that time was occupied with the probe's descent to the surface, so there was little time available for sending back data. The probe managed to transmit data from the surface for 90 minutes. The *Cassini* orbiter was well out of transmission range at this point, so the data was received by large radio telescopes back on Earth, which were monitoring the probe's faint signals. Scientists couldn't receive images from *Huygens,* but they could confirm that the probe was still alive and transmitting. It's not like they missed out on much though. Because *Huygens*'s cameras couldn't move, the probe sent back the same view of Titan's surface over and over again. Talk about déjà vu!

**Figure 17-4:** The surface of Titan as seen by *Huygens;* a composite image from Apollo, at a similar scale, is on the right.

*Courtesy of ESA/NASA/JPL/ University of Arizona*

## Fast facts about Cassini-Huygens

Following are the basic facts about the Cassini-Huygens mission:

- **Launch date and site:** October 15, 1997; Cape Canaveral Air Force Station, Florida
- **Spacecraft names and mass at launch:** *Cassini* (orbiter), 4,739 lb (2,150 kg); *Huygens* (probe), 770 lb (350 kg)
- **Launch vehicle:** Titan IV-B/Centaur
- **Number of Saturn orbits:** 100, as of this writing
- **Length of time in Saturn orbit:** 11 years to date

# Part IV
# Current Space Exploration

## In this part . . .

Space is a busier place than ever before, and the rewards are great, as you find out in this part on current space exploration. Telescopes in space take amazing views of the cosmos, and the International Space Station epitomizes the spirit of international cooperation that was absent during the days of the Space Race, because it's built and operated by crews from many nations. Mars exploration also has been a highlight of recent operations, which include the highly successful and popular Mars Exploration Rovers.

# Chapter 18

# What a View! Telescopes in Space

### In This Chapter

- Seeing through the eyes of Hubble
- Observing gamma rays with Compton
- X-raying the universe with Chandra
- Heating up with Spitzer
- Identifying the universe's oldest galaxies with Webb

Telescopes have revolutionized society's understanding of the cosmos. However, you can see only so much from the ground, because the atmosphere on Earth blocks the view of distant stars and galaxies at some wavelengths and blurs the view in others. For this reason, most observatories are located on the tops of tall mountains, where the air is thinner. But for an even better view, why not turn to space? Telescopes in space don't have to peer through any atmosphere at all, which means they can view the cosmos with unprecedented clarity.

To take advantage of this fact, NASA has established the Great Observatories Program, which consists of a series of large space telescopes. The Hubble Space Telescope observes in visible wavelengths; the Compton Gamma Ray Observatory and the Chandra X-Ray Observatory study the high-energy universe; and the Spitzer Space Telescope and the planned James Webb Space Telescope observe longer-wavelength infrared emissions. We describe all of these telescopes in detail in this chapter.

# The Hubble Space Telescope Clears the Way (1990–Present)

The Hubble Space Telescope, the world's biggest telescope in space, has not only made amazing discoveries about the nature of the universe but it has also greatly helped popularize astronomy. Hubble has taken extremely clear photographs of the cosmos, and these images have graced the covers of calendars, textbooks, and many other forms of media. In fact, they've become some of the most iconic, and reproduced, astronomical images ever.

Hubble images have revealed aspects of the universe that were largely unknown in previous years. For instance, Hubble data has suggested that black holes are likely to exist in the centers of all galaxies, and it has even helped scientists estimate their masses. More recently, observations from Hubble have provided direct images of extrasolar planets for the first time and revealed young stars with planet-forming disks of dust and debris around them.

In addition to the pictures, the scientific discoveries made by Hubble are magnificently varied. For example, Hubble has recorded optical phenomena related to gamma-ray bursts that give crucial information about how stars are formed. Additionally, in 2001, research teams were able to use Hubble data to make a measurement of the rate of expansion of the universe, which can be used to determine both the age and the ultimate fate of the universe.

In the following sections, we share how Hubble went from being an impressive idea to an amazing reality, explain the technology used by Hubble, and describe the repairs Hubble has needed over the years.

## How Hubble came to be

First suggested in the 1940s by Lyman Spitzer, the Hubble Space Telescope took many years to reach fruition as the world's biggest telescope in space. A deal with the European Space Agency (ESA) in the late 1970s resulted in the ESA supplying solar panels and one instrument for the telescope in return for a percentage of observing time. Work began on the telescope in 1978 with an initial planned launch date in 1983, but the launch was postponed due to construction delays.

Hubble was finally completed in the mid-1980s and was scheduled for launch in late 1986. However, the *Challenger* disaster of early 1986 (see Chapter 5) grounded all Space Shuttle flights until 1988. As a result, the telescope didn't launch until April 1990 when Space Shuttle *Discovery* carried it into orbit (flip to Chapter 14 for details on the launch).

The telescope is in a low Earth orbit so it can be reached by the Space Shuttle for servicing missions. Hubble (shown in Figure 18-1) orbits the Earth every 97 minutes and moves at about 18,000 miles per hour.

**Figure 18-1:** The Hubble Space Telescope in orbit.

*Courtesy of NASA*

The Hubble Space Telescope was named after Edwin Hubble, a pioneering American astronomer. After a brief career in law, Hubble studied astronomy and created a system for listing and classifying the galaxies that he observed. He was one of the first modern-era scientists to map the cosmos. He also came up with a method, known as Hubble's Law, for discerning the nature of the ever-expanding universe.

## The telescope's technology

Hubble is composed of three main elements: the spacecraft itself, the instruments that collect and store data, and the optics that make it all happen. The telescope-spacecraft is a solar-powered, insulated structure that serves as a home for the instruments, which in turn are driven by optical technology designed to collect light and precisely reflect it through mirrors.

The light-gathering capability of telescopes on Earth is usually measured by the size of their mirrors. Hubble's 7.9-foot (2.4-meter) diameter mirror is small compared to some ground-based telescopes. However, the lack of atmosphere, the cool operating temperatures in space, and the telescope's very large detector (the chip that detects light coming from the mirror) mean that Hubble can take stunning images of unprecedented clarity. (Head to the color section to see an image taken by Hubble.)

The images produced by Hubble are created using one of several instrument systems that take images ranging from the near-infrared to the ultraviolet. The instruments have been upgraded and changed out over the years; the current complement in orbit includes cameras that observe in visible and ultraviolet light, take infrared images, capture wide field and planetary images, and measure temperature and composition.

Many of the current instruments have failed and are awaiting repair or replacement on the next Hubble servicing mission (see the later section "Keeping Hubble up to snuff: Other servicing missions" for the scoop).

## A few fixes for Hubble

Hubble was the first space observatory designed to be serviced in space by astronauts. We describe the events of Hubble's servicing missions in the following sections. (Check out an image of a servicing mission in the color section.)

### Providing clarity to blurry images: The first repair

Shortly after Hubble's launch, NASA officials realized that the telescope's main mirror had a slight manufacturing defect and needed to be repaired. Although the mirror had been built to exacting specifications, its edges were just ever-so-slightly too flat. The flaw was only a matter of 0.00009 inches (0.00002 centimeters), but it was serious enough that instead of bringing images to a sharp focus, the light was smeared out over the image, resulting in pictures that looked blurry and unfocused.

After spending decades — and billions of dollars — on the telescope, the out-of-focus images opened NASA up to public ridicule. Hubble was widely regarded as a dud by the popular media, even though it was able to carry out a portion of its observing duties despite the blurry images.

Fortunately, NASA didn't have to abandon the telescope completely. Because the problem was a stable, optical issue, NASA engineers were able to design an adaptive mirror called COSTAR. Similar to a pair of corrective eyeglasses, COSTAR fit into the telescope to correct the images Hubble was capturing.

Hubble was restored to its intended glory during the first servicing mission in 1993. Astronauts on mission STS-61 matched the orbit of the Space Shuttle *Endeavour* to Hubble and then reached out with the Shuttle's robotic arm to grab the telescope and bring it into the Shuttle's *payload* (cargo) bay. In addition to installing the COSTAR corrector, they installed a new camera with built-in optical correction, replaced the solar panels, and replaced other components of the telescope's electronics and pointing systems. The ten-day mission, which involved five long *spacewalks* (the more-common term for *extravehicular activities* [EVAs] done in space) that were covered live on national television, was grueling but widely regarded as a complete success. Astronomers proudly showed off crystal-clear images of the cosmos, and today, 18 years after its launch, Hubble is thought of as one of the gems of NASA's astronomical program, with its blurry early days largely forgotten.

### Keeping Hubble up to snuff: Other servicing missions

Three other servicing missions have kept Hubble going over the years:

- Servicing Mission 2, which took place aboard Space Shuttle *Discovery* on flight STS-82 in 1997, upgraded the telescope to take pictures at longer wavelengths, better study distant galaxies, and search for black holes.
- Servicing Missions 3A and 3B involved further maintenance upgrades and repairs:
    - Mission 3A, once again aboard Space Shuttle *Discovery* but this time on flight STS-103 in 1999, replaced four failed gyroscopes responsible for keeping the telescope at a steady orientation so it can perform observations. This mission also installed a new computer, bringing Hubble's onboard memory to a whopping 2 megabytes of RAM.
    - Mission 3B, aboard Space Shuttle *Columbia* on mission STS-109 in 2002, installed a new camera (one that could produce even higher-quality images), repaired the near-infrared imaging device, and replaced solar panels responsible for powering the telescope.

## Fast facts about the Hubble Space Telescope

Following are the basic facts about the Hubble Space Telescope:

- **Launch date and site:** April 24, 1990; Kennedy Space Center, Florida
- **Spacecraft name and mass launch:** Hubble Space Telescope; 24,250 lb (11,110 kg)
- **Launch vehicle:** Space Shuttle *Discovery*
- **Number of Earth orbits:** More than 110,000, as of this writing
- **Length of time in Earth orbit:** 18 years to date

## Expanding the rainbow: The electromagnetic spectrum

If you've ever seen a rainbow, you know that sunlight can be divided into different colors. What's actually going on is that visible white light is made up of a combination of different colors, each with a different wavelength. Think of light as a wave that, like the waves at the beach, has a certain distance between peaks. This distance is called a *wavelength* — in the rainbow of visible light, red light has a longer wavelength, and violet light has a shorter wavelength.

The idea of wavelengths of light can be expanded beyond what your eye can actually see. Visible light is just a small part of what's called the *electromagnetic spectrum,* which refers to a huge range of energy that's given off by stars like the Sun and radiates though space. The wavelength of electromagnetic radiation is related to its energy — shorter-wavelength radiation, such as ultraviolet waves, X-rays, and gamma rays, has increasingly higher energy than visible light. On the longer wavelength side of things, past the red side of the visible spectrum, are infrared radiation, microwaves, and radio waves, with increasingly longer wavelengths and lower energies.

Because celestial objects such as stars and galaxies give off radiation all along the electromagnetic spectrum, NASA's four great observatories were designed to collect data in the gamma ray, X-ray, visible, and infrared parts of the spectrum. Radio waves are long enough to pass through the Earth's atmosphere almost unaltered, so radio telescopes can easily be located on the ground without the need for a dedicated space telescope.

To keep Hubble operational, a fourth servicing mission is crucial. Although it was initially canceled due to the few remaining Shuttle flights being designated solely for International Space Station construction (see Chapter 19 for details on the ISS), the fourth servicing mission was reinstated. Originally scheduled to launch in October 2008 aboard Space Shuttle *Atlantis,* the mission was postponed when an essential computer component aboard the telescope failed. NASA is delaying Servicing Mission 4 until 2009 to allow astronauts to train to replace that component. Ultimately, this mission will breathe new life into Hubble, fixing or upgrading all the scientific instruments and giving the telescope an orbital boost that's expected to keep it alive and well until at least 2014, and perhaps as long as 2020.

If you want to stay in the know about Hubble's fourth servicing mission, visit `www.nasa.gov/mission_pages/hubble` and click on the Servicing Mission 4 link in the left-hand navigation bar.

# The Compton Gamma Ray Observatory's High-Energy Mission (1991–2000)

Gamma rays are a very high-energy, short-wavelength form of electromagnetic radiation given off by some of the most energetic, and rarest, celestial phenomena such as *supernovas* (star explosions), black holes that are in the process of pulling in mass, solar flares, and *pulsars* (rapidly spinning stars). They're also difficult to study from the ground, which is why NASA's second Great Observatory, the Compton Gamma Ray Observatory (CGRO), was launched into orbit by Space Shuttle *Atlantis* in 1991. (Figure 18-2 shows CGRO at launch.)

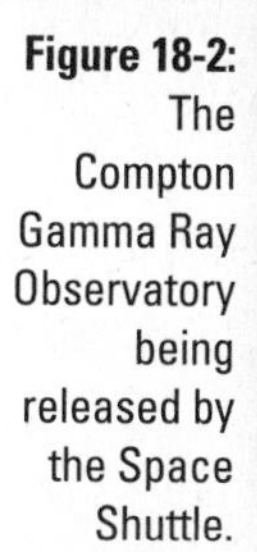

**Figure 18-2:** The Compton Gamma Ray Observatory being released by the Space Shuttle.

Courtesy of NASA

The CGRO had several notable achievements. It

- Discovered a "young" (nearly 700-years-old) supernova that had gone undetected by previous telescopes
- Helped astronomers develop a full data base of gamma ray bursts, thereby enabling the discovery that these rare events come from outside the Milky Way Galaxy
- Supplied valuable information on the source of gamma ray bursts, including radio galaxies, newly discovered pulsars, and solar flares

## Fast facts about the Compton Gamma Ray Observatory

Following are the basic facts about the Compton Gamma Ray Observatory:

- **Launch date and site:** April 5, 1991; Kennedy Space Center, Florida
- **Spacecraft name and mass at launch:** Compton Gamma Ray Observatory; 37,500 lb (17,000 kg)
- **Launch vehicle:** Space Shuttle *Atlantis*
- **Number of Earth orbits:** About 53,000
- **Length of time in Earth orbit:** 9 years, 2 months

CGRO didn't look like a telescope in the traditional sense of the word. Instead of having a small primary mirror and light-gathering arrangements, CGRO's instruments had to be large to detect rare gamma-ray photons. In fact, at the time of its launch, CGRO was the heaviest satellite devoted to astrophysics ever flown. (The Chandra X-Ray Observatory, which we cover in the next section, later beat CGRO to become the heaviest payload launched by the Space Shuttle.)

CGRO takes its name, but not its looks, from Dr. Arthur H. Compton, a Nobel Prize–winning scientist who made great strides in the field of high-energy physics.

At the end of its planned lifetime, one of the gyroscopes that helped control the spacecraft's orientation in space failed. Unlike the Hubble Space Telescope, CGRO wasn't designed to be serviced by astronauts, so the failed device couldn't be replaced. Instead of waiting for a second gyroscope to fizzle out and leave the spacecraft unmaneuverable, NASA decided to bring CGRO out of orbit in a controlled crash. Most of the spacecraft burned up in the atmosphere on June 4, 2000, with the rest falling safely into the Pacific Ocean.

# The Chandra Observatory X-rays the Universe (1999–Present)

The third of the four Great Observatories is the Chandra X-Ray Observatory, launched in 1999 by Space Shuttle *Columbia.* X-ray observations are particularly well-suited to space telescopes because they're absorbed by the Earth's atmosphere almost completely. Chandra was designed to be able to detect X-ray emissions that were 100 times fainter than anything seen before by previous satellites.

## Fast facts about the Chandra X-Ray Observatory

Following are the basic facts about the Chandra X-Ray Observatory:

- **Launch date and site:** July 23, 1999; Kennedy Space Center, Florida
- **Spacecraft name and mass at launch:** Chandra X-Ray Observatory; 50,200 lb (22,700 kg)
- **Launch vehicle:** Space Shuttle *Columbia;*
- **Number of Earth orbits:** About 1,200, as of this writing
- **Length of time in Earth orbit:** 9 years to date

X-rays have a lower energy than gamma rays, which means a satellite observing an object can actually image that object by using X-rays. That's what makes this X-ray–focused satellite so special: Because X-rays are given off by energetic sources in the Milky Way Galaxy and in other galaxies, Chandra has been able to observe remnants from supernovas and the centers of galaxies. It has also observed neutron stars and black holes and helped astronomers understand dark matter. Chandra has also detected cool gas spiraling into the center of a nearby galaxy and caught X-rays being given off as material in a planet-forming disk fell back onto its central star, helping astronomers understand the role of gas and dust in planet formation.

To detect high-energy X-rays, Chandra possesses a series of four nested mirrors that are specially angled to reflect X-rays. Observations are made with one of two instruments, either the Advanced CCD Imaging Spectrometer or the High Resolution Camera. The spacecraft has a highly elliptical orbit that takes it out to one-third of the distance to the Moon at its farthest point. This orbit allows it to spend much of its time outside the Van Allen radiation belts that protect the Earth from harmful radiation. However, this orbit also means Chandra can't be visited by the Space Shuttle.

Chandra was officially designed as a five-year mission, but it's still going strong nine years after launch. Astronomers estimate that the observatory could last for up to 15 years, giving it another 6 years to contribute to humans' understanding of the universe.

You can see some great images and find out all about Chandra's discoveries here:

- chandra.harvard.edu
- www.nasa.gov/mission_pages/chandra/main

The observatory was named after Indian astrophysicist Subrahmanyan Chandrasekhar. Frequently called by the less-formal moniker Chandra, he was widely regarded as a preeminent physicist who was one of the first to apply the principles of physics to astronomy. His studies of the stars earned him a Nobel Prize in Physics in 1983.

# The Spitzer Space Telescope Heats Things Up (2003–2009)

Infrared telescopes, such as the Spitzer Space Telescope, are a great addition to the study of space because they detect and measure the heat *(infrared energy)* given off by objects within certain wavelengths. The view of telescopes that operate at visible wavelengths (such as the Hubble Space Telescope, which we cover earlier in this chapter) is frequently blocked by gas, dust, and other obstructions; infrared telescopes like Spitzer, on the other hand, can see through the haze and yield brand-new information from deep galaxies, new stars, and other objects that can't be seen through conventional means.

*Infrared radiation,* or radiation that falls along the electromagnetic spectrum between visible light and microwave radiation, may be best known for its use in night-vision goggles. With that in mind, thinking of Spitzer as the world's biggest and most-expensive set of night-vision goggles isn't entirely off-base!

Spitzer (shown in Figure 18-3) was launched aboard a Delta II rocket in 2003 and is expected to function through mid-2009. In the following sections, we explain how Spitzer works and why it needs to keep cool in order to function.

**Figure 18-3:** The Spitzer Space Telescope.

Courtesy of NASA/JPL-Caltech

The Spitzer Space Telescope was named after Lyman Spitzer, who first proposed the idea of a space telescope in the 1940s.

## Capturing heated images of the cosmos

The Spitzer Space Telescope is equipped with three main instruments that are sensitive enough to take the most-detailed and farthest-reaching infrared images ever seen. The instruments themselves provide additional data, including broad views of the sky at a series of different near-infrared and mid-infrared wavelengths; mid-infrared observations in narrow wavelength ranges (for determining the chemical composition of the object being observed); and measurements of dust distribution in the galaxy.

Spitzer has taken a variety of stunningly beautiful images of the cosmos, which are also scientifically indispensible (check out an image for yourself in Figure 18-4). As a complement to Hubble, Spitzer has revealed the details of star-forming regions and extrasolar planets. It has also provided proof of supermassive black holes, found evidence of stars that could've formed shortly after the Big Bang, and provided new details about the structure of the center of the Milky Way Galaxy.

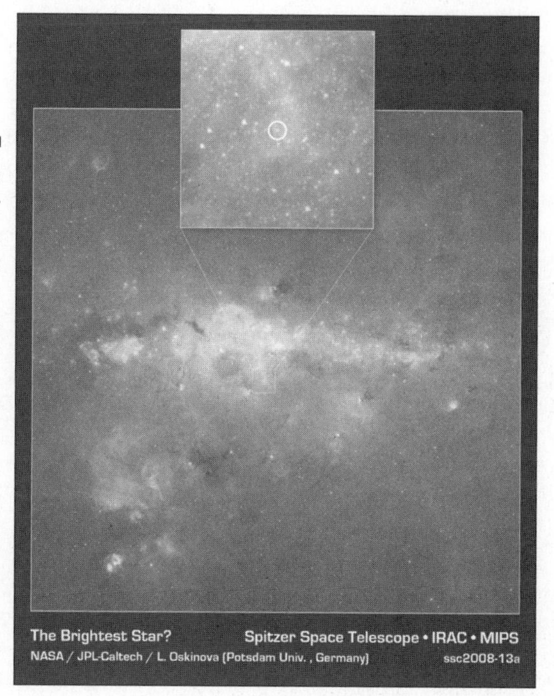


**Figure 18-4:** This Spitzer image combines data from all three instruments to show a bright star (inside the circle in the box) that's almost completely obscured by a cloud of dust.

*Courtesy of NASA/JPL-Caltech/Potsdam Univ.*

### Fast facts about the Spitzer Space Telescope

Following are the basic facts about the Spitzer Space Telescope:

- **Launch date and site:** August 25, 2003; Cape Canaveral Air Force Station, Florida
- **Spacecraft name and mass at launch:** Spitzer Space Telescope; 2,090 lb (950 kg)
- **Launch vehicle:** Delta II rocket
- **Number of solar orbits:** 5, as of this writing
- **Mission duration:** 5 years to date

### Keeping Spitzer cool

Because an infrared telescope detects heat, the telescope itself must remain as cold as possible to perform its observations, and even in the cold, dark vacuum of space, extra cooling is required. Spitzer keeps itself cool with the aid of a *cryostat,* a device that uses a helium tank to lower the telescope detector's temperature. The cryostat is capable of keeping the telescope cool for the expected life span of Spitzer's mission. Solar shields on the spacecraft also help, but it's a good idea to keep the telescope as far away from the hot, bright Earth as possible (while still leaving it able to do the work it needs to do).

Scientists sent Spitzer into an *Earth-trailing heliocentric orbit,* which means that the telescope follows Earth in its orbit around the Sun instead of orbiting around the Earth itself. A natural cooling effect is one benefit of this orbit, because the spacecraft is relatively far away from the heat-emanating Earth; as a result, the mission was able to reduce the amount of coolant it had to carry. The helium onboard Spitzer is expected to be used up sometime in 2009, at which time the telescope will only be able to operate in a "warm" mode with just one of its instruments.

## Looking to the Future: The James Webb Space Telescope

An infrared project that's planned but not yet launched is the James Webb Space Telescope (JWST). Scheduled for liftoff in 2013, the mission goals of JWST are to use infrared technology to help locate the oldest galaxies in the universe. Such information would help scientists connect the dots between the beginning of the universe, the Big Bang theory, and the Milky Way Galaxy. JWST will be capable of studying early to recent solar systems, and ultimately, it'll be able to help demonstrate how solar systems like ours were formed.

The design for the telescope is quite grand: The main mirror measures 21 feet (6.5 meters) in diameter (although the overall mass of JWST will be smaller than Hubble). The sheer size of the mirror means JWST must be launched in a folded state and opened up after the telescope makes it into orbit. Several other new technologies will help JWST gather significant data, and like the Spitzer infrared telescope described in the previous section, JWST will require cooling mechanisms in order to maintain its effectiveness.

The JWST mission is officially being planned by NASA, the Canadian Space Agency, and the European Space Agency, but its data will be valuable to scientists worldwide. For more info, visit `www.jwst.nasa.gov`.

# Chapter 19

# A Symbol of Global Cooperation: The International Space Station

### In This Chapter

- Making preparations for the International Space Station
- Scoping out the station's parts
- Constructing the station and conducting experiments in space
- Guessing how long the International Space Station will last

The world's preeminent space research facility, the International Space Station (ISS), is a modularized structure that orbits about 217 miles (350 kilometers) above the Earth. At its conception, the ISS's main goals were to establish the world's first continuous human presence in space and provide a permanent lab for conducting space research. It's a mammoth example of global cooperation because many nations worked together to make it possible.

The ISS continues the research begun by earlier space stations, such as the various Salyut stations, Skylab, and Mir (see Chapters 12 and 13 for details on these early space stations). It's one of the most technologically advanced — and most expensive — construction projects in human history. Although budget realities have scaled back the station many times, the current incarnation of the ISS features several lab modules and an astronaut habitat.

In this chapter, we walk you through the beginnings and the future of the ISS. We also describe the station's layout and the important work being done inside it.

To discover even more about the ISS, check out this Web site: www.nasa.gov/mission_pages/station/main/index.html.

# Planning the International Space Station

Space Shuttle missions can fulfill short-term scientific, technical, and political goals (as you find out in Chapters 13 and 14), but to really establish a presence in space, NASA engineers and scientists realized that a destination was crucial. During the Apollo years of the 1960s and 1970s, the Moon was the destination of choice. In the 1970s and 1980s, Skylab and the various Salyut stations were short-term destinations for Soviet and American crews. Finally, Mir became the site of joint missions in the mid-1980s and into the 1990s.

The International Space Station (ISS), however, was conceived as a first step toward a continuous human presence in space and a step toward the future of humans as a space-faring race. It was established early on that a permanent orbital base would greatly facilitate exploration of the rest of the solar system (largely by serving as a "way station" for other destinations), in addition to providing a way to discover more about human survival in space. Following experience gained from Mir and Skylab, both Russia and the U.S. began planning more-advanced, modular space stations that could support larger crews with enhanced research opportunities; both countries, however, suffered from design and financial setbacks. With Cold War secrecy and competition no longer an obstacle, Russia and the U.S. (along with Europe, Japan, and Canada) agreed to pool economic and intellectual resources toward the creation of a common space station project, eventually named the International Space Station.

ISS partner countries include

- Belgium
- Brazil
- Canada
- Denmark
- France
- Germany
- Italy
- Japan
- The Netherlands
- Norway
- Russia
- Spain
- Sweden
- Switzerland
- United Kingdom
- United States

Saying you're going to work together with multiple nations to construct a space station is great, but actually constructing that station requires a lot of prep work, as the ISS partner countries soon found out. Here are some of the factors the various nations had to consider as they worked to make the ISS a reality:

- **Protection:** National security concerns were foremost, but proprietary engineering documents and other intellectual property also needed to be protected.
- **Compatibility of parts:** The countries had to work very closely together to make sure their modules and equipment would all fit together and work smoothly, despite country-to-country differences in typical electrical voltages and even units of measurement.
- **Ownership:** The partner nations had an important question to answer: Would astronauts from different countries be allowed onboard the ISS only in direct proportion to how much their country contributed to the construction costs? Political tact wound up playing a large role in the granting of credit and privileges. Russia, for example, demanded major access to the completed ISS because its scientists and engineers contributed substantial knowledge and experience in the area of human spaceflight.
- **The watchful public eye:** The ISS provided an unimaginable wealth of worldwide publicity possibilities, and the people in charge of the station's design were acutely aware of the potential that every decision — good or bad — could make front-page news.

# Touring the International Space Station

Crews aboard the International Space Station (ISS) have everything they need to live and work in their home away from home. In the following sections, we describe the overall layout of the ISS and the latest 'n' largest addition to the station: Japan's Kibo lab. (Be sure to check out an image of the ISS in all its glory in the color section.)

## The piece-by-piece construction (1998–present)

Since 1998, a mix of Soyuz rockets, Proton rockets, and the Space Shuttle has launched various components of the ISS into orbit. In fact, ISS assembly missions have become the Space Shuttle's primary mission. Space Shuttle flights have been largely devoted to ISS construction since 2000, and the end of the Shuttle program in 2010 is currently arranged to coincide with the anticipated completion of the ISS in 2011 (which includes a final launch of some components on an unmanned rocket). Afterward, the ISS will be serviced by Russian Soyuz capsules and automated supply ships. (Check out the Space Shuttle approaching the ISS in Figure 19-1.)

When completed, the ISS will be composed of 15 modules joined together. The modules currently in orbit are described in Table 19-1.

**Figure 19-1:** The Space Shuttle *Endeavour* approaches the International Space Station. In its cargo bay is an Italian-built logistics module full of equipment and supplies.

Courtesy of NASA

**Table 19-1 Current Modules of the International Space Station**

| *Module Name* | *Country of Origin* | *Launch Date* | *Launch Vehicle* | *Mass* | *Function* |
|---|---|---|---|---|---|
| Zarya | Russia & U.S. | November 20, 1998 | Proton-K | 42,600 lb (19.325 kg) | Source of electrical power, propulsion, and other essential functions |
| Unity | U.S. | December 4, 1998 | Space Shuttle *Endeavour* | 25,600 lb (11,610 kg) | Berthing bay for other modules |
| Zvezda | Russia | July 12, 2000 | Proton-K | 42,000 lb (19,050 kg) | Service module |
| Destiny | U.S. | February 7, 2001 | Space Shuttle *Atlantis* | 32,000 lb (14, 515 kg) | Research laboratory |
| Quest | U.S. | July 12, 2001 | Space Shuttle *Atlantis* | 13,370 lb (6,065 kg) | Airlock |
| Pirs | Russia | September 14, 2001 | Soyuz-U | 7,900 lb (3,580 kg) | Docking ports |
| Harmony | Europe & U.S. | October 23, 2007 | Space Shuttle *Discovery* | 31,500 lb (14, 290 kg) | Utility hub for the entire ISS |
| Columbus | Europe | February 7, 2008 | Space Shuttle *Atlantis* | 28,000 lb (12,800 kg) | Research laboratory |

| Module Name | Country of Origin | Launch Date | Launch Vehicle | Mass | Function |
|---|---|---|---|---|---|
| Experiment Logistics Modules (part of Kibo) | Japan | March 11, 2008 | Space Shuttle *Endeavour* | 18,490 lb (8, 335 kg) | Transportation and storage |
| Kibo's Pressurized Module | Japan | May 31, 2008 | Space Shuttle *Discovery* | 33,000 lb (14, 800 kg) | Research laboratory |

Planned additions include the following:

- Mini-Research Module 2 (launch planned for 2009), Russia, docking and storage
- Node 3 (launch planned for 2010), Europe/U.S., life-support system and recycling
- Cupola (launch planned for 2010), Europe/U.S., observatory module
- Mini-Research Module 1 (launch planned for 2010), Russia, storage and docking
- Multipurpose Laboratory Module (launch planned for 2011), Russia, research module

Figure 19-2 shows the layout of the ISS as of June 2008. The *photovoltaic arrays* (solar panels) power the station. Zarya and Zvezda provide power and other service functions, whereas Kibo and Columbus serve as research facilities. Harmony serves as a utility hub, and the trussed segment provides power and a physical connection between the disparate modules and elements.

## The newest (and biggest) addition: Japan's Kibo lab

Various countries have supplied astronauts, equipment, and modules to the ISS to help further scientific research and exploration in space. One of the more recent successes in the area of international cooperation is the Japanese Experiment Module, called Kibo in Japanese (see Figure 19-3). This space-borne laboratory will be brought to the ISS in three parts, two of which are already in orbit.

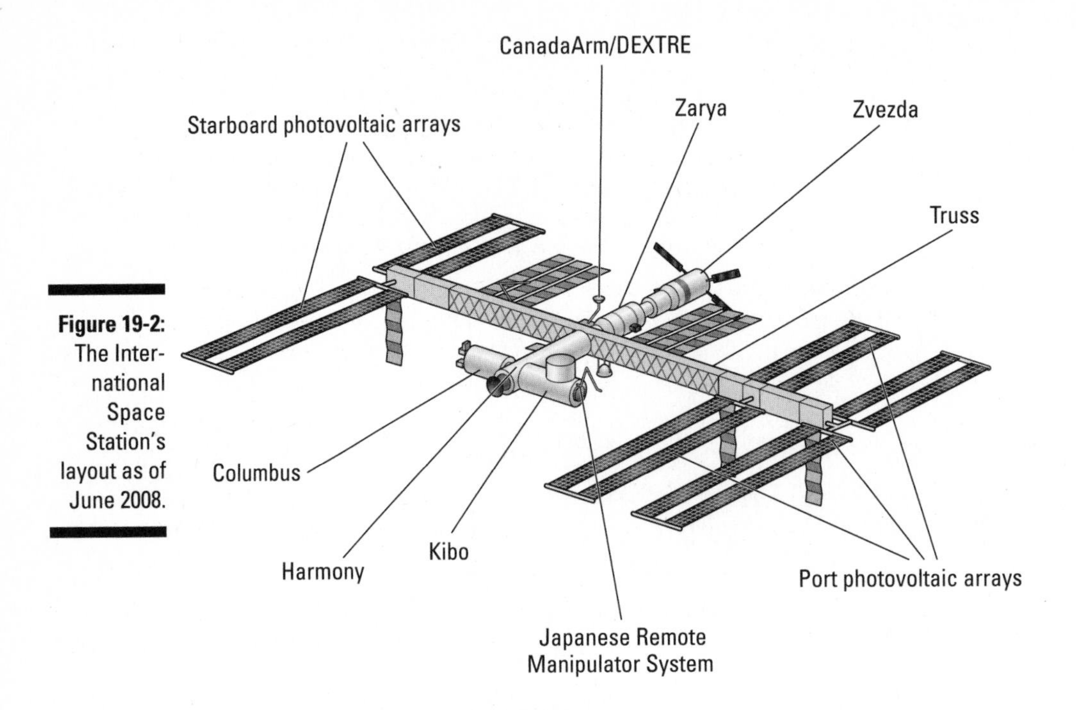


**Figure 19-2:** The International Space Station's layout as of June 2008.

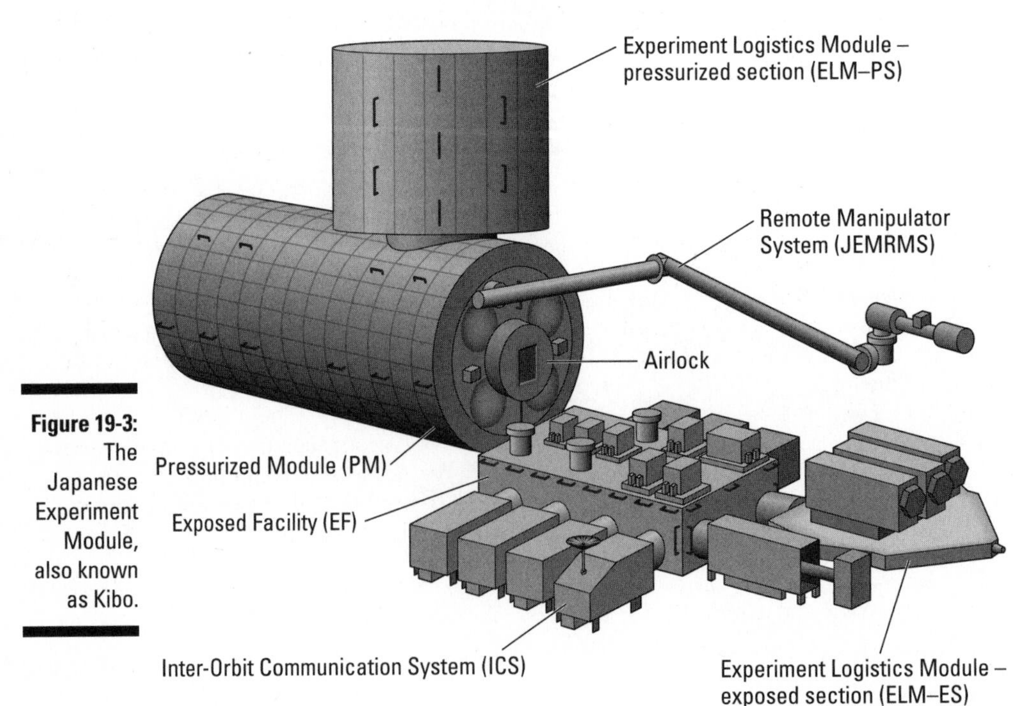


**Figure 19-3:** The Japanese Experiment Module, also known as Kibo.

In keeping with the theme of modular space station development, Kibo has four main parts:

- **Pressurized Module:** The place where astronauts work and perform experiments. Supplies, equipment, power sources, thermal controls, and other requirements for effective research are housed here. The Pressurized Module also features an Experiment Logistics Module, which provides additional storage for experiment racks.
- **Exposed Facility:** An area for work outside of the space station. Astronauts can send and receive equipment and info straight to the Pressurized Module through an airlock. An Experiment Logistics Module also is located in this section.
- **Remote Manipulator System:** A set of two robotic arms that can move heavy payloads between modules. It's controlled from inside the Pressurized Module.
- **Inter-Orbit Communication System:** A system that coordinates the exchange of data between Kibo and Japanese scientists.

Kibo was built in Japan and shipped to Florida to await its launch. Three Space Shuttle missions are required to completely transport and assemble it:

- The first part of Kibo was launched to the ISS aboard Space Shuttle *Endeavour* on mission STS-123 on March 11, 2008; the second part was launched aboard Space Shuttle *Discovery* on mission STS-124 on May 21, 2008. These two launches brought parts of the Pressurized Module to the ISS.
- The remainder of Kibo is scheduled for launch (likely aboard Space Shuttle *Endeavour*) on mission STS-127, currently planned for 2009.

When fully assembled, the Kibo module will be the biggest component of the ISS. Its Pressurized Module (which is the main "indoors" working area) measures 36.7 feet (11.2 meters) by 14.4 feet (4.4 meters). Within it, astronauts can conduct experiments in areas such as X-ray astronomy, biology, and communications. Kibo experiments are run from the Tsukuba Space Center in Japan.

Great hopes are placed on Kibo — and not just because the word *kibo* means "hope" in Japanese — because it's the newest lab, one that contains significantly improved equipment for the ISS researchers and represents a major milestone for Japanese achievements in space research.

## Working aboard the International Space Station

The International Space Station (ISS) has been home to at least one crew (or *expedition,* as ISS crews are officially known) since 2000. The first crew consisted of Commander Bill Shepherd, Soyuz Commander Yuri Gidzenko, and Flight Engineer Sergei Krikalev. Due to the station's current size, ISS crews consist of just three people, but eventually the ISS will be able to support six individuals, thanks to extra living space and life-support systems installed during a 2008 Space Shuttle mission. (Check out a team of astronauts aboard the ISS in Figure 19-4.)

**Figure 19-4:** Astronauts inside the International Space Station.

*Courtesy of NASA*

The Expedition 1 crew traveled to the ISS in a Russian Soyuz rocket and remained in orbit for 135 days. Subsequent teams have journeyed to the ISS via the Space Shuttle or a Soyuz rocket and spent increasing amounts of time in space. The most-recent crew, Expedition 18, reached the ISS in the fall of 2008 and will remain aboard it until the spring of 2009.

In the following sections, we describe a few of the major tasks that astronauts handle aboard the ISS — namely, constructing and maintaining the station, taking care of their health, and conducting scientific experiments.

## Maintaining the station's and the astronauts' health

The ISS is the most health conscious place in all of space. After all, the men and women aboard it must constantly check on and maintain the operational health of the station (as well as their own health). The ISS astronauts spend considerable time doing construction work and maintenance on their habitat. An estimated 960 hours per year are required to build the station in space and keep it running; about 160 of those hours are spent on *spacewalks* (*extravehicular activities,* or EVAs, performed in space). Because spacewalks require two crew members, that comes out to about 2,000 hours of crew work.

When they're not performing upkeep on the station (or conducting experiments, as described in the next section), astronauts aboard the ISS spend time caring for themselves. They aim to sleep a reasonable amount each evening and fit in exercise sessions throughout the day to try and mitigate the loss of muscle tone and bone mass that comes in space. Treadmills and specially designed resistance training equipment (such as the Resistive Exercise Device, a machine that uses vacuum cylinders rather than weights) are expected to be used for about two hours daily by each astronaut.

Although astronauts can conceivably become ill aboard the ISS, they're carefully screened for infections before making the journey. And sure, bacteria and fungi are bound to grow wherever humans are involved, but the ISS astronauts use advanced tests to locate the areas that require extra cleaning. In addition to cleaning the station, personal hygiene also occupies more time in orbit than it does on the ground. Fortunately for the crew, taking showers on the ISS is a lot easier than on early space stations (long-duration crews on the early Salyut stations had an entire day devoted to bathing, which included heating up the water one small pot at a time).

One of the more interesting aspects of living in orbit has to do with *microgravity,* the very low-gravity environment found in orbit around the Earth. Thanks to this lovely phenomenon, even ordinary tasks, such as writing with a pen or walking across the room, take on an entirely different meaning (like when the pen floats away and walking is really more like flying). It takes time for astronauts to get used to the fact that they're no longer weighted down by gravity. Imagine getting up in the morning and, rather than standing for a stretch, whacking your head on the ceiling!

## Waste not, want not

Because the limited confines of the Space Shuttle and the Soyuz spacecraft prevent them from transporting waste from the ISS back to Earth, ISS waste products are recycled whenever possible. In late 2008, astronauts installed a new high-tech filter system that recycles waste water, including urine, into clean drinking water. All nonreusable supplies, including water, must be launched from the ground, so being able to recycle water frees up cargo space for other essential items.

In cases where objects (such as old equipment, food containers, used clothes, and human waste) can't be reused in any capacity, they must be discarded to make room inside the modules for the crew and equipment. Too much stuff inside the ISS means more work for the astronauts because they have to spend time and precious energy hustling trash rather than working.

In between dockings, the station's trash must be compressed and stored for eventual transport. It must be tempting just to throw all that trash overboard, but ejecting garbage from the ISS is done quite carefully, considering any objects that head back toward the ISS could become dangerous missiles. Although astronauts aboard the former Mir space station ejected bags full of trash, ISS crew members generally pack unmanned Progress supply ships with garbage after removing the new supplies. The Progress ship is then sent on an Earth-bound trajectory where much of the module burns up in orbit; any remaining debris is targeted to fall into a designated dumping zone in the middle of the Pacific Ocean, far from land.

## *Conducting science experiments*

Three of the ISS's ten currently active modules are laboratory modules, each of which is capable of supporting studies in biology, physics, meteorology, and astronomy. Most of the lab science at the ISS currently takes place in the NASA-made Destiny module, which has been in space since 2000. This lab measures 14 feet (4.3 meters) by 28 feet (8.5 meters) and is pressurized to allow for full human occupation. Racks provide space and equipment for running ducts, wires, pipes, and other necessities throughout the lab.

Biology experiments take a high priority on the ISS, because how the human body performs long-term in space is of the utmost interest for future missions to the Moon, Mars, or beyond. Astronauts study bone density, muscle tone, and other aspects of the human anatomy as they change in microgravity. Other experiments on the physics side of things help solve the mysteries of how fluids behave in space, as well as in examining the effect of low gravity on fluids and fluid combinations.

### Controversy in space: The U.S. nearly backed out of the International Space Station

As the ultimate symbol of international agreement, it would seem a foregone conclusion that all countries involved and committed to the International Space Station would remain so. Not necessarily! In 2005, it appeared that the U.S. was on the verge of gradually withdrawing from the ISS project, due in part to its increasing commitments to other areas of space exploration. In addition, there was a great deal of skepticism about the longevity of the Space Shuttle program following the disintegration of Space Shuttle *Columbia* in 2003 (as explained in Chapter 5).

However, the U.S. kept up its end of the bargain. Why?

- There was ultimately little choice, given that Space Shuttle flights are the only way to deliver essential ISS components and modules. No other spacecraft are large or powerful enough to lift certain space station components into orbit.
- America made a commitment to other nations in terms of money, time, effort, manpower, and heart. Backing out of supporting the ISS could've been interpreted as a flagrant violation of this partnership, and other space agencies (the Japan Aerospace Exploration Agency and the European Space Agency in particular) have invested quite a bit of money in modules, cargo, and other supplies that are scheduled for future launch to the ISS.
- Future goodwill is another consideration — America's quest to return to the Moon will undoubtedly require cooperation from other nations, and burning that goodwill now can only damage America's credibility later.

Crystal growth in microgravity is studied for its ability to teach scientists about the physical properties of particular chemical solutions. The value of growing these tiny crystals in space is that gravity doesn't disturb or impact their growth, meaning their structure can be observed in a purer fashion. The crystals are then brought back to Earth for study in laboratories. Growing pure crystals allows their physical properties to be studied in detail in Earth-bound laboratories, which may lead to the development of new materials and better medicines.

## Predicting the Future of the International Space Station

The International Space Station (ISS) is scheduled for completion by 2010, the year the last of the current Space Shuttles will fly to the ISS with the necessary parts (although an unmanned Russian Proton rocket may bring some remaining components to the ISS in late 2011). After the Space Shuttle is retired, the Russian Soyuz spacecraft will serve as the only access to the ISS until NASA's new Orion spacecraft (which we tell you all about in Chapter 21) takes flight several years later.

How long the station will last is a question on the minds of many. Generally speaking, each module in the ISS can last as long as its parts and equipment are functional and safe. Certain pieces can be replaced as needed, but the overall life span of the ISS will depend on how long its components remain in operational condition. Factors affecting that life span include radiation damage, extreme temperatures, and other conditions unique to space that may make the metal components wear faster than they would on Earth.

Officially, the ISS is expected to last no longer than 2015. Practically speaking, many items used in space have lasted long past their anticipated expiration dates, but remember that the ISS has many different ages. The first modules, in place since 1998, were originally rated with a 15-year life span, but more-recent modules were added 10 years later (and came with similar life spans). As of this writing, no nation has officially committed to the ISS beyond 2015, and no specific plans have been made public as to what will happen to the ISS after that time.

Though regarded as an embarrassing waste of money by some, the completed ISS will be a capable orbital laboratory. However, its future may depend more on political, rather than engineering or scientific, concerns. In 2004, President George W. Bush urged NASA to focus on a human return to the Moon with future missions to Mars — a move that largely supplanted the ISS because it moved the space agency's focus beyond Earth orbit. Regardless, the ISS is a tremendous technological and political achievement that continues to bring the space agencies of many different countries together to build a very complex piece of technology. Lessons learned from collaborating and operating the ISS will be very useful for the next phase of space exploration, whatever that stage may be. Keeping the ISS in working order for as long as possible will only further those goals.

# Chapter 20

# New Views of Mars in the New Millennium

### *In This Chapter*

- Venturing on a *Mars Odyssey* in search of water
- Seeing in three dimensions with *Mars Express*
- Roving the Red Planet: *Spirit* and *Opportunity*
- Mapping Mars via the *Reconnaissance Orbiter*
- Studying polar ice with *Phoenix*

Although Mars exploration has had its more-than-fair share of failures in the past, missions from the first decade of the 21st century have been largely successful. In fact, current robotic exploration of Mars is providing a better view of the Red Planet than ever before. While rovers are traveling across Mars's surface and digging in the icy dirt near its poles, orbiters are making compositional measurements and even taking pictures of the landers from orbit! In this chapter, we keep you up to date on the latest 'n' greatest missions to Mars.

For a complete, interactive timeline of Mars exploration, we recommend visiting `phoenix.lpl.arizona.edu/timeline.php`.

## *Mars Odyssey*: Searching for Water and Sending Signals (2001–Present)

*Mars Odyssey,* which reached Mars orbit in 2001, was the first robotic spacecraft to successfully reach Mars in the 21st century. Its purpose? To start looking for evidence of volcanic activity and water on the surface of Mars. Perhaps the spacecraft's most-significant finding is the existence of places on the Martian surface that may hold chloride minerals such as salt — the current presence of salt may indicate the previous presence of liquid water. Additionally, data gathered by *Mars Odyssey* has revealed concentrations of hydrogen that are thought to indicate near-surface water, which could indicate the more-recent presence of habitable locations on the Red Planet.

A repeat of many experiments on the failed *Mars Climate Orbiter* (described in Chapter 15), *Mars Odyssey* was also designed to relay communications from future Mars landers. Because landers typically rely on an orbiting spacecraft to relay their info all the way back to Earth, having *Mars Odyssey* around for this purpose is critical.

*Mars Odyssey* uses three main instruments to take its measurements. THEMIS (Thermal Emission Imaging System) takes near-infrared pictures to study surface composition; MARIE (Mars Radiation Environment Experiment) studies the Red Planet's radiation environment; and GRS (Gamma Ray Spectrometer) looks for hydrogen on the surface, which could indicate the presence of water (or ice) below the surface.

The *Mars Odyssey* spacecraft fulfilled its mission goals as planned in 2001 and helped relay the vast majority of images from three different Mars landers. Because the spacecraft and its equipment were still functioning (and because *Mars Odyssey* was already out in space) NASA officials decided to extend the Mars Odyssey mission through 2010 to allow the spacecraft to study climate phenomena, such as ice and dust, and continue relaying information from other Mars missions.

For more information about *Mars Odyssey,* go to `mars.jpl.nasa.gov/odyssey` or `www.nasa.gov/mission_pages/odyssey/index.html`.

## Fast facts about Mars Odyssey

Following are the basic facts about the Mars Odyssey mission:

- **Launch date and site:** April 7, 2001; Cape Canaveral Air Force Station, Florida
- **Spacecraft name and mass at launch:** *Mars Odyssey* (officially the *2001 Mars Odyssey* orbiter); 1,600 lb (725 kg)
- **Launch vehicle:** Delta II rocket
- **Date Mars orbit achieved:** October 24, 2001
- **Number of Mars orbits:** About 30,000, as of this writing
- **Length of time in Mars orbit:** 7 years to date

# Mars Express: A Bittersweet European Effort (2003–Present)

In 2003, the European Space Agency (ESA) got into the act of Mars exploration with its ambitious Mars Express mission. Intended to send both a lander and an orbiter to study the geology and biology of Mars, Mars Express was also charged with taking climactic and atmospheric measurements and investigating the Martian surface and subsurface. (It also served as the basis for Venus Express, an ESA mission that launched successfully in 2005, as you can read about in Chapter 16).

Like other spacecraft that traveled to Mars, *Mars Express* didn't go empty-handed. It had a hefty science *payload* (cargo, such as experiments and equipment) that included infrared and visible spectrometers for measuring thermal emissions, a radar altimeter for measuring surface height, and a camera for capturing stereo images of the Martian surface to make three-dimensional views.

The spacecraft left Earth on June 2, 2003, via a Soyuz/Fregat launch vehicle (thanks to an agreement with the Russian Federal Space Agency). It arrived at Mars on December 25 and successfully launched the lander, called *Beagle 2* after Charles Darwin's famous ship. *Beagle 2* managed to descend into the Martian atmosphere, but it stopped communicating with Earth immediately afterward and was officially declared lost. The landing site chosen for *Beagle 2* was the Isidis Planitia, an area where scientists had been hoping to search for signs of life. The site wasn't a particularly dangerous one, as far as planetary landings go, and scientists speculate that *Beagle 2* may have been irreparably damaged during landing if the atmosphere on Mars was thinner than expected due to recent dust storms.

Despite the failure of *Beagle 2* to land on Mars, the *Mars Express* orbiter reached orbit successfully and has been returning valuable scientific images and measurements since 2004, including new information about the types and abundances of minerals on the Martian surface. Two of the mission's most-amazing scientific results to date are its beautiful three-dimensional stereo views of Mars's surface (see Figure 20-1) and its unexpected detection of methane on the Red Planet's surface. This detection was unexpected because on Earth, methane usually comes from biological sources; as far as we know, there is no life on Mars, so finding methane there was something of a surprise! Methane is unstable at the surface of Mars, which means it doesn't last very long. The discovery of methane on Mars means that a source must be currently supplying it, which in turn implies that there's some sort of activity, geologic or otherwise, on Mars right this very moment.

**Figure 20-1:** A three-dimensional view of Mars taken by *Mars Express.*

Courtesy of ESA/DLR/FU Berlin (G. Neukum)

Originally scheduled to end in 2005, ESA has extended the Mars Express mission until at least mid-2009. To keep up with *Mars Express,* visit `marsprogram.jpl.nasa.gov/express`.

## Fast facts about Mars Express

Following are the basic facts about the Mars Express mission:

- **Launch date and site:** June 2, 2003; Baikonur Cosmodrome, Kazakhstan
- **Spacecraft name and combined mass at launch:** *Mars Express* (orbiter) and *Beagle 2* (lander); 2,470 lb (1,120 kg)
- **Launch vehicle:** Soyuz rocket with Fregat booster
- **Date Mars orbit achieved:** December 25, 2003
- **Number of Mars orbits:** More than 5,000, as of this writing
- **Length of time in Mars orbit:** 5 years to date

# Roving with Spirit and Opportunity (2003–Present)

The ability to move around on the surface of Mars has revolutionized people's understanding of the wide variety of environments present on the Red Planet. Lessons learned from the tiny yet successful *Sojourner* rover (see Chapter 15) helped NASA engineers design the two Mars Exploration Rovers (MER): *Spirit* and *Opportunity.* These rovers are fully mobile, solar-powered labs capable of sending their data to an orbiting spacecraft for relay back to Earth. If necessary, they can even contact Earth directly from the Martian surface. (See Figure 20-2 for a close-up glimpse of a Mars Exploration Rover.)

**Figure 20-2:** An artist's conception of a Mars Exploration Rover.

*Courtesy of NASA/JPL-Caltech*

The initial goals of the MER mission were to explore and start defining the surface characteristics of Mars. NASA wanted information on the rocks, minerals, and soil found at the rovers' targeted landing sites so scientists could determine how these geologic features may have been created. Of particular interest was the search for specific rock types and compositions that could only have been formed in the presence of liquid water. The significance of that? Where there once was water, there could once have been life too.

*Spirit* (also known by the far more boring name of MER-A) launched on a Delta II rocket on June 10, 2003, followed by *Opportunity* (also known as MER-B) on July 7, 2003. (*Opportunity* left Earth on a more heavy-duty version of the Delta II because it launched a month later than *Spirit,* and the Earth-Sun-Mars alignment required more from the rocket to make it to Mars.) *Spirit* landed on Mars first, hitting the large Gusev Crater on January 4, 2004. *Opportunity* followed on January 24, 2004, landing in the Meridiani Planum.

The rovers are designed to do just about everything a geologist would do if he or she were to walk the surface of Mars. Cameras mounted to them take 360-degree panoramic images of the Red Planet's terrain, and robotic arms extend instruments in order to touch and study rocks up close. Both rovers are capable of grinding the surfaces of rocks to expose fresh, new material that can then be scanned for thermal emissions and bombarded with X-rays to determine the rocks' composition.

From orbit, *Spirit*'s landing site in Gusev Crater looked like a geologic wonderland of features formed by water. However, once on the ground, *Spirit* found rock after rock that had been formed by volcanic, not *fluvial* (flowing liquid), processes. It turned out that any water-related features in Gusev Crater had long ago been covered over by lava flows. Fortunately, *Spirit*'s ability to travel long distances meant that it could be directed toward some interesting-looking hills in the distance (dubbed the Columbia Hills after the lost Space Shuttle; see Chapter 5). *Spirit* has spent the past several years traveling to and exploring a series of cliffs that reveal different rock types, including silica-rich deposits that could come from ancient hot springs, driving almost five miles in the process.

*Opportunity*'s landing site, by contrast, had geologic features that seemed less interesting from orbit, but the appearance of large deposits of hematite was too good to pass up because hematite usually only forms in the presence of liquid water on Earth. After landing on the other side of Mars from its twin lander, *Opportunity*'s airbags rolled it into a small crater with fascinating layers that included small round spheres dubbed "blueberries" that turned out to be the mysterious hematite. Scientists used their studies of the fine details of these and other rock types to determine that the Meridiani Planum was once covered with water, a fascinating discovery that improves the chances of Mars once supporting life. *Opportunity* has spent the last five years driving more than eight miles across the surface of Mars, investigating a large crater called Victoria Crater (which you can see in Figure 20-3), among other sites.

## Fast facts about the Mars Exploration Rovers

Following are the basic facts about the Mars Exploration Rover mission:

- **Launch date and site:** June 10, 2003 *(Spirit)* and July 7, 2003 *(Opportunity)*; Cape Canaveral Air Force Station, Florida
- **Spacecraft names and mass at launch:** *Spirit* (MER-A) and *Opportunity* (MER-B); 2,343 lb (1,063 kg) each
- **Launch vehicle:** Delta II 7925 rocket *(Spirit)*; Delta II 7925 H rocket *(Opportunity)*
- **Number of Mars orbits:** 0 (The rovers landed directly on Mars and didn't linger in the planet's orbit.)
- **Mars landing date:** January 4, 2004 *(Spirit)*; January 24, 2004 *(Opportunity)*
- **Length of time on Mars surface:** 5 years to date

**Figure 20-3:** The surface of Mars as seen from *Opportunity,* showing tracks as the rover climbs out of Victoria Crater.

Courtesy of NASA/JPL-Caltech

Although their initial mission was only supposed to last 90 days, both *Spirit* and *Opportunity* are still going strong after five years on the Martian surface, leading NASA to extend their mission into 2009. If you want to follow the rovers' progress, check out `marsrovers.jpl.nasa.gov/home/index.html` or `www.nasa.gov/mission_pages/mer/index.html`.

# The Mars Reconnaissance Orbiter in Action (2005–Present)

Many images of Mars have been taken from orbit, but none boast the impressive degree of detail provided by the images taken from *Mars Reconnaissance Orbiter (MRO). MRO*'s mission goals were to view the surface of Mars at as high a resolution as possible from orbit to help NASA better understand how the landing sites for the rovers *Spirit* and *Opportunity* fit into the overall picture of Mars (see the previous section for details on the rovers), as well as to help select sites to which future missions could be directed.

*MRO* left Earth on August 12, 2005, and entered orbit around Mars about seven months later with the help of *aerobraking* (using the atmosphere of a planet to change the orbit of a spacecraft). MRO was intended as a follow-up to the Mars Global Surveyor mission of 1996 (see Chapter 15), and it has a wide variety of state-of-the-art instruments aboard, including several cameras. The HiRISE (High-Resolution Imaging Science Experiment) camera takes extremely high-resolution pictures of the surface that reveal tiny details; the CTX (Context Imager) camera helps scientists understand the geologic features seen around the supersmall HiRISE pictures; and MARCI (Mars Color Imager) adds lower-resolution color images to help understand Mars's surface composition.

## Fast facts about the Mars Reconnaissance Orbiter

Following are the basic facts about the Mars Reconnaissance Orbiter mission:

- **Launch date and site:** August 12, 2005; Cape Canaveral Air Force Station, Florida
- **Spacecraft name and mass at launch:** *Mars Reconnaissance Orbiter;* 4,800 lb (2,180 kg)
- **Launch vehicle:** Atlas V-401 rocket
- **Date Mars orbit achieved:** March 2006
- **Number of Mars orbits:** More than 10,000, as of this writing
- **Length of time in Mars orbit:** 3 years to date

*MRO* isn't just a camera machine, however. It also features instruments designed to perform specific scientific observations such as searching for subsurface water with radio waves, studying the composition of the Red Planet with an imaging near-infrared spectrometer, and measuring the temperature and water vapor levels in the Martian atmosphere.

Although all of *MRO*'s instruments have returned important scientific results, the HiRISE images are the most stunning (see one for yourself in the color section). The HiRISE camera has revealed layered deposits near Mars's North Pole that may hold data on how the planet's climate has changed over the last few million years. Images of salt deposits — a key indicator of areas where liquid water could've evaporated over time — near the southern highlands have also been sent back. Additionally, the HiRISE camera has observed *Spirit* and *Opportunity* from orbit, as well as both Viking landers (see Chapter 11), and captured a dramatic image of the *Phoenix* Mars Lander while it was landing (see the next section for more about *Phoenix*).

The latest updates on *MRO,* whose primary mission lasts until the end of 2010, are available at `marsprogram.jpl.nasa.gov/mro`. For amazing pictures taken by the HiRISE camera, check out `hirise.lpl.arizona.edu`.

# The Phoenix Mars Lander Visits the North Pole (2007–2008)

The most recent visitor to the Red Planet was the *Phoenix* Mars Lander, a robotic spacecraft sent to Mars to find definitive evidence of water. And find water it did! In June 2008, the lander dug a trench that was filled with a bright material scientists thought could've been ice. Several days later, the bright clumps had vaporized, suggesting that they were probably water ice. Additional evidence of water was gleaned from water vapor measurements sent back in late July 2008.

All previous Mars landers have set down near the Martian equator, but *Phoenix* was targeted for terrain near the North Pole. Like the poles on Earth, both poles of Mars have ice caps, though the ice on Mars is primarily made of carbon dioxide rather than water. The North Pole was chosen for *Phoenix's* landing site because the northern polar cap is larger than the southern polar cap (and because more water ice is thought to exist there).

The primary mission goals for *Phoenix* were to find water, examine the history of water on Mars, and, in the process, discover more about the mechanisms of climate change. A related goal was to determine how habitable Mars was, or had been, as a result of the presence (or absence) of liquid water at various points in its history.

The spacecraft was dubbed *Phoenix* because it was said to rise from the ashes of the failed *Mars Polar Lander* spacecraft (see Chapter 15). This rising was largely figurative but partly literal — some spare instruments from *Polar Lander* were used in *Phoenix*.

The Phoenix Mars Mission launched successfully on August 4, 2007. The lander made it to Mars on May 25, 2008, setting down in Vastitas Borealis (one of the areas that promised a high concentration of water ice). Initial pictures showed a flat surface scattered with pebbles and polygonal cracks thought to be due to the freezing and thawing of ice just below the surface.

Unlike the previous few Mars missions, *Phoenix* was a stationary lander rather than a rover. Because the lander itself couldn't move, one of its most important tools was a robotic arm that could dig about 1.5 feet (0.5 meters) into the surface of Mars, bringing up soil and ice samples that could then be analyzed and studied. A scoop on the end of the robotic arm (see Figure 20-4) delivered soil samples into a variety of chemical analysis tools aboard the spacecraft, which wetted and heated the soil to perform various compositional analyses.

## Fast facts about Phoenix Mars Lander

Following are the basic facts about the Phoenix Mars Mission:

- **Launch date and site:** August 4, 2007; Cape Canaveral Air Force Station, Florida
- **Spacecraft name and mass at launch:** *Phoenix* Mars Lander; 772 lb (350 kg)
- **Launch vehicle:** Delta II 7925 rocket
- **Date of Mars landing:** May 25, 2008
- **Number of Mars orbits:** 0 (*Phoenix* landed directly on Mars, so it didn't spend time orbiting the planet.)
- **Length of time on Mars surface:** 5 months

**Figure 20-4:** *Phoenix's* scoop for bringing soil samples aboard.

Courtesy of NASA/JPL-Caltech/University of Arizona

*Phoenix*'s wet chemistry lab, MECA (the Microscopy, Electrochemistry, and Conductivity Analyzer), provided the means for soil samples to be baked, imaged, and doused with chemicals — all in the interest of determining more about the soil's composition. For example, it found that the soil contained minerals such as potassium, sodium, and magnesium.

*Phoenix* was intended to last for 90 days following its landing on Mars, but it managed to survive an extra two months before diminishing light levels in the region reduced the amount of available sunlight below operating levels for the spacecraft's solar panels. NASA's last contact with *Phoenix* came on November 2, 2008. It's possible that the spacecraft can survive the cold Martian winter and emerge from its safe mode during Martian spring, in 2009, but because the spacecraft will be covered by frozen carbon dioxide *(dry ice)* for many months, mission designers think that another resurrection for *Phoenix* is unlikely.

For all the results and images from *Phoenix,* go to `phoenix.lpl.arizona.edu`.

# Part V
# The Future of Space Exploration

The 5th Wave By Rich Tennant

"It's another deep space probe from Earth seeking contact with extraterrestrials. I wish they'd just include an email address."

## In this part . . .

Part V deals with the future of space exploration. A whole fleet of exciting robotic missions is in the pipeline from NASA and other countries' space agencies. These missions include advanced journeys to Mars; a return to Jupiter; and treks to Pluto, Mercury, and the asteroids. On the astronaut side o' things, the venerable Space Shuttle is headed on its last few missions and will soon be replaced with Project Constellation's Orion, a capsule-based system that will allow astronauts to journey to the International Space Station and eventually return to the Moon and travel to Mars.

You may be surprised to find out that space travel is no longer the sole province of astronauts. Space tourism has allowed private citizens to fly in space — for a price. Commercial space travel will undoubtedly take on an important role in the future.

We conclude this part with a look at the reasons for space exploration, from a philosophical view of the frontier to the necessity of expansion given the finite resources here on Earth.

# Chapter 21

# More in Store: Upcoming Space Missions

### *In This Chapter*

- Designing a new space vehicle for human exploration
- Sending robotic spacecraft to the Moon, Mars, and beyond
- Encountering asteroids and Pluto, among other notable missions

In earlier parts of this book, we describe a great number of past and present missions, but space exploration is far from finished. Many exciting developments are in store for the future of both human and robotic exploration of the solar system:

- On the human exploration side, President George W. Bush's vision for space exploration guided NASA to plan for a return to the Moon in preparation for human exploration of Mars. To meet these goals, the Space Shuttle will be retired, and the new Orion spacecraft will take over.
- On the robotic side, the upcoming car-sized Mars Science Lab will perform a modern search for life on Mars, and a new Outer Planets Flagship Mission will travel to the Jupiter system. A number of robotic missions to the Moon are also either underway or planned for launch in the next few years.

Other interesting missions are also on their way courtesy of the United States and other nations. We give you the goods on what to watch for in this chapter.

## Recommitting to Human Exploration with NASA's Project Constellation

The Space Shuttle has had a long and celebrated career as the first reusable spacecraft able to bring astronauts and cargo to Earth orbit (check out Chapters 13 and 14 for details). However, with the International Space Station nearing completion and targets beyond Earth beckoning, the development of a new human spacecraft is long overdue.

With an American return to the Moon, and an eventual destination of Mars, in mind, NASA engineers were faced with the challenge of developing a new launch vehicle and spacecraft for human space exploration. The resulting effort, called Project Constellation, includes the Orion spacecraft (formerly known as the Crew Exploration Vehicle) and the Ares I rocket, as well as Ares V, a heavy-lift launcher that will bring nonhuman *payloads* (cargo) into orbit. We describe all three components of Project Constellation (as well as the end of the Shuttle program and the basics of how second-generation lunar exploration will work) in the following sections.

## Transitioning from the Space Shuttle

To make way for the development of a new spacecraft for astronauts, the venerable Space Shuttle program will be shut down. The final Shuttle flight is currently scheduled for 2010, but the first Orion flight won't take place until 2014 at the earliest. The gap between the two programs is unfortunate, but given scheduling constraints and funding considerations, keeping the Shuttle flying until Orion is ready to take over is very unlikely. Each Shuttle flight is extremely expensive, and that money — plus the money for continued ground support, maintenance, repairs, and more for the Shuttle — can instead be freed up so NASA can focus solely on Orion and Ares I.

Fortunately, the International Space Station (ISS) shouldn't be negatively affected by the Space Shuttle's retirement. A plan is already in place for the ISS to be visited solely by Soyuz capsules (for crew exchanges) and robotic Progress ships (for supplies) until Orion is ready for its first trip to the ISS in 2014 or later. (Flip to Chapter 19 for details about the ISS.)

## Introducing Orion, the next-generation spacecraft for human exploration

Unlike the Space Shuttle, which launches like a rocket but lands like a glider, the Orion spacecraft looks much more like the Apollo spacecraft (which we describe in Chapters 9 and 10). In fact, Orion has been called "Apollo on steroids" because it's larger and more capable than its predecessor. The resemblance is indeed striking — astronauts will ride into space in a small capsule shaped like a rounded cone, although Orion will be able to carry up to six astronauts compared to Apollo's maximum capacity of three (see Figure 21-1).

Figure 21-1: Introducing the Space Shuttle's replacement: Orion.

Courtesy of NASA

Orion will launch with the help of the Ares I rocket (see the next section for details on Ares I). Orion's capsule will either land on solid ground with parachutes and airbags, similar to the Russian Soyuz spacecraft that we describe in Chapter 7, or it'll return to Earth via Apollo-style water landings. Ground landings are more technically difficult, but they also don't require the support of an entire Navy fleet to retrieve the astronauts. NASA is currently leaning toward a water landing for Orion, but the final decision has yet to be made as of this writing.

The Orion spacecraft will be completely reusable for up to ten spaceflights per capsule, whereas the Ares launch vehicle will be only partially reusable. Orion will have a crew module (more than twice the size of Apollo's Command Module) in which the astronauts will ride for takeoff and landing. This capsule will return to Earth after each mission. Similar to Apollo, Orion will also have a service module for supplies and equipment, but the Orion version will deploy solar panels when it reaches space so as to reduce the amount of fuel the spacecraft has to carry.

## Lifting off with the Ares I and Ares V rockets

The U.S. space program has used various rockets for propelling spacecraft into space over its 50-year history, but none have ever been as powerful and efficient as the new Ares models.

Ares I (shown on the right in Figure 21-2) consists of two stages:

- The first stage is a solid rocket booster similar to that found on the Space Shuttle, but it's more powerful and able to burn longer. This solid rocket booster stage is reusable.
- The second stage is a J2-X rocket engine that uses liquid hydrogen and liquid oxygen as its propellant. This engine is similar to the one used on Project Apollo's Saturn V Moon rocket (see Chapter 9). The big difference is that it'll be able to start both in vacuum and in a planet or satellite's atmosphere.

**Figure 21-2:** Designs for the Ares I (right) and Ares V (left) rockets.

*Courtesy of NASA*

The Apollo spacecraft and the Space Shuttle were singular spacecraft that brought both cargo and astronauts into space. Now that model's a way of the past. As part of Project Constellation, NASA officials have decided to split the cargo capability off from the human spaceflight system. Orion will carry only crew and necessary life-support systems into space. Cargo, such as supplies and possible future ISS modules, will be brought into space aboard two-stage Ares V rockets (shown on the left in Figure 21-2). Splitting functions between crew and cargo allows the design of each spacecraft to better address the needs of its occupants, in addition to reducing some of the hardware costs. The cargo rocket will also launch lunar modules and spacecraft to take future astronauts from Earth orbit to the Moon. (Check out Figure 3-1 in Chapter 3 to see the size of the Ares I and Ares V rockets as compared to the Space Shuttle, Saturn V, and other rockets.)

# Flying to the Moon (again)

People want to go back to the Moon. That's just fine and dandy, you may think, but is there a plan in place? Never fear, reader dear! NASA's current plans for a human return to the Moon begin with the launch of astronauts aboard Orion via the Ares I rocket. Separately, an Ares V rocket will launch a payload into space that includes the *Earth Departure Stage* (EDS; a powerful rocket stage designed to be launched separately from the Orion spacecraft only to be united with it in orbit in order to propel the spacecraft toward the Moon), as well as a brand-new lunar module (the Altair Lunar Lander). Once in orbit, Orion will dock with the EDS and use the EDS's rocket engines to send the whole spacecraft on its journey to the Moon. After the crew arrives at the Moon, it'll either send the EDS off into solar orbit or intentionally send it to crash into the Moon in order to study the seismic waves raised by the impact.

Astronauts will use the Altair Lunar Lander to physically land on the Moon. Altair will be larger and more capable than Apollo's Lunar Module. In one possible mission configuration, a four-person crew will be able to land on the surface, leaving the Orion spacecraft operating all by its lonesome in orbit as it awaits the return of Altair.

Other differences between Altair and Apollo's Lunar Module include

- An airlock (for keeping pesky Moon dust out of the rest of the spacecraft and sparing the astronauts from having to depressurize the entire lander before a moonwalk)
- A toilet similar to those on the International Space Station
- The ability to heat up food for hungry moonwalkers

Altair will be able to support a mission to the lunar surface of up to a week's duration, after which it'll blast off from the lunar surface and dock with Orion for the journey back to Earth.

Orion's first Moon landing is currently set for 2018, but this date will likely slip farther into the future if technological, political, and budgetary constraints arise. After 21st century Moon landings have been thoroughly established, the next step will be to send astronauts on a much longer trip to Mars and/ or a near-Earth asteroid. Lessons learned from humanity's return to the Moon will likely result in changes to Orion, but by that time, being confined to Earth orbit will be a mere memory.

For more details about Project Constellation and Orion, head to NASA's Orion Web site at www.nasa.gov/mission_pages/constellation/orion.

# Revving Up Future Robotic Exploration

Humans may be headed back to the Moon, but robotic explorers of the not-so-distant future will travel both near (to the Moon) and far (to Pluto, the celestial body formerly known as the ninth planet). NASA's planned robotic spacecraft will be joined by an innovative fleet from a number of other countries, some of which are launching their own solar system explorers for the first time.

Robotic missions serve as both an essential precursor to future human exploration, in the case of the Moon and Mars, and humanity's emissaries to locations that are too far away, or too dangerous, to send people to anytime soon.

## Rekindling interest in the Moon

The Moon has long been a fascinating space exploration destination for Americans and Europeans alike. Now Asian nations are delving into lunar science as well. More and more countries are making the foray into lunar exploration for several reasons, including the fact that significant international prestige comes with sending spacecraft to the Moon and the knowledge and technology required to make such trips no longer belongs solely to the United States and Russia. Plus, the Moon is an exotic target that's also close by, a perfect destination for a country that's just testing the space exploration waters. Recent and current robotic Moon missions include the following:

- **SMART-1:** Launched in 2003 by the European Space Agency, SMART-1 (an acronym for Small Missions for Advanced Research in Technology) was a small, lightweight, low-cost mission that studied the Moon's surface composition. The spacecraft was crashed into the Moon in September 2006 to look for water in the impact plume.

  To find out more about SMART-1, visit www.esa.int/SPECIALS/SMART-1.

- **SELENE:** Japan's Selenological and Engineering Explorer mission, nicknamed "Kaguya" and launched in 2007, is Japan's first major Moon mission. It's currently studying the Moon with many instruments, including a high-definition television camera that has relayed stunning images. *SELENE* is scheduled to be crashed into the Moon in 2009.

  Check out www.jaxa.jp/projects/sat/selene/index_e.html for the official word on *SELENE*.

- **Chang'e:** The first major Chinese planetary spacecraft, *Chang'e* (launched in 2007) is observing the Moon's surface geology and composition with a focus on future human exploration. The one-year mission has been extended to allow mapping of the lunar poles.

  Here's a Web site with more news about *Chang'e,* just be sure to click on the English link at the bottom-right corner of the page (unless of course you're fluent in Chinese): `www.clep.org.cn`.

- **Chandrayaan-1:** The first major Indian planetary spacecraft, *Chandrayaan-1* (launched in 2008) is studying the Moon's topography, geology, and surface composition. *Chandrayaan-1* launched a lunar impact probe, whose purpose was to take pictures of the Moon's surface at increasingly high resolution, as it made its way toward a hard surface landing. The mission is still operational as of this writing.

  Want to keep up with *Chandrayaan-1?* Then visit `www.isro.org/Chandrayaan`.

On the other side of the globe, American interest in returning to the Moon grew following President George W. Bush's declaration of new priorities for the U.S. space program in January 2004. But in order to return an American to the Moon, NASA officials determined some additional robotic research was worthwhile. Upcoming American Moon missions include the following:

- **Lunar Reconnaissance Orbiter:** NASA will launch this spacecraft in 2009 to study resources on the Moon and prepare for future human exploration by observing potential landing sites.

  Head to `lunar.gsfc.nasa.gov` for updates on this mission.

- **Lunar Crater Observation and Sensing Satellite:** This NASA satellite will be launched from the *Lunar Reconnaissance Orbiter* in 2009 and will look for water vapor sent into orbit after part of the spacecraft impacts a polar crater.

  If this particular mission sounds interesting to you, keep up with it by going to `lcross.arc.nasa.gov`.

- **Lunar Atmosphere and Dust Environment Explorer:** To be launched in 2011, NASA's *LADEE* will orbit the Moon and study the thin lunar atmosphere and the properties of Moon dust, which will affect future human exploration.

  *LADEE*'s mission page (`nasascience.nasa.gov/missions/ladee`) is a good resource for further info.

- **Gravity Recovery and Interior Laboratory:** Also scheduled for a 2011 launch, NASA's *GRAIL* will orbit the Moon with two satellites to map the lunar gravitational field in detail in order to understand the satellite's subsurface structure and assist with future Moon landings.

  Check out `moon.mit.edu` for *GRAIL*-related updates.

The Russian Federal Space Agency is also planning renewed efforts in robotic Moon exploration. The Luna-Glob 1 mission, scheduled to launch in 2012, will mark Russia's first return to the Moon since the Apollo-era Luna Program (see Chapter 8). The orbiting spacecraft will include four penetrators to study seismic signals and determine the Moon's subsurface structure.

With recent precedents for international cooperation in space exploration, such as the International Space Station, the number of redundant missions being launched from different countries may seem strange. However, for new space-faring nations, such as India, China, and Japan, launching their own missions and using their own launch vehicles and mission control centers is a matter of national pride. Such missions can be a way to prove to themselves and the world that they're ready to be taken seriously as contenders in space exploration. There are signs of cooperation though. For example, Russia plans to collaborate with both China and India on future spacecraft in the Luna-Glob series. And NASA recently announced plans to build up the International Lunar Network, a group of lunar landers that will be able to work together to gather combined data about the Moon's subsurface.

## Looking for signs of life directly on Mars

Although many recent Mars landers (described in Chapter 20) have looked indirectly for signs of life on the Red Planet, by searching for the presence of water and studying the chemical composition of the soil, no direct experiments have been done since NASA's Viking landers searched for life in the 1970s (see Chapter 11). Given scientists' new understanding of the surface properties of Mars — and the new discoveries regarding extreme types of life on Earth, such as microbes that can live deep in the ocean or at extremely high temperatures — the time is right for a repeat mission to explicitly search for Martian life (and we're not talking about little green men!). We describe a few important upcoming Mars missions in the following sections.

### The Mars Science Laboratory

Mars exploration will take another leap forward when NASA's upcoming *Mars Science Laboratory (MSL)* rover lands on the Red Planet's surface. Originally scheduled for a 2009 launch, this car-sized rover, which is twice the size of and three times as heavy as Mars Exploration Rovers *Spirit* and *Opportunity* (see Chapter 20), simply wasn't ready for prime time yet. Consequently, its launch was postponed until 2011. However, its expected mission duration on Mars (at least one year) hasn't changed.

Whereas *Spirit* and *Opportunity* can perform remote chemical analysis, *MSL* will actually be able to perform onboard chemistry of both soils and ground rock powders, similar to the wet chemistry lab on the *Phoenix* Mars Lander. The suite of chemical-analysis instruments will specifically look for organic materials that could relate to life, including carbon compounds such as methane and chemical elements such as nitrogen and oxygen.

Another interesting instrument on *MSL* will be a laser capable of vaporizing a small portion of a rock from a distance of up to 43 feet (13 meters) and measuring the chemical properties of the vaporized material to learn about the rock's composition.

*MSL* will be powered by a radioisotope thermoelectric generator (RTG), rather than solar panels, which will increase the lifetime of the mission as well as the amount of power available for instruments like the vaporizing laser. In addition, because *MSL* is too heavy to land daintily via airbags, it'll use a new landing method (called the *skycrane*) in which the spacecraft is lowered from a rocket on a long tether to a soft landing. When the wheels touch the ground, the tether is cut, and the rocket carrier flies off to a crash landing at a safe distance. *MSL*'s landing will mark the first use of a skycrane.

### The Astrobiology Field Laboratory

A true *astrobiology mission* to Mars (one that will search for not just organic compounds but actual evidence of past or present life on Mars) will have to wait until NASA's planned *Astrobiology Field Laboratory* (*AFL;* see Figure 21-3). This rover, currently planned for a 2016 launch, will be similar to *MSL* (described in the preceding section), but it'll have instruments that are specifically designed to search for evidence of life on Mars.

*AFL* will be able to look for thousands of carbon compounds that form the basis of life here on Earth. Any evidence of life would probably be simple single-celled organisms, but even the discovery of microscopic life would result in a huge scientific revolution. The mission will likely be directed to land either in an icy polar region or at a site of a former hot spring. The mission is initially planned to last for two years, and the spacecraft will be powered by an RTG (like *MSL*).

**Figure 21-3:** The *Astrobiology Field Laboratory* will search for signs of life on Mars.

*Courtesy of NASA/JPL-Caltech*

### Planning to return a sample to Earth

Another big goal of Martian exploration is to bring a sample of Martian rocks to Earth. Selection of the rock samples will be key. One current suggestion for simplifying the selection process is to have a precursor mission choose interesting rocks along its path and either mark them or carry them to a specified location so that the sample return mission can bring a variety of well-selected samples back to Earth. Of course, planetary protection concerns will be in full swing whenever a sample from Mars is finally brought to Earth (after all, no one wants dangerous Martian germs or viruses infecting the Earth), but scientists will also be eagerly anticipating the myriad scientific advances that will be in store. NASA is thinking about launching a sample return mission in 2018 or later, perhaps as a joint mission with ESA, but detailed planning has yet to occur.

## Sending the best to the outer planets

As a follow-up to the success of NASA's Galileo mission and the joint NASA-ESA Cassini-Huygens mission, NASA and ESA are currently planning a series of Outer Planet Flagship Missions. The first such mission will be a return to the *Jovian system* (Jupiter and its moons); a later mission will revisit Saturn and its satellites. The Europa Jupiter System Mission will launch in 2020 and arrive at Jupiter in 2026, with an anticipated mission duration of at least three years.

### Verifying Europa's ocean

Following the demise of the *Galileo* spacecraft (see Chapter 17), NASA has been planning a follow-up mission to help confirm the presence of a liquid water ocean under the surface of Jupiter's moon Europa. If such an ocean exists relatively near the moon's surface, Europa becomes a prime target in the search for life in outer space. (The rest of the system, including Jupiter and its other moons, is also an interesting target for further exploration.)

The current mission design for a trip to Europa is called the Europa Jupiter System Mission, and it actually includes two separate spacecraft. First, a dedicated NASA-built spacecraft, currently called *Jupiter Europa,* would make multiple passes through the Jupiter system before entering orbit around Europa. The Europa orbit phase is essential to make detailed maps of Europa's surface and obtain precise gravitational and magnetic field measurements that can be used to help determine Europa's subsurface structure. The spacecraft will also likely bring a radar instrument to sound beneath the satellite's surface in order to directly detect the ocean layer.

The Europa Jupiter System Mission will also include an ESA-built spacecraft that will orbit Ganymede, another of Jupiter's moons. Additionally, the Russian Federal Space Agency is considering building a Europa lander.

Although an orbital mission at Europa will certainly be intriguing, it will also be relatively short-lived. Jupiter's strong magnetic field means that the radiation environment at Europa is extremely harsh, and a mission in Europa's orbit will likely only be able to survive for a few months at most.

### Imaging Titan's surface

NASA's *Cassini* spacecraft is currently studying Saturn and its moons, in particular the large moon Titan and the active moon Enceladus (you can read more about *Cassini* in Chapter 17). A follow-up mission named the Titan Saturn System Mission will be the second joint NASA-ESA Outer Planets Flagship Mission, after the Europa Jupiter System Mission. The Titan mission will again involve two spacecraft, one built by each agency. The NASA-built contribution will begin by orbiting Saturn. The mission will then enter into orbit around Titan, equipped with a suite of instruments specially optimized to see through the moon's clouds in order to map and image its surface.

The radiation environment in the Saturn system is much more benign than the Jupiter system, allowing the spacecraft to survive in Titan's orbit for much longer — at least a year and a half after its two years in Saturn orbit. In addition, Titan's thick atmosphere makes landing easy: A lander can simply use a parachute to slow down instead of requiring the tricky retrorockets needed to land on airless worlds such as the Moon or Europa.

The Titan Saturn System Mission is currently planned to include a Titan lander built by ESA. This lander will be a traditional lander and/or a Titan balloon that could navigate through Titan's thick atmosphere to view many different regions of the surface (flip to the color section for an image of the balloon concept).

## Keeping an Eye Open for Other Exciting Missions

In addition to the new missions planned to the Moon, Mars, and beyond (which we describe earlier in this chapter), other missions are either already in place or in the works to other solar system targets. These missions will study asteroids, search for extrasolar planets, and examine the two extremes of the solar system: Mercury and Pluto. In keeping with the increasingly international nature of space exploration, many of these missions will be operated by countries other than the United States.

## Hayabusa looks at an asteroid

Most people probably hope that an asteroid never lands in their backyards, but that doesn't dissuade scientists from wanting the opportunity to study one up close. The Hayabusa mission, conducted by the Japanese Aerospace Exploration Agency (JAXA), involves a robotic spacecraft designed to bring a sample from near-Earth asteroid 25143 Itokawa back home to Earth. The spacecraft, also called *Hayabusa,* launched in 2003 and transmitted extensive data about the size, shape, topography, and other features of the asteroid.

The mission plan called for *Hayabusa* to actually touch the asteroid's surface with a sampling device. It appears that the spacecraft did land successfully on the asteroid and gathered samples for about 30 minutes, but no one knows whether the sampling procedure was performed correctly. Japanese scientists think that some dust from the asteroid's surface made it into the collection device, so they told the spacecraft to seal the receptacle and then sent the spacecraft on a return trajectory to Earth. *Hayabusa* also carried a tiny "hopper" lander called *MINERVA* that would've been able to use the asteroid's extremely low gravity to bounce around, but *MINERVA* was accidentally deployed at too high an altitude and drifted away from the asteroid instead of landing on it.

The spacecraft, along with its sample-return capsule, is currently on its way back to Earth, with an arrival date of mid-2010. After the capsule has safely landed on Earth, scientists will be able to find out whether it managed to return the first-ever sample from an asteroid.

To find out more about *Hayabusa* straight from the horse's mouth, check out `www.jaxa.jp/projects/sat/muses_c/index_e.html`. For additional perspective on the mission's contribution to near-Earth asteroid research, visit `neo.jpl.nasa.gov/missions/hayabusa.html`.

## MESSENGER revisits Mercury

In 1973, *Mariner 10* gave scientists a glimpse at one side of Mercury; they had to wait more than 30 years until the MESSENGER (MErcury Surface, Space ENvironment, GEochemistry, and Ranging) mission launched to see the other half of the planet closest to the Sun. *MESSENGER* was launched in 2004 with a plan of doing several Mercury flybys and eventually entering the planet's orbit. The end goal of this mission is to completely map Mercury's surface. As an added bonus, if MESSENGER is successful, it'll be the first mission to orbit Mercury.

Because of the gravity-assisted trajectories required to make it to Mercury, *MESSENGER* first did an Earth flyby and two Venus flybys before continuing along to the target planet. January 2008 marked *MESSENGER*'s first flyby of Mercury, during which it took a series of high-resolution images that show the other side of the planet for the first time (see Figure 21-4). The spacecraft is due to enter Mercury's orbit in early 2011.

**Figure 21-4:** A view of Mercury's other half from *MESSENGER.*

*Courtesy of ASA/Johns Hopkins University Applied Physics Laboratory/ Carnegie Institution of Washington*

*MESSENGER* had to be lightweight (to counterbalance the amount of fuel required for the trip to Mercury), highly heat-resistant due to its proximity to the Sun, self-powering, communicative for data relays and remote programming, and able to successfully maneuver into orbit. NASA addressed these design requirements one by one. Heat from the Sun is counteracted by *MESSENGER*'s Nextel sunshade, and solar panels provide a reusable source of power. Massive thrusters use the spacecraft's fuel and oxidizer to meet propulsion needs, and electrically controlled high- and low-gain antennae send and receive data. A series of scientific instruments, including a high-resolution camera, a laser altimeter, and instruments to study fields and particles, will characterize Mercury and its space environment.

You can find out more about *MESSENGER,* and see the latest pictures, by going to `messenger.jhuapl.edu` or `www.nasa.gov/mission_pages/messenger/main`.

## New Horizons encounters Pluto

NASA's New Horizons mission was launched in 2006 to keep with the U.S. space program's theme of exploring the heavenly bodies about which little is known. It's the first mission sent to Pluto and its three moons: Charon, Hydra, and Nix.

The New Horizons mission was designed and launched back when Pluto was still considered a planet. (If the fact that Pluto is no longer called a planet is news to you, flip to Chapter 2.) Even though Pluto has since been reclassified as a dwarf planet, the mission's goals remain valid:

- To study Pluto's geology and atmosphere via high-resolution imaging
- To better understand the chemical makeup of the surfaces of Pluto and its moons
- To look for new discoveries such as additional moons or rings

The *New Horizons* spacecraft itself is triangular in shape and stocked full of scientific instruments (including cameras and a dust counter) that will provide significant new information about Pluto and its environment. Because Pluto is so far from the Sun, the spacecraft relies on radioactive energy rather than solar power to keep it functioning.

After being launched directly on a path that will take it to Pluto and beyond, *New Horizons* passed by the orbit of Mars and flew through the asteroid belt. It received a gravitational assist from Jupiter in 2007 and took images of the planet and its satellites during a relatively close flyby. The spacecraft is currently well on its way toward its Pluto flyby in 2015, after which it will continue on to the Kuiper Belt to fly past one or more Kuiper Belt Objects on its way out of the solar system.

For the latest updates on this long-lasting project, and to find out where *New Horizons* is right now, head to `pluto.jhuapl.edu`.

## Kepler tries to find Earth-like planets

Planet detection is one of the most important ways in which society's knowledge of the universe is expanded. Locating *extrasolar terrestrial planets* (Earth-sized planets that orbit other stars) is of particular interest to scientists looking for signs of liquid water and life. NASA's *Kepler* spacecraft is geared toward precisely this type of exploration and will study the Milky Way Galaxy looking for Earth-like planets. The launch of *Kepler* took place in March 2009, and the mission is expected to last at least three-and-a-half years.

### In the habitable zone

The *habitable zone* is the distance from a star at which water could exist in liquid form on a planet's surface. Scientists place great importance on the habitable zone because it's an indicator of that planet's potential for life.

If a planet is too close to its star, the surface would be too hot for liquid water to be stable. On the other hand, if the planet is too far away, the planet would be too cold to support liquid water. The habitable zone is a range in which life-supporting planets could exist. In our solar system, Earth is the only planet currently in the habitable zone — Venus is a bit too close to the Sun, whereas Mars is a bit too far out from it. Finding an Earth-sized extrasolar planet in the habitable zone of its star would greatly increase that planet's chances of supporting life.

The spacecraft will use a *photometer* (a detector very sensitive to changes in light levels) to monitor the sky for changes in star brightness. A change in brightness could indicate the passing of a planet in front of its star due to its orbit, and the amount of dimming of the star can be used to calculate the size of the planet.

This Web site is your number one resource on all things *Kepler*: `kepler.nasa.gov`.

## MAVEN plans to study the upper atmosphere of Mars

Among the many missions in the works for the Red Planet is MAVEN, scheduled for launch in 2013. MAVEN (which stands for Mars Atmosphere and Volatile Evolution) was selected from a field of 11 missions proposed by NASA scientists. It'll study the upper atmosphere of Mars, with a primary focus on how the atmosphere interacts with the Sun and how much atmospheric material is lost over time. The spacecraft will arrive at Mars in 2014 and then enter orbit around the Red Planet for a mission that's expected to last at least two years.

Stay up to date on MAVEN news by visiting `lasp.colorado.edu/maven`.

## BepiColumbo heads to Mercury

Mercury has historically been a difficult planet to study because of its close distance to the Sun. That fact has left scientists with many unanswered questions about Mercury for quite some time, but thanks to the European Space Agency's (ESA) BepiColombo mission, those questions may soon be answered. Scheduled for launch in 2013, the BepiColombo mission will last for at least a year and perhaps much longer. It'll consist of two orbiting spacecraft: ESA's *Mercury Planetary Orbiter* and Japan's *Mercury Magnetospheric Orbiter.* A range of instruments will help provide critical information, such as data about Mercury's magnetic field, environment, and composition, that may help better describe the origins and evolution of Mercury. This data will complement the observations currently being made by *MESSENGER* (described earlier in this chapter).

Want the full scoop on the BepiColumbo mission? Be sure to bookmark these Web sites: `sci.esa.int/science-e/www/area/index.cfm?fareaid=30` and `www.stp.isas.jaxa.jp/mercury`.

## A few other notable space missions

In addition to the missions described in the previous sections, a variety of other space missions are also looming on the horizon:

- **Phobos-Grunt:** This Russian mission, planned to launch in 2009, will study Phobos, a moon of Mars, and return a sample of it to Earth.
- **Yinghuo-1:** Set to launch in 2009 on the same rocket as *Phobos-Grunt,* this Chinese mission will orbit Mars and study its space environment.
- **MetNet:** This mission planned by Finland will launch between 2009 and 2011. It'll establish a network of small landers on Mars to study the Red Planet's atmosphere and climate.
- **Hayabusa 2:** This Japanese mission, planned to launch between 2011 and 2012, will be a follow-up mission to Hayabusa (described earlier in this chapter). It'll target another asteroid and attempt to deploy the *MINERVA* lander a second time.
- **Juno:** Set to launch in 2011, this American mission will embark on a polar orbit of Jupiter to study the planet's composition, magnetic field, and formation.
- **Aditya:** This Indian mission, planned to launch in 2012, will study the Sun's corona.

# Chapter 22

# Increasing Access to Space

### *In This Chapter*

- Taking the bronze: China's Shenzhou program
- Opening space travel to nonastronauts
- Understanding what it takes to be a space tourist

Although space travel started out as a venture operated by huge national space agencies (namely, those of the United States and the Soviet Union), modern innovation is beginning to open up access to space, as you discover in this chapter. For example,

- NASA and the Russian Federal Space Agency have new company in space in the form of the Chinese Shenzhou program, which sent the first Chinese taikonaut into space in 2003.
- Russia has flown space tourists on its Soyuz trips to the International Space Station. At $20 million per trip, the cash-strapped Russian Federal Space Agency used these flights as a crucial source of funding.
- Innovation continued in the form of the Ansari X Prize, a prize offered by a nonprofit foundation for the first private suborbital flight.

## The Emergence of China as a Space Power

NASA and the Russian Federal Space Agency may be two of the biggest names in the history of space exploration, but they aren't the only game in town. Many other nations have space agencies; some, such as China, Japan, India, and Europe, have launched satellites into orbit and even sent robotic spacecraft to explore the solar system. Until recently, however, other countries had to partner with either the Russian or the American space program in order to get their astronauts into space. But in the last few years, China's space program, run by the China National Space Administration, has emerged as a third major player in the world of international space exploration — and one with keen eyes toward the future.

The next couple sections get you up to speed on the history of China's space program and where it's headed.

## The history of Chinese spaceflight

The Chinese have a long-standing interest in rocketry and flight, beginning with their use of rockets in warfare in the 12th century CE. Their space program originally had a military bent when it was formed in the early 1950s as an initiative of the People's Republic of China. Naturally, the program's early emphasis was on missiles and warheads. However, interest began turning toward space exploration and satellites as the 1960s drew to a close. The Space Race between the Soviet Union and the U.S. was likely a factor in China's interest, although at that point China wasn't technologically advanced enough to attempt to join in.

The 1970s saw the development of China's first astronaut program, and a series of rocket launch tests followed (though human spaceflight was put on hold). China successfully launched a recoverable satellite in the 1970s and made steady progress in the development of launch vehicles in the ensuing years. As China matured into a technological and economic power on the world stage, a human space exploration program became an important milestone with both engineering and political implications.

China's serious push to join the "astronaut club" began in the early 1990s with the development of the *Shenzhou program,* China's foray into manned spaceflight, and the formation of the China National Space Administration. Although China had been working to put a human in space starting in the 1960s, it wasn't until 2003 that the country achieved its goal. Following a series of unmanned test missions, the Shenzhou 5 mission carried Yang Liwei into orbit around the Earth, making China just the third country to put a person into orbit. China launched Shenzhou 6 two years later, on October 12, 2005, and Flight Commander Fei Junlong and Flight Engineer Jie Haisheng spent almost five days in space, making 75 orbits of the Earth and testing out the orbital module of their spacecraft.

Due to cooperative agreements between Russia and China in the 1990s, the Shenzhou spacecraft are very similar to Soyuz spacecraft (described in Chapter 7). However, the former are larger than Soyuz and feature a number of completely redesigned key systems. The Shenzhou spacecraft are based on a three-part design, and they contain an orbital module that provides space for living, working, and conducting experiments. (Unlike Soyuz, the orbital module can actually fly on its own apart from the rest of the spacecraft.) Shenzhou spacecraft also have a reentry capsule with space for the crew to return to Earth and a service module that contains power supplies, life-support, solar panels, and other items essential to manned spaceflight. The reentry capsule is the only part of the Shenzhou spacecraft designed to reenter the Earth's atmosphere intact and make a ground landing in Mongolia via parachutes.

The third Shenzhou flight, Shenzhou 7, took place on September 25, 2008, when three *taikonauts* (the Chinese term for astronauts) were launched into space. Commander Zhai Zhigang performed China's first spacewalk, testing out a new Chinese-designed spacesuit while crew member Liu Boming, wearing a Russian spacesuit, stayed nearby in the airlock in case of any problems.

## Future plans for Chinese space exploration

China plans to continue developing its space program, following a similar plan to that taken during NASA's Project Apollo (although without the political impetus of the Space Race, the Chinese timetable is much slower). The next step is for China to launch an orbiting laboratory to practice docking two spacecraft in orbit, leading up to the creation and launch of a small Chinese space station. Because the Shenzhou spacecraft are based on Soyuz designs, they're compatible with docking facilities on the International Space Station (ISS), but as of this writing, no plans have been discussed to allow them to dock at the ISS.

Chinese officials have stated that they also have plans to send people to the Moon. Although such plans are in the future and their timetables are subject to change, one published plan calls for a robotic spacecraft to orbit the Moon (*Chang'e* is currently in orbit; see Chapter 21 for details), followed by a lunar lander and a sample-return mission. In this plan, development of a human lunar mission would begin around 2017, with taikonauts landing on the Moon sometime after that.

The success of the Shenzhou program has shown that China is a force to be reckoned with in the arena of space exploration. In fact, the possibility of the Chinese reaching the Moon has spurred NASA officials to take the prospect of NASA astronauts returning to the Moon more seriously. (Hey, sometimes a little competition is good to spur technological development!)

### An Asian Space Race?

China may not be the only Asian country planning on challenging the American and Russian domination of space. In 2009, India, following its successful robotic *Chandrayaan-1* Moon mission (described in Chapter 21), announced plans to launch its own astronauts into space as soon as 2014. Russia will serve as a partner in this plan, which is thought by many to be a response to China's budding space program. And India's not alone in wanting "in" on the space club — Iran has also announced plans to launch astronauts into space by 2021.

# A Brief History of Space Tourism

When you think of tourism, do images of Hawaiian luaus or European tours come to mind? Well, now you can add outer space to the list! *Space tourism,* or the possibility of nonastronauts voyaging to the limits of Earth's atmosphere and beyond, is a relatively recent phenomenon. It has hefty technological, social, and financial implications — some more beneficial than others. Space travel by nonastronauts can occur in several different ways. Journeying aboard a Russian Soyuz capsule is the way to go today, but other methods are likely to become available in the future.

Check out the following sections to see how space tourism came to be, from the first space tourists to the technology developments that will soon give birth to commercial spaceflight.

## The first space tourists

The earliest space tourists owe a debt of thanks to the cash-strapped Russian space program. Although it met with international controversy at the time, the Russian Federal Space Agency decided in 2001 to allow wealthy individuals to purchase seats on the Soyuz flights it operated to and from the International Space Station (ISS) in order to raise some extra dough to keep the Russian space program flying through tough economic times.

Legally, Russia's decision was on questionable ground. Russia was an international partner in the ISS, but as the sole operator of the Soyuz flights (the only way other than the Space Shuttle to reach the ISS), lawyers eventually determined that Russia had the legal right to bring paying passengers along. To soothe fellow ISS partners anxious about the prospect of tourists, the Russians restricted their tourists to Russian-built ISS modules. (The tourists could only visit modules built by other countries if an escort from that nation was present.) NASA made additional legal requests of the Russians: Space tourists had to acknowledge that they were responsible for repair costs if they broke anything during their trip. They also had to agree that NASA was free of any liability if they were injured or worse while in space. Despite these caveats, and the huge price tag, a sizable group of interested candidates appeared almost immediately.

The first official space tourist was Dennis Tito, a former NASA rocket scientist who made millions post–NASA as an investment fund manager. Tito paid $20 million to the Russian Federal Space Agency for his flight and trained for months outside of Moscow alongside the two cosmonauts who were to accompany him on the flight. Both Tito and the cosmonauts were turned away from NASA's Johnson Space Center when they arrived there for further training on NASA's space station modules. (NASA's administrator at the time, Daniel Goldin, reportedly spoke out strongly against Tito's flight, formally expressing NASA's mixed feelings on the subject.)

Of course, NASA's opposition to Tito's flight didn't really matter. Tito went up into space in 2001 and spent close to eight days there during his flight to and from the ISS. He conducted a number of his own scientific experiments on the space station, in addition to having responsibilities assisting the Russian cosmonauts.

As a private citizen purchasing a ticket for spaceflight, Tito was followed by Mark Shuttleworth (South Africa) in 2002, Gregory Olsen (U.S.) in 2005, Anousheh Ansarai (Iran) in 2006, Charles Simonyi (Hungary) in 2007, and Richard Garriot (U.S.) in 2008.

Russia has recently announced that 2009 will be the last year for such flights because the permanent ISS crew will increase from three people to six. Lack of space on the ISS for tourists, coupled with increased demand for Soyuz flights when the Space Shuttle is retired in 2010, mean that Charles Simonyi's scheduled 2009 repeat Soyuz trip is planned to be the last of the Russian space tourist program. Space Adventures, the U.S. company that has organized these commercial Soyuz trips, is reportedly attempting to charter an entire Soyuz flight as a final adventure.

## Recent innovations in commercial spaceflight

One of the main thorns in the side of the space tourism industry is that venturing into space is, at this point, only possible for people of extreme wealth. Perhaps as a way of reducing the cost of commercial space travel, organizations are planning the development of commercial space vehicles that could transport more people at a time (and, presumably, lower the per-person cost of space travel in the process).

The first step in developing a commercial spaceflight industry is the creation of a commercial suborbital vehicle. Such a vehicle would bring passengers to the edge of space, where they could experience weightlessness but not actually go into orbit — the equivalent of Alan Shepard's first flight in 1961 (see Chapter 7). To encourage development of the commercial spaceflight industry, the Ansari X Prize was established in 1996. The winner of the prize, *SpaceShipOne,* is serving as a model for future commercial space vehicles.

### Racing to win the Ansari X Prize

A little healthy competition, and a significant payday, can be good incentives toward achieving greatness. The *Ansari X Prize,* a competition to develop a manned, reusable spacecraft without the use of government funds, provided an abundance of both. It offered a $10 million prize to the first private team to develop a spacecraft that could fly to a height of 62 miles (100 kilometers) and be launched into space twice in two weeks (meaning the spacecraft had to be largely reusable and couldn't burn up or disintegrate for a grand finale).

The overall goal? To increase the development of space technology in the private sector.

The Ansari X Prize was originally called the X Prize; it was renamed as the Ansari X Prize in 2004 after a sizable donation from the Ansari family. The prize was initiated by the X Prize Foundation, a nonprofit institute whose goal is to promote public competitions in areas that provide the greatest and longest-term benefits to the human race. X Prizes have been offered in areas of spaceflight, ultra-efficient automobiles, and lunar landings. Funding comes from different sources (for example, the Google Lunar X Prize is paid for by, you guessed it, Google).

Competition for the $10 million prize was steep, with 26 teams participating. Although some teams were large and possessed substantial financial backing, others were smaller and had more-limited resources. Ultimately, the competition was won by Tier One, a Northrop Grumman subsidiary that developed a spacecraft and launcher designed by aerospace engineer Burt Rutan. Microsoft's Paul Allen contributed substantially (more than twice the amount of the prize money, in fact) to the Tier One team's efforts.

The winning spacecraft, called *SpaceShipOne* (see Figure 22-1), was rocket-propelled and capable of carrying three people to a height of at least 62 miles (100 kilometers). Its cabin was appropriately pressurized, and a nearby truck performed Mission Control duties. The spacecraft was carried aloft aboard a carrier airplane; then it ignited its rocket engine for a trip to the edge of space. *SpaceShipOne* then glided to the ground, landing like the Space Shuttle.

**Figure 22-1:** *SpaceShipOne* (lower center) aboard the *White Knight* carrier aircraft, which lifts *SpaceShipOne* off the ground for an air launch.

*SpaceShipOne* made a number of suborbital flights, including two back-to-back flights on September 29 and October 4, 2004, qualifying it to win the Ansari X Prize. The first flight was piloted by Mike Melvill; the second was piloted by Brian Binnie. Both flights carried an extra *payload* (cargo) of 400 pounds (180 kilograms) to simulate the presence of two passengers. The spacecraft was retired after winning the competition; it now hangs in the Smithsonian Air and Space Museum as the first privately funded manned spacecraft.

### Developing models for future commercial space travel

Because *SpaceShipOne* could carry just three passengers at maximum, it was really built as a demonstration of the potential for privately funded spaceflight rather than an actual model for future commercial flights. However, much of the technology used in *SpaceShipOne* may make its way into a new spacecraft, called *SpaceShipTwo,* also to be designed by Burt Rutan. This project is being funded by Virgin Galactic, a space-faring offshoot of the company that also operates Virgin Airlines.

Virgin Galactic plans to begin offering commercial suborbital flights as early as 2010. These flights will take paying passengers to the edge of space, which is usually officially marked as an altitude of 62 miles (100 kilometers), where they'll experience weightlessness for about six minutes before returning to Earth. Each flight will be able to carry eight people (specifically two pilots and six passengers).

The first 100 seats on Virgin Galactic's initial flights were priced at a hefty $200,000 apiece, but the company reports that more than 200 seats have already been reserved. The price is scheduled to drop to $20,000 per seat after the first 500 seats have been reserved — still a pricy excursion, but much more within the reach of the average person who isn't a millionaire!

Following is the quick scoop on some other companies developing commercial spaceflight endeavors:

- **Rocketplane:** This company plans to begin suborbital flights in its spacecraft in 2010, provided all tests go well. Visit `www.rocketplane.com` for more details.
- **Rocket Racing and Armadillo Aerospace:** These two groups are planning to develop a suborbital spacecraft. Seats would sell for under $100,000 beginning in 2010.
- **Space Adventures:** The company that organized Soyuz flights to the ISS is also planning suborbital flights beginning in 2011 from spaceports in the United Arab Emirates and Singapore. Interested? Check out `www.spaceadventures.com`.
- **Xcor Aerospace:** This company is currently selling seats on its two-person ship for $95,000 apiece. Flights will go to a height of 38 miles (61 kilometers) and are planned to start in 2010. For more info, head to `www.rocketshiptours.com`.

# The Requirements and Realities of Current (And Future) Space Tourism

Commercial space travel is in its infancy, and despite high interest, it's still fairly primitive, risky, and costly. Oh yeah, and it requires specialized training, much like conditions during the early days of air travel. However, with continued technological development and financial investment, commercial space travel may become a more comfortable luxury excursion — and eventually a cheap transportation method for the masses (similar to today's airplanes). The following sections show you what it takes to be a space tourist today and what the space tourism industry will look like down the road.

## The necessary criteria for "regular people" to travel to space

Since the dawn of the Space Age, the only way to get into space was to be an astronaut. That's not an easy profession for most folks to enter given that the requirements for becoming an astronaut are many, the selection process is rigorous and competitive, and years of training are required before you can even qualify for a space mission. However, with new options for space travel now available (as described earlier in this chapter), "regular people" too can experience the thrill of weightlessness and see Earth from high above the atmosphere.

To plan a trip into space as a private citizen, you have to go through a broker such as Space Adventures to negotiate and plan out your trip. But that's not all. You also need to

- **Have a great deal of money:** Space travel isn't cheap. If you want to go into space without being an official astronaut, it helps to be a millionaire. Flights to the International Space Station (ISS) aboard a Soyuz spacecraft currently cost between $20 and $30 million, which places them squarely out of range for the vast majority of people interested in space travel. As commercial space travel becomes more popular and more common, costs are expected to drop substantially (under $100,000 or so for brief flights on a suborbital spacecraft). The odds are good, however, that space travel will remain considerably more expensive than other forms of transportation for some time to come.
- **Meet specific physical criteria:** Spacecraft are designed to hold people of certain heights and weights, therefore all astronauts — including the recreational ones — must be able to meet these specific physical criteria. On the health end of things, all space travelers must undergo rigorous screening to ensure their fitness can withstand the rigors of space travel.

- **Be able to handle stress:** Between the cramped quarters, space-hygiene systems, and microgravity, traveling in space is a stressful experience. Having a high tolerance for stress is a requirement for anyone considering the voyage.

- **Complete specialized training:** Space tourists must undergo a subset of the same training as astronauts in order to make sure they're qualified and able to do what's required of them in space. On Soyuz flights to the ISS, for example, space tourists occupy a valuable seat in the space capsule and need to be able to fulfill certain duties. Consequently, they're trained on how to perform the tasks assigned to them, how to move about and function in the ISS, how to use appropriate spacecraft controls, and how to maneuver inside the space capsule. They're also given a basic course in Russian so they can better communicate with the other people aboard the spacecraft. For a less-advanced trip, such as one on a suborbital flight, the required training will be shorter, but it'll likely involve at least a weekend of training and preparation (as well as medical clearance).

Space tourists currently don't need to have pilot credentials or any of the specialized abilities that professional astronauts do. As with astronauts, there's no fixed upper limit on astronaut age, though young children (or very small adults) would be prohibited from traveling under current guidelines.

All recreational astronauts need to realize that they're taking a risk beyond that of the spacecraft maintaining its integrity on takeoff and landing. Exposure to radiation while in space is high, and other medical issues can result from extended exposure to microgravity. Therefore, traveling in space, although incredibly exciting, isn't something to be taken lightly.

## What to expect from the future of space tourism

Say you're all signed up for a space excursion in the near future. What should you expect? First off, don't think heading into orbit means you're embarking on a luxury vacation with all the amenities of a cruise ship. On the contrary, space capsules and the ISS were designed for efficiency and specific scientific goals, not necessarily human comfort. For example: The toilet, bathing, and kitchen facilities on the ISS aren't quite what you're used to. Because gravity doesn't exist in space, expect the toilet to have a vacuumlike attachment to help everything get where it should be going. You can also expect to have some trouble getting wet (and then getting dry again!) in a shower with floating drops of water. Although astronauts can put up with facilities that are usable but perhaps not comfortable or familiar, tourists will expect more of the comforts of home, requiring commercial space hotels of the future to develop innovative new solutions.

That's right: space hotels. As commercial space travel becomes more and more viable, it's not crazy to imagine that space hotels will one day be developed to support human visitors. Unlike the ISS, commercial space property will be designed with a new purpose in mind: that of ensuring the safety, well-being, and physical comfort of its occupants. For example, the ISS is designed as a research station, filled with scientific equipment. A space hotel, on the other hand, would be expected to have big windows and observation equipment to take advantage of the breathtaking views of the Earth below. Instead of mundane fitness equipment meant solely to keep the ISS astronauts strong, a space hotel might have fitness classes in microgravity gymnastics and floating ballet. And rather than the small berths on the ISS and sleeping options that include attaching your sleeping bag to the wall with Velcro, a space hotel would probably be able to develop innovative guest rooms with incredibly fluffy beds on the ceiling!

# Chapter 23

# Why Continue to Explore Space?

### In This Chapter

- Quenching society's thirst for answers
- Looking ahead to the end, when space travel will be a necessity
- Realizing how you benefit from space science

Justifying space exploration can be difficult from a purely rational point of view. With all the problems that abound on Earth, from financial woes to disease to protecting the environment, what possible justification can there be for keeping our eyes on the stars rather than the ground beneath our feet? In this chapter, we present several reasons for continuing space exploration, including the satisfaction of human curiosity, the possibility of needing space travel to save the human race, and the ways in which space research has benefitted (and will continue to benefit) the general public.

## Satisfying an Innate Curiosity about Space

Perhaps the single most compelling reason for space exploration is simple human curiosity. Living beings are curious by nature. When you think about it, curiosity is a truly fundamental emotion; it spurs people to investigate and explore, learn and accomplish, observe and act. Without curiosity, you wouldn't just be boring — you wouldn't feel a need to create or do much of anything!

Curiosity is particularly important to scientists and astronomers, who tend to be compulsive about understanding that which can't be seen, heard, smelled, tasted, or touched. The child who looks up at the night sky in wonder becomes the astronaut of tomorrow, and she didn't choose this profession because it was the only job she could get. Space exploration is fueled by curiosity about the universe, and that curiosity is fundamental to society's continued investment in space science. The urge to push forward and explore the unknown is an insatiable aspect of human character, one that propels space scientists in their quest for knowledge.

When asked in 1924 why he wanted to climb Mount Everest, British climber George Mallory replied simply, "Because it is there." He disappeared a few months later on his way to the top, but the allure of Mount Everest remained until it was finally climbed more than 25 years later. Similarly, space exploration remains the next great adventure — space is there, so why not go see it?

The following sections illustrate just what makes space so curiosity-inspiring.

## Investigating the true "final frontier"

Humanity has always had a penchant for extending its borders past the edges of the known frontier — a character trait that gives human curiosity its sometimes nationalistic tone. Just consider the mid-19th century to see what we mean. American politics at the time focused on the idea that the United States had a duty — make that a national obligation — to expand its borders from the eastern seaboard to the west coast of the continent. This idea was embodied under the concept of Manifest Destiny, and although it was never an official government policy, the power of Manifest Destiny and the allure of the frontier drove the American population inexorably westward. From the first intrepid explorers to the settlers who staked out homesteads on the vast prairies or pushed onward to reach the western coast, the western frontier became a symbol of the nation's promise.

Fast-forward a few generations and you can see some iffy parallels with modern space exploration. Although no one's saying the U.S. should seek to discover and dominate the rest of the universe, there is, perhaps, a connection to that 19th-century spirit of exploration and the allure of the frontier. If space truly is the final frontier, then we're still in the days of Lewis and Clark — the initial exploration of the vast unknown. The actual familiarization of that unknown, and perhaps the future colonization of the solar system, is still generations off.

Some scholars argue that humanity, as a species, *needs* a frontier — that without a physical place to look ahead to, to explore, and to eventually expand into, humanity will stagnate and turn inward instead of continuing to evolve and strive onward and upward. Although today some may suggest that cyberspace takes the place of that physical frontier, the idea of one day living exclusively in cyberspace is questionable. Space, on the other hand, is a physical location that can be visited — a location that may one day be mankind's new home. Even if your generation never takes up residence in the heavens, the simple idea that they're there for the visiting opens up worlds of possibility.

## Looking "there" for answers "here"

The poet T.S. Eliot wrote: *"We shall not cease from exploration. And the end of all our exploring will be to arrive where we started and know the place for the first time."* A case for exploring and attempting to understand the solar system and galaxy, and humanity's place in it, is that doing so helps us understand our own planet.

Just as learning a foreign language can often illuminate grammar rules and other fundamental underpinnings of your native tongue, seeing your own planet from an outside perspective can also be rewarding. The Earth is no longer an isolated planet but rather a member of a larger solar system. Scientists can compare the history and development of Earth to what happened on Venus and Mars, discovering previously unknown aspects of Earth's past and clues about its future.

Scientists can also study other solar systems. Over the last ten years, space exploration has revealed planets around other stars, as well as how they differ from our own solar system. More than 300 other solar systems have been discovered to date; scientists can use the details of these systems' formations and structures to better understand the forces that shaped *this* solar system.

# The Future Necessity of Space Travel

In addition to pure curiosity and the exploration of the frontier, more down-to-Earth reasons exist for the continued exploration of space. For example, population pressure or a worldwide disaster of either human or natural origin may eventually propel humanity into the vast reaches of space. Additionally, the Sun will reach the end of its lifetime some billions of years in the future, and future humans will need to escape to a different solar system to ensure the survival of the species.

## Presenting the real deal with population overflow

Earth and its natural resources are finite in size — there's only so much land, water, food, and air to go around. The human population, however, keeps growing, and even with industrialized countries attempting to cut consumption, the billions of people in up-and-coming countries such as China and India will continue to significantly increase demand on commodities that will grow ever scarcer.

Attempts to control population growth have met with limited success, partially because the bulk of population increases comes from developing countries where access to resources such as fertility control (not to mention prevailing religious or cultural attitudes) present barriers. Policies such as China's one-child edict have received outrage from a human rights perspective, but they can still be acknowledged as an effective way to slow population growth.

How many people are considered to be too many? Overpopulation is based not on any particular number of individuals but on the ability of their host (in this case, Earth) to meet the needs of those individuals. Complicating matters is the fact that in many countries, birth rates continue to increase while death rates decrease due to better healthcare, nutrition, and sanitation. Practically, this means that the Earth has more people to sustain (and probably will for the foreseeable future).

The real price of attempting to sustain a population for which the Earth isn't equipped will potentially be paid in many different ways, such as

- Climate change
- Food shortages
- Increased cost of available resources
- More human diseases due to crowded living conditions
- Destruction of native species and habitats

Regardless of how life on Earth devolves, there's little dispute that at some point in the future, if things continue as they are now, the Earth will no longer be capable of supporting future human life. When that tipping point is reached, population pressures may provide the impetus for a mass migration into space — to habitats in orbit, on the Moon, or on other planets. It's this possible future that space scientists have in the back of their minds when considering the future goals of space exploration.

## Looking at space as a place for escaping disaster

Concern is growing that a worldwide disaster may change space exploration from a nice-to-have intellectual exercise into a full-blown requirement for preserving human life. Stephen Hawking, the well-known astrophysicist, gave a speech in 2006 in which he mentioned that the possibility of nuclear war, an extreme collision of an asteroid with Earth, or a similar event could force space exploration into the driver's seat of a new frontier.

In the wake of some cataclysmic event that would render the Earth's surface uninhabitable, several options exist for humanity's survival. Subterranean settlements and underwater cities are viable possibilities in some parts of the world, but they'd require substantial research and development in order to be made feasible. Colonization of a potentially hospitable place, such as Mars or the Moon, would also require significantly more research into space exploration than is currently planned. However, colonization just may offer the best hope for placement and growth because such a colony could expand to one day include a whole new world as a new home base for humanity.

## Considering the future of the solar system

Even if humans manage not to blow up the Earth or each other, this planet (and the entire solar system) still has a finite lifetime. Although the Sun currently provides enough heat and light to sustain life on Earth, keep in mind that it actually loses 8.8 billion pounds (4 billion kilograms) of mass every second as it converts hydrogen into helium fuel. The Sun is large enough, of course, that it can sustain this rate of mass loss for a long time, but in about 5 billion years, the Sun will run out of hydrogen.

### Examining the drive for new transportation

In order to reach distant worlds, scientists must come up with a way to traverse the enormous distances between planets in a relatively quick manner. At society's current levels of technology, a trip to the Moon takes days and a trip to Mars takes months. Even if more-efficient propulsion technologies are developed in the future that allow people to travel the solar system with ease, actually leaving the solar system is a whole different problem.

The distances between the stars are vast — in fact, such distances are measured in light-years, where one *light-year* is the distance that light travels in one year. A fundamental law of physics states that the speed of light in a vacuum is a fixed constant, meaning light always travels at the same speed. Another law of physics states that nothing can travel faster than the speed of light. So as far as we know, traveling that fast is a physical impossibility in our universe.

Of course, modern ships can't come anywhere near the speed of light, so bumping up against this cosmic speed limit anytime soon is unrealistic. However, even if at some point in the future engineers manage to construct a spacecraft that can travel at some sizable fraction of the speed of light, such a ship would still take years to travel to the nearest star system, Alpha Centauri, which is 4.3 light-years away.

Science fiction has a number of ways around this inconvenient scientific truth, from transporters to warp speed to wormholes. Perhaps the most mundane of these innovations is also the most technologically likely: generation ships, or sleeper ships. Acknowledging the reality of travel times that are measured in decades or centuries, colony ships of the future could carry a manifest of passengers in suspended animation, with the ship itself run by a small crew or even fully automated. Or such a ship could simply carry families whose children's children may one day reach their final destination.

Such ideas are still obviously beyond society's current technological capabilities, but they don't require fundamental revolutions in the laws of physics to become reality. A voyage of this type would necessarily be one-way, requiring passengers to leave behind everyone and everything they know. These space pioneers will be like past generations who gave up everything to walk for miles as part of a wagon train to settle the West.

Running out of gas in your car is an inconvenience, but running out of fuel for the Sun is guaranteed to be cataclysmic. Even before the Sun is completely out of fuel, perhaps in just a billion years or so, the mass loss from the Sun will cause its temperature to increase, a step that's required in order for the Sun to maintain a constant pressure in its core. This temperature increase will translate into increased solar radiation on Earth, resulting in a heating of the surface that could cause a runaway greenhouse effect like that found on Venus. Surface temperatures on Earth will become uncomfortably warm as solar radiation is trapped by thickening clouds. Eventually the oceans will evaporate, and the Earth will become extremely inhospitable.

Assuming that humans somehow manage to survive this phase, the next phase in solar evolution will be even worse. When the Sun runs out of hydrogen fuel, it'll switch to burning helium in its core. Before this switch can take place, the outer layers of the Sun will expand hugely while its core contracts. This is called the *red giant phase* because the Sun will cool as it expands, changing its color to red (a lower-energy wavelength). This change will spell the end for the inner solar system — the red giant Sun will spread so large that first Mercury and Venus, and then probably Earth, will be engulfed. (Figure 23-1 shows the Sun's red giant phase.)

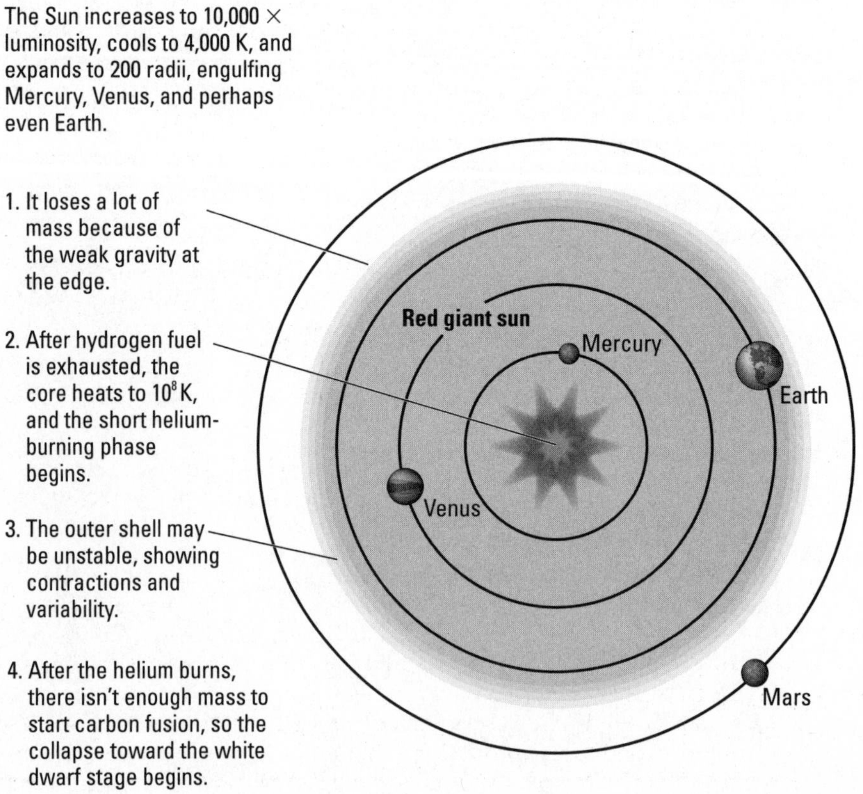

The Sun increases to 10,000 × luminosity, cools to 4,000 K, and expands to 200 radii, engulfing Mercury, Venus, and perhaps even Earth.

1. It loses a lot of mass because of the weak gravity at the edge.
2. After hydrogen fuel is exhausted, the core heats to $10^8$ K, and the short helium-burning phase begins.
3. The outer shell may be unstable, showing contractions and variability.
4. After the helium burns, there isn't enough mass to start carbon fusion, so the collapse toward the white dwarf stage begins.

Red giant sun
Mercury
Earth
Venus
Mars

**Figure 23-1:** The red giant phase of the Sun.

Although the inner solar system will become uninhabitable during this phase, the outer solar system may undergo a brief period of habitability. When the Sun's intensity increases roughly 1 billion years from now, Jupiter's moon Europa and Saturn's moon Titan may become a refuge for the remnants of humanity escaping a sweltering Earth. Such environments will only be transient, however, because the Sun's energy output will decrease when it enters the red giant phase. Eventually, the Sun will use up all of its fuel sources and burn out, ejecting its outer layers into a bright planetary nebula while the center collapses into a tiny white dwarf star.

Given the finite nature of the solar system itself, space exploration is becoming even more of a necessity for the long-term survival of humanity. Sure, colonization of the galaxy may sound like science fiction, but real scientific reasons exist to push for expansion beyond Earth — and even beyond this solar system. If, a billion years from now, humans are still confined to this solar system, then the extinction of the human race is a distinct possibility.

# Space Lends a Hand to the Common Man

Unlike most other fields of study, space science manages to evoke a childlike sense of wonder while also providing practical applications that can make life better for much of the general public. Granted, the end goals of space exploration may be lofty, but the mechanics of getting into space require down-to-earth innovations in everything from rocket mechanics and solar cells to water-filtration systems.

Astronomers, astronauts, and aeronautical engineers seek to explore space not just for their own personal edification but also for the benefits space exploration brings to the larger experience and understanding of society. Spinoffs of space science have improved just about every aspect of humanity's existence, and what scientists find out from space exploration has greatly informed mankind's ability to live and work here on Earth.

## Applying space technology to everyday life on Earth

Some of science's best and brightest minds are focused on the tools, spacecraft, and equipment required for space missions. Fortunately, many of the resulting space innovations and inventions have a direct applicability to everyday life. For example, did you know that the CCD (Charge Coupled Device) chips responsible for detecting breast tumors were initially developed for the Hubble Space Telescope? How about the fact that NASA engineers contributed to the development of the Speedo racing swimsuits worn by most gold-medal Olympic swimmers? Or that the DHA (docosahexaenoic acid, an omega-3 fatty acid) and ARA (arachidonic acid, an omega 6-fatty acid) found in most infant formulas today (and a range of other foods) are a direct result of NASA's research into astronaut life-support?

Part of the allure of basic space research is that it can produce innovative results that are initially wholly unexpected. Many other types of devices and advances owe their origins to the space program — some of which you may not expect! You can dive into the details of them in Chapter 26.

## Finding new sources of natural resources in space

Although space will eventually be a destination for human habitation, in the nearer term it just may be a source of raw materials for humans on Earth to use. Commercial development of space may soon become a reality, especially if it can be made profitable. From the mundane to the exotic, many materials in space may be used either to help support space exploration or be brought back to Earth. Here are a couple examples:

- Many asteroids are composed primarily of valuable metals such as iron and nickel. A small asteroid could provide enough metal to meet the worldwide demand for several years if that iron and nickel could be mined out and brought back to Earth. A number of commercial companies are seriously investigating plans for asteroid mining, particularly of near-Earth asteroids that are in accessible orbits.
- Another component common in the solar system is water. Water ice is found in many comets and asteroids. It probably wouldn't make financial sense to bring water from space back to Earth, but such water would be a precious resource to supply the crews of mining operations, space stations, and spaceships. Water can also be broken down to provide hydrogen fuel for rockets.

After a viable commercial plan is developed, private companies will likely go far beyond the space-tourism model to push the envelope in space exploration. Wide-eyed curiosity has always gone hand-in-hand with profiteering, and space exploration may get a boost far beyond what can be done solely by government-sponsored expeditions if (or perhaps when) a new Space Age–style gold rush begins.

# Part VI
# The Part of Tens

"Paul, turn off your flashlight. There's a real interesting star cluster I'm trying to get a picture of."

## In this part . . .

The Part of Tens takes a lighter view of space exploration. We start off with ten places to look for life beyond Earth — some are, of course, more likely than others! If all this talk about space travel has made you want to dig out some classic movies, make sure to check out the ten reasons that space travel isn't like the movies (or TV!) so you can know what to look out for (or complain to your friends about). And finally, if you aren't already convinced about how exciting and necessary space travel is, this book's final chapter covers all the amazing types of items that owe their existence to the space program. Happy reading!

# Chapter 24

# Ten Places to Look for Life beyond Earth

### In This Chapter

- Pondering life on Venus and Mars
- Looking for life on other planets' moons
- Creating life from comets
- Spreading the search to planets that orbit other stars

Water. Certain *biogenic* elements (ones that are essential for living organisms), such as carbon and nitrogen. An energy source. What do all three have in common? They're the requirements for life as we know it. The Earth is currently the only place in the solar system known to meet all three requirements. Consequently, it's the only place that we know of where life currently exists. Of all the other places in the universe that could support current life, or could've supported life at some point in the past, some are more likely hosts than others. Read this chapter to explore ten places to look for life beyond Earth (in order of their distance from Earth). Some may seem likely to you, such as Mars and Jupiter's moon Europa; others may seem less likely but not unthinkable, such as comets and *extrasolar planets* (those planets found outside of our solar system).

## Venus: (Potential) Life in the Clouds

Surface temperatures on Venus measure around 800 to 900 degrees Fahrenheit (427 to 482 degrees Celsius), and the atmosphere is heavy and acidic. Venus is about 30 percent closer to the Sun than the Earth, but that's not the only reason that modern Venus is a very hot place. The high surface temperature comes primarily from the thick atmosphere, which insulates the surface instead of allowing the Sun's heat to escape back to space. All of these characteristics make for rather inhospitable conditions for most forms of life.

However, the surface isn't the only place to look for life on Venus. Some terrestrial bacteria and other microbes are able to survive in very harsh environments, and scientists speculate that the dense clouds surrounding Venus may be able to harbor these types of life-forms. Both temperature and pressure are lower in the high-atmosphere clouds, and droplets of water could exist at some levels, perhaps providing a stable environment for organisms escaping the harsh surface conditions.

Four billion years ago, when the Sun was a lot cooler and the Earth was likely a ball of ice, Venus may have been the right temperature to support a wider variety of life-forms. After all, carbon and water are essential elements for life as we know it, and Venus has an abundance of carbon dioxide in its atmosphere. Some theories actually suggest that the planet's surface may have been covered with liquid oceans in cooler times. Those oceans likely evaporated as the planet grew hotter and hotter. The European Space Agency's Venus Express mission is currently gathering data that may help fill in the gaps of scientists' current Venus knowledge; flip to Chapter 16 for details.

## Mars: Once a Warm and Wet Planet

Mars is one of the places that's most likely to have sustained life, because it has one of the greatest abundances of life-supporting materials. Thanks to measurements taken by the *Phoenix* Mars Lander (which we describe in Chapter 20), scientists know for sure that the Red Planet currently has water in the form of water ice, which is frozen in ice caps at its North and South Poles. In addition, planetary geologists have studied images of the Martian surface and seen features that were once carved by flowing water (much like river channels on Earth). This fact means that Mars was once much warmer and wetter than it is today.

During this warm and wet period, about 4 billion years ago, the conditions on Mars would've been a lot like those on Earth today. Scientists think there's a good possibility life could've formed on the Red Planet at this time. Although life on Mars could've been possible way back then, it's much less likely to be able to survive there today — the surface of Mars is too cold, and the atmosphere is too thin, for liquid water to be stable aboveground. Perhaps a future space mission will find signs of fossilized Martian organisms, but current life would be a huge surprise.

## Europa: An Ocean beneath the Surface

Jupiter's moon Europa is about the same size as Earth's Moon, but its surface is covered with a bright layer of water ice crisscrossed by cracks and fractures. Scientists think this surface ice layer may be hiding an ocean that contains more water than all of Earth's oceans combined. Water is one of the three requirements for life as we know it, so the likely presence of liquid water makes Europa a prime candidate for life.

In addition to liquid water, Europa also likely contains the right mix of chemical elements (think carbon and nitrogen) to support life. If sufficient energy is available, such as from hydrothermal systems at the bottom of the ocean, then life could very well exist on Europa right now, making it one of the prime solar system targets in the search for life beyond Earth. A future mission to Europa may well explore an active ecosystem beneath its icy surface.

## Ganymede: An Ocean Buried between Layers of Ice

Magnetic field measurements strongly suggest a layer of liquid water deep inside Ganymede, the largest of Jupiter's moons. This layer, which is buried more deeply than Europa's liquid water layer, is likely sandwiched between two layers of ice. The top layer is believed to be regular water ice, whereas the bottom layer is thought to be a special form of high-pressure ice.

Water that's on top of a layer of rock, like in Europa's ocean, can circulate through the rock and dissolve interesting kinds of chemicals that can react with each other in the rock. Ganymede's ocean, however, just has another ice layer to circulate through underneath it. Consequently, the kinds of chemicals the water can pick up are limited. So although Ganymede's ocean is still a possible place for life, the water itself has less potential for chemical reactions and energy sources that could help support an ecosystem.

## Callisto: An Even Deeper Layer of Water

Surprisingly, even Callisto, a more distant moon of Jupiter, could have a liquid water layer like Europa and Ganymede (covered earlier in this chapter). Callisto's old, battered surface looks much more like the Moon than an active world like Europa. However, magnetic field results suggest that a liquid water layer could exist below Callisto's surface ice layer.

As is the case for Ganymede, this water layer is deeply buried and likely also located between two ice layers rather than on top of a rock layer. In fact, scientists aren't even sure that Callisto's rock is concentrated in the center of the satellite — bits of rock and dust could be scattered throughout the surface ice, making Callisto a much more primitive body than Europa and Ganymede. Again, although a water layer makes life a possibility, the extreme depth of such a layer reduces possible energy sources and communication with the surface. Chemical reactions are also reduced by the water layer being on top of ice rather than rock.

# Titan: Swimming in Methane

Titan, the largest of Saturn's moons, has an extremely thick atmosphere that makes liquid hydrocarbons such as ethane and methane possible on its surface. Organic compounds such as nitrogen are present in abundance as well. In fact, nitrogen makes up the majority of Titan's atmosphere.

Speaking of Titan's atmosphere, scientists believe that it's similar to the atmosphere surrounding Earth when life first formed there roughly 3.5 billion years ago. Note, though, that Earth is a warm planet. Titan has a surface temperature of about –290 degrees Fahrenheit (–180 degrees Celsius), meaning any water present on the surface would be in the form of ice. The cold temperature alone isn't a good sign in the search for life, but Titan's surface is covered with features carved by running liquid (in this case, methane), and the *Cassini* spacecraft has even found methane lakes and signs of rain (check out Chapter 17 for the scoop on *Cassini*).

The continued presence of methane in Titan's atmosphere suggests that something else may be generating it. On Earth, methane is produced by life, and some scientists have suggested that Titan's methane could also be related to some sort of biological process.

Interestingly, methane is an unstable component of Titan's atmosphere — it quickly decays in sunlight, producing nitrogen compounds that fall out onto the surface.

# Enceladus: Tiny but Active

Enceladus, a tiny moon of Saturn, was known to be anomalously bright and suspiciously smooth according to images captured by the *Voyager 1* spacecraft in 1981 (see Chapter 11 for details on the Voyager missions). *Cassini* images from 2005 revealed a stunning explanation for Enceladus's appearance in the form of giant plumes of material jetting off the satellite's South Pole. Water vapor, nitrogen, and carbon dioxide have all been detected in the plume, and liquid water is likely present just beneath the surface. Methane and other organic molecules are also possibilities.

Liquid water always raises the chances for life, and the geysers of Enceladus are a dramatic example of how that water can reach the surface. One unanswered question, however, is how stable the liquid water environment is — the rate at which Enceladus is currently ejecting material isn't sustainable over the long term. A future mission, the Titan Saturn System Mission, will return to Titan and Enceladus, although this mission is still in the early planning stages (check out Chapter 21 for more info).

# Triton: Venting Plumes of Nitrogen

Triton, a large moon of Neptune, is an icy body very far from the Sun. Its surface is covered primarily with nitrogen ice, but frozen carbon dioxide and water ice are also present. Triton has a young surface, and plumes of nitrogen have been observed being vented off the surface, making Triton one of just a few satellites in the solar system known to be currently geologically active (the others are Jupiter's Io, Saturn's Enceladus, and perhaps Saturn's Titan).

Triton orbits Neptune in a *retrograde orbit,* meaning it moves backward from the other satellites in the system. This and other factors suggest that Triton formed elsewhere in the solar system, such as in the Kuiper Belt, and was captured by Neptune. The capture event could've contributed enough heat to melt Triton's interior thoroughly, which in turn could help support Triton's current activity.

Like Jupiter's Europa, the upper layer of Triton's subsurface is composed primarily of water ice on top of rock. Also like Europa, evidence shows there may even be sufficient heat to keep the base of the ice layer melted, creating a deep subsurface ocean that has the benefit of lying on top of a rocky foundation, thereby making life a possibility.

# Comets: Filled with Life's Building Blocks

*Comets,* leftover ice balls from the early days of the formation of the solar system, hold the building blocks of life because they're composed primarily of water ice, with bits of dark, carbon-containing organic dust mixed in. Comets could've been very important to the formation of life on Earth. One theory suggests that much of Earth's water actually came from comet impacts early in the solar system's history. Other theories suggest that comets could've carried the seeds of life to Earth, where it thrived. How can such a thing be possible? Well, scientists know, for example, that meteorites can actually exchange material between Earth and other planets —pieces from the Moon and even from Mars have been found on the Earth. It's even possible that microbes could survive the trip from Mars to Earth, or vice versa.

But can comets actually hold life? Because they're primarily small, primitive bodies, it's unlikely that they possess enough of an energy source to actually drive the chemical reactions needed to form life. However, one possibility is that the interiors of comets may hold small amounts of liquid water, perhaps just individual droplets around grains of ice and other materials. This tiny amount of water could nevertheless be sufficient to carry life from one place to another, allowing comets to transport *prebiotic materials* (simple compounds that may be used in the formation of life), or perhaps even simple life-forms, throughout the solar system.

## Extrasolar Planets: Life around Other Suns

The search for planets that orbit other stars has heated up in the last ten years, with hundreds of planets now known to circle other suns. Scientists have yet to find planets similar in size to Earth, however. All known extrasolar planets are large bodies the size of Jupiter or Saturn, or larger. Future missions (such as Kepler, covered in Chapter 21) will be able to find Earth-sized planets. If such planets exist in the *habitable zone,* the distance around a star at which liquid water would be stable on the surface of a planet, then the possibilities for life become much greater. But how can humans detect life on another planet if actually traveling there isn't possible?

One option for detecting life on far-away worlds is to look at the color of the planets and study the gases present in their atmospheres and on their surfaces. Plants convert water and carbon dioxide into glucose and oxygen through photosynthesis; oxygen, combined with methane, is unstable unless continually supplied through some chemical reaction. Thus, the discovery of an oxygen- or nitrogen-containing atmosphere on an extrasolar planet would send up an immediate red flare to scientists looking for life.

Another way to search for life is to look for the telltale color of chlorophyll, which makes plants on Earth appear green. This pigment takes sunlight and uses it to produce energy and fuel, but the process is somewhat imperfect because green plants can't absorb 100 percent of the available sunlight. Scientists have speculated that plant life on extrasolar planets, if it exists, may be black rather than green. Black vegetation would be a more-perfect absorber of solar energy because all light would be absorbed — but even that would depend on the spectrum of light given off by the planet's star.

If an Earth-sized planet is ever discovered in the habitable zone of a far-off star, then scientists will regard it as a prime possibility for the existence of life. However, even huge Jupiter-sized planets may have inhabitable moons, extending the definition of the habitable zone far outside what's currently considered a limiting factor for life. Although receiving an actual signal from an intelligent civilization would be the best, and most exciting, confirmation of life beyond Earth, scientists will continue to look for simpler forms of life on extrasolar planets as technology improves to allow for direct imaging of these worlds.

# Chapter 25

# Ten Ways Space Travel Isn't Like Television or the Movies

### In This Chapter

- Limiting the speed and realm of space travel
- Knowing how space affects humans
- Debunking ideas about aliens

Although science fiction movies and TV shows shape people's ideas of how space exploration should look and work, reality, like with so much on the big (or small) screen, is often a completely different story. From humanoids with funny noses to whooshing spaceships, many science fiction films and shows make basic errors in biology and physics. Whether these errors exist for the sake of plot convenience or aesthetics, they're errors nonetheless. This chapter sets the record straight on ten of those mistakes.

## A Cosmic Speed Limit Exists

The phrase "Move to light speed and beyond!" sounds great when you're in a hurry, but the reality is that anything with mass is constrained to the speed of light. Einstein's theory of special relativity is largely to blame. Einstein first began with the observation that the speed of light in a vacuum is always constant. By combining that idea with the concept that the laws of physics are the same for all observers, he showed that things get really weird as you approach the speed of light: Time stretches out, lengths get shorter, and really the only thing that remains the same is the speed of light! Because of these weird effects, it's theoretically possible to travel at speeds that approach the speed of light, but it's scientifically *im*possible to go any faster.

If Einstein's theories continue to be borne out in the future (and so far they've passed every test), then it would be possible for a spaceship to travel, say, at 99.99 percent of the speed of light (186,000 miles, or 300 million meters, per second), although that would require huge amounts of energy. Traveling any faster, however, would fall square into the realm of science fiction, not science fact.

Various movies and television shows have relied on things like *warp speed* and *hyperdrive* to refer to faster-than-light space travel. Because distances in the galaxy are vast, it'd still take more than four years to reach the nearest star system when traveling at the speed of light — a fact that makes for not-so-entertaining entertainment! Unless scientists of the future come up with a way to change the laws of physics, it's unlikely that the future of space cruisers and starships warping between stars in a matter of hours or days will ever become reality.

## Wormholes Aren't On Call

Can a stable wormhole be used as a quick way to travel between two distant locations? Well, in the 1997 movie *Event Horizon,* a spaceship with the ability to create wormholes flew into one and resurfaced years later with a supernatural visitor. A stable wormhole also featured prominently in the TV series *Star Trek: Deep Space Nine* as an instant portal between two far-away parts of the galaxy. If it's on the big screen or the boob tube, it must be true, right? Wrong, scientifically speaking.

*Wormholes* are hypothetical ways of circumventing the normal laws of space and time; they're an unproven phenomenon that can't be summoned on demand. Also, although general relativity does provide for the possibility of wormholes connecting universes, scientists think they'd consist of a tunnel-like boundary. In the movies and on television, wormholes can lead anywhere at any time. Furthermore, what scientists know about wormholes suggest that they're incredibly tiny and exist only for a very brief amount of time — less than a second — suggesting that sending anything as large as a spaceship through one for an extended period of time is sheer folly (but, we admit, great entertainment).

## Space Isn't Noisy

From *Star Wars* to the arcade down the street, most popular portrayals of space depict it as a very noisy place. Between the whooshing of starships and the sounds of laser guns shooting audibly (directly into space, mind you), space appears about as quiet as a New York City street during rush hour.

Explosions are fiery, loud events, and one spaceship crashing into another results in audible, audience-pleasing sounds.

The fundamental problem with space noise is that sound waves don't transmit in a vacuum. A *vacuum* is defined as matter-free space, or space with a gaseous pressure significantly lower than its atmospheric pressure. The parts of outer space that lie between planets and stars (in other words, most of it) are considered a vacuum. Sound waves are mechanical vibrations that require a molecular substance, such as water or air, to travel through. Vacuums are devoid of those molecules, so there's no way sound can transmit through them.

The classic 1968 movie *2001: A Space Odyssey* got this one right. During the scenes in which an astronaut travels outside his ship in a spacesuit, all you can hear is his breathing. Unfortunately for scientific accuracy, audiences find silence boring. They much prefer special effects, so expect on-screen spaceships to continue whooshing by.

# Asteroids Aren't Packed Together

Many movies, from the *Star Wars* saga on, show ships flying through a tightly packed region of giant rocks the size of small moons, narrowly avoiding collisions. Some depictions of asteroid belts have rocks moving in all directions; others have the rocks packed very closely together. Sure, this makes for exciting cinema, but it doesn't represent reality. In our solar system, the asteroid belt located between Mars and Jupiter does in fact contain millions of objects. However, they're spread over such a large amount of space that hitting one without aiming precisely for it would be very difficult. In fact, if you were located right next to one asteroid, you likely wouldn't even be able to see any others.

Similarly, the rings of material surrounding Saturn are spectacular when seen from a distance. Up close, you can see that they're composed mostly of tiny particles of ice and dust. The rings themselves are thousands of kilometers wide with an average thickness of only 33 feet (10 meters)! In fact, when seen edge-on, the rings are so thin that they seem to disappear completely.

# Humans Can't Breathe "Air" on Another Planet

If films like *Rocketship X-M* are correct, people would be able to land on, say, Mars, get out, take a stroll, and (if they're lucky enough) interact with some very humanlike Martians. The sad fact is that oxygen-breathing beings — like us or you or your favorite Fido — can't breathe the atmosphere on Mars or any other planet. Humans evolved on a planet with a specific mixture of oxygen, nitrogen, and other atmospheric gases. There's no reason to think that such a mixture would be replicated elsewhere just through random chance.

In fact, it's very unlikely that Earthlings could breathe the air on a different planet without some sort of mask or filter (think of the clunky spacesuits worn by Armstrong and Aldrin during the Apollo 11 Moon landing). Real space exploration just isn't quite as fashionable (or easy!) as *Star Trek, Stargate SG-1,* and other science fiction sagas make it out to be.

# You Won't Explode in Space

The vacuum of space clearly isn't a place where you'd want to travel unprotected, but movies generally exaggerate the effects of exposure to it. For example, in the 2000 movie *Mission to Mars,* an astronaut removes the helmet of his spacesuit, while in space, and promptly freezes and dies instantly. Other movies, such as the 1981 film *Outland,* show a person exploding when exposed to the vacuum of space.

True, exposure to vacuum is never a good idea, but you aren't going to die instantly if you find yourself in such a scenario. Both theory and actual tests show that you can probably survive exposure to vacuum for between 60 and 90 seconds without lasting effects. (You'd only remain conscious for about 10 seconds, though, so you should make sure to get to safety as soon as possible.) You won't explode because your skin is strong enough to keep your body intact and your blood from boiling. And although you'd eventually freeze, it'd take hours for your body to lose all of its heat. The actual cause of death would be a lack of oxygen.

Our props to *2001: A Space Odyssey* (1968) for actually getting this one right. In the film, a main character must travel through space from a ship to an airlock for a few seconds without his helmet. He survives without too much incident. So did a NASA astronaut-in-training from the 1960s. His spacesuit sprung a leak, and he lost consciousness after 14 seconds. When the chamber was repressurized, he regained consciousness and was just fine.

# People Float, Not Walk, in Space

On the deck of the starship *Enterprise* or in the bays of the *Millennium Falcon,* people walk as freely as they do on Earth. Check for yourself the next time you watch an episode of *Star Trek* or one of the original *Star Wars* films. We bet you won't spot Captain Kirk or Han Solo floating to the top of the ceiling! If the actors were really on a ship in outer space, though, they most certainly would be. Think about astronauts on the International Space Station — they float around because the station is in free-fall as it orbits the Earth. All parts of the space station and everything in it, including the inhabitants, are falling at the same speed. Thus, they're effectively in a zero gravity environment. Whenever space travel extends to interplanetary or interstellar space, travelers really will be floating in zero gravity.

Because long-term exposure to microgravity environments can have negative consequences for the strength of an astronaut's bones, inhabitants of the International Space Station must exercise daily. One way to simulate gravity in space is by spinning a space station or spaceship. The motion results in a centripetal force that pushes the contents of the spacecraft toward the outside, simulating gravity. Such a spinning space station was correctly depicted in the 1968 film *2001: A Space Odyssey*. However, most spacecraft you encounter onscreen don't spin. Some sort of artificial gravity device would therefore be necessary to allow the ships' inhabitants to walk around inside.

# Aliens Don't Look Like Humans

The media-driven perception of alien life is completely derived from the human experience, which makes little sense because alien life is, well, alien. Human biases are likely reflected in the appearance of aliens as humanoid, with two arms, two legs, and a physical structure basically similar to that of Earthlings. In addition, dressing up a sci-fi show's alien-of-the-week actors in humanoid costumes, perhaps with funny noses or green skin, is much easier than inventing truly unique beings. The reality is that life on other planets wouldn't look human. There's just no reason for it to.

Some shows have attempted to create beings that are completely different, from silicon-based, rocklike creatures to sentient energy clouds. These efforts acknowledge the reality that if humans ever do find life elsewhere, it likely won't be at all recognizable!

# Aliens Don't Speak English

According to just about any science fiction movie or TV show that features intelligent alien life, extraterrestrials speak a language. Sometimes it's English; sometimes it's a similar language. The first problem with this situation in real life is figuring out how to understand an alien language. Without *Star Trek*'s universal translator, a *Star Wars* protocol droid, or even the Babel Fish from the *Hitchhiker's Guide to the Galaxy* series, such a feat would be difficult.

Secondly, almost all of these languages are verbal, which is a direct extrapolation of mankind's perception of what defines language. Studies of communication have shown that species such as Humpback whales and honeybees actually have complex languages that can communicate vast amounts of information very efficiently. The honeybees' communication in particular is completely nonverbal. If an alien civilization ever does try to contact Earth, or if astronauts ever come into direct contact with one, its style of communication might be completely different from anything humans have ever known.

# Space Invaders Aren't Among Us

Movies such as *Starman* and *Close Encounters of the Third Kind* bring extraterrestrial invaders to Earth, and numerous television shows have suggested that UFOs are common features in Earth's skies. However, there's no scientific evidence that UFOs have visited our planet. If alien ships were really landing on Earth and abducting people, wouldn't some sort of evidence exist that could be studied by scientists, such as traces of material not found on Earth or clear pictures of an alien spacecraft?

Television shows such as *The X-Files* are fond of portraying aliens masquerading as humans. The idea is that these nonhuman humans can take over the body of your friends, your teachers, or your parents. Not so, say scientists. If such aliens ever made it to Earth, the odds of their being able to survive our climate, atmosphere, and food are slim.

Although chameleons can change their color and some can even change their gender, no precedent exists in either the natural or technological world for the type of transformation that an alien invasion of a human body would require. Not a good enough reason for you? Just consider this: If so many people aren't happy with their bodies half the time, why would a superior life-form want to be human in the first place?

# Chapter 26

# Ten Types of Everyday Items Brought to You Courtesy of NASA

### *In This Chapter*

- Improving medical procedures and technology, as well as your health
- Making life easier, from transport and tools to food and entertainment

Exploring space, discovering new worlds, understanding Earth — that's all well and good, but how do your space-centric tax dollars actually help you? If you've ever used a laptop without burning your lap, had a CAT scan, used wireless headphones, driven a car, or fed jarred food to a baby, chances are you've used NASA-created technology. In fact, going an entire day without running into conveniences and necessities that were spun off of NASA research is pretty hard to do. Read this chapter to find out about ten types of items brought to you by NASA.

## *Medical Diagnostics and Procedures*

Diagnostics are one key to preventing and treating a host of medical conditions, and NASA technology has greatly improved this area of medicine. Here are a couple examples:

- **CAT scan:** The computer-aided tomography scan (aren't you glad they shortened the name?) allows doctors to "look inside" a person without surgery or patient discomfort. The core functionality of this scanning technique came from NASA's methods of enhancing lunar pictures taken during the Apollo missions (which we cover in Chapters 9 and 10).
- **Ocular screening:** This technology has helped hundreds of thousands of children see more clearly. Developed by NASA to process the images being sent by space probes, ocular screens reflect light into a child's eyes and create an accurate image, allowing ophthalmologists to study and detect problems earlier than they were previously able to.

Various medical procedures have also benefitted from your friendly U.S. space agency. *Laser angioplasty,* a technique that allows doctors to unclog arteries without doing invasive surgery, uses technology that was originally developed for satellites to study the atmosphere.

Need another example? LASIK vision correction allows military and laypeople alike the chance to achieve their dreams without being halted by vision problems or encumbered by eyeglasses. The technique, which involves reshaping the cornea using lasers, is based on techniques developed by NASA contractors for spacecraft rendezvous and docking maneuvers.

Interestingly enough, NASA only recently allowed astronaut candidates to undergo laser eye surgery to meet the eyesight requirements for astronaut qualification. Originally, NASA officials were concerned about how laser-corrected eyes might be affected by the changes in pressure and gravity that come with spaceflight.

## Medical Devices

A wealth of medical devices owes its origins to NASA research. One of the most-vital medical devices to have originated in the space program is the programmable pacemaker. This mechanism, which helps regulate an irregular heartbeat, was created using technology that originated in the electrical communication design for spacecraft.

The voice-activated wheelchair, a device that can greatly improve the quality of life for people who are completely paralyzed, was also developed via NASA technology. Research into robotics and remotely operating instruments in space was later used to create these wheelchairs, which can be directed to perform a number of motions by using single-word commands.

Recovering from an accident is no easy task, but the odds of success can be improved with the *SAmbulation Module* (SAM), a rehabilitative device for physical therapy patients who need to practice walking. The technology for SAM was developed during NASA's robotics research and work on assembly of sounding rockets.

Even the cochlear implant was partially developed by a NASA engineer! This device allows deaf people to hear by translating sounds into digital signals sent to a speech processor. It was first designed by an electronics instrumentation engineer who had undergone a number of failed surgeries to try to restore his hearing.

## Personal Health and Safety

A number of other spinoffs of NASA technology may seem even further afield from their original purposes, but that doesn't make them any less useful. For example, NASA clean rooms use a special airflow technique to keep contaminants off spacecraft during their assembly. The same technique has been used at tollbooths located on highways and bridges to keep the toll collectors from inhaling high levels of hazardous engine exhaust.

NASA technology has also been used to produce a small radiation detector that can sense harmful levels of microwave radiation. For people who work in hazardous areas, such a device sounds an alarm when microwave radiation levels become too high.

Another device, an alarm system that can be used by the elderly, prison guards, and others who need to send a quick message, uses ultrasonic transmitters to send an inaudible signal to a base station, which can indicate exactly where the transmitter is located. This technology was originally developed by NASA to determine the exact position of a spacecraft in space.

The life rafts developed for Project Apollo, which can inflate in just 12 seconds and turn themselves upright automatically, have also found commercial applications for emergency use on airplanes and other places where a quick-inflating raft comes in handy. Additionally, NASA technology has helped create lightweight air tanks for firefighters and aided in the development of a device that can give a 30-minute warning of an impending lightning strike.

# Transportation Aids

The next time you travel from here to there, consider NASA's role in getting you safely to your destination, particularly if you're flying. The impact of space research on commercial airplanes has been enormous. NASA's techniques for rendezvous and docking in space greatly influenced modern Air Traffic Control and collision-avoidance systems. Cabin pressure monitoring was developed from NASA spaceflight research, as were more-advanced parachute recovery systems, engine combustors, airplane materials, wind shear management, and even ways to prevent jet lag.

And by the way, if you drive a car, you should know that today's spiffier, sturdier tires are Martian in origin. NASA's development work with materials that could help rovers rock and roll on Mars went on to improve radial tires sold locally today.

Many other aspects of automotive engineering were improved with NASA research. Chemical sensors, high-performance lubricants, stud-free snow tires, more safely designed cars — all these areas, and more, emerged from research that went into the Space Shuttle and other space missions. Even rescue tools designed to help extricate injured people from crushed cars came from NASA instruments that were originally intended to separate connected modules and other devices on the Space Shuttle.

## Power Tools

From aerospace companies that build spacecraft to astronauts who do repairs in space, NASA has required special tools to perform its missions. Some of these tools have actually turned out to have interesting applications back on Earth:

- Cordless drills are ubiquitous in households today, and they can thank the Apollo missions for their existence. The Apollo astronauts needed a quick, portable way to drill into the Moon's crust to take core samples, and cordless drills were the answer. Today, cordless tools abound in size, variety, and application; the cordless vacuum cleaner is a similar spinoff.
- A special welding torch that was originally developed to help join together the special materials used to construct the Space Shuttle's fuel tanks is now used by commercial companies to weld sheet metal.
- Microlasers are commercially used to melt, drill, or cut materials extremely precisely, but this technology was originally developed by NASA as an alternate way to communicate with spacecraft over the long distances between the planets. Such communication is commonly done via radio waves, but laser communication systems may be used on upcoming spaceflights.

## Computing Power

Laptop computers have been around since the early 1980s, but they weren't always as lap friendly as they are today. NASA engineering helped keep laptops from scorching their users by inventing the fans required to help keep them cool. Portable computers were used in many space navigation systems and are found throughout the world today.

Other breakthroughs in the world of computers that had their start in NASA research include

- A more compact and efficient three-dimensional package of semiconductors, which are stacked in a cube shape to allow computers to have faster processing speeds while also being smaller and lighter.
- Virtual reality software, which allows a user to wear a headset and seem to be immersed in a three-dimensional, virtual world (such software was used for astronaut training).
- Software for structural analysis; first used to help design spacecraft, this software is currently used to design cars, tools, and other complicated systems.

# Foods and Nutrition

"What's for dinner?" is a universal question, even if the questioner is orbiting Earth! Advances in space food keep today's astronauts well-fed — and help make meals easier and healthier for those of us on the ground as well.

- Modern DHA (docosahexaenoic acid)-fortified baby foods can thank NASA for their presence on grocery store shelves. NASA research into long-term survival in space led to the isolation of two fatty acids with many protective properties. At first found only in breast milk, and not in processed infant formulas and baby food, these substances are now routinely added to infant formulas and meals.
- Although freeze-dried Apollo space food may have left something to be desired, its principles have found ready use today. Freeze-dried meals and snacks are shelf-stable, require little effort to reheat, and are tastier than their predecessors. They're commonly used by hikers who want lightweight, long-lasting foods.
- The entire method of displaying refrigerated products in the grocery store owes thanks to NASA scientists. Work that went into the development of airflow and refrigeration coils found use in the display system that most grocers use to simultaneously show off their cold merchandise and keep it fresh and safe.

# Intelligent Clothing

Would you believe that NASA has even contributed to clothing? An astronaut's "cool suit," which helps regulate his or her body temperature despite the heat and weight of a bulky spacesuit, found a quick application in patients with multiple sclerosis. People with this degenerative neurological disease have seen rapid improvement in their symptoms when wearing a version of the suit, which provides cooling liquid through flexible tubing.

Don't forget your feet! It's hard to imagine participating in any sport without shock-absorbing shoes. Yes, that's right; NASA made those possible, too. The original material developed for the "Moon Boots" worn by the Apollo astronauts was repurposed for athletic shoes, allowing both athletes and astronauts to play and work harder without ruining their feet.

# Advanced Fabrics and Plastics

Do you remember those ribbed swimsuits that helped the U.S. team achieve victory in the 2008 Olympics? They were developed using NASA techniques for laser-closed seams and fabric that actually repels water. (And yes, we're still talking about swimwear here.) They were even tested in NASA wind tunnels.

The flame-retardant fabric that makes up a large portion of infant and toddler sleepwear is also a NASA spinoff. Treating fabrics with chemicals to make them fire-resistant was a technique originally developed by NASA to protect astronauts from accidental fires, in space or on the ground. Today, such fabrics are standard equipment for babies and children.

Over time, other fabrics and plastics have made their way from NASA's drawing board into consumer life. Check out the following:

- Racecar insulation material originated as part of the Space Shuttle's thermal system.
- Rust- and scratch-resistant metal coatings were developed for the Space Shuttle.
- Scratch-free plastic eyeglass lenses were derived from the NASA technique of creating polymers that are light, strong, and resilient (and impervious to trampling by school children).

# Entertainment Necessities

Space technology isn't used just for health and work. Check out these entertainment necessities developed directly from space technology:

- Wireless earphones and headsets, used for listening to music or conducting a conference call, are the direct result of technology developed for NASA that allowed the Apollo astronauts to communicate with Earth from the Moon.

- You can thank NASA for your eye-protecting, UV-blocking sunglasses. The filtering technology came from work done to protect astronauts and equipment from damaging solar radiation.
- Those creamy cosmetics that people apply to their skin are the result of lunar research. The same image processing and analysis tools that went into studying the Moon's surface have been used by major cosmetics manufacturers to make their products more effective.
- Fore! The metallic glass alloy commonly used in golf clubs is a spinoff of NASA's Space Shuttle metals research (flip to Chapters 13 and 14 for general Shuttle information). This strong yet flexible material has many potential uses in automotives, defense, sports, and other industries.

# Index

## • A •

## • F •



## • G •



## • H •

• L •



• M •

## • R •

## • U •



## • V •



## • W •

• Z •

## BUSINESS, CAREERS & PERSONAL FINANCE

**Accounting For Dummies, 4th Edition***
978-0-470-24600-9

**Bookkeeping Workbook For Dummies†**
978-0-470-16983-4

**Commodities For Dummies**
978-0-470-04928-0

**Doing Business in China For Dummies**
978-0-470-04929-7

**E-Mail Marketing For Dummies**
978-0-470-19087-6

**Job Interviews For Dummies, 3rd Edition*†**
978-0-470-17748-8

**Personal Finance Workbook For Dummies*†**
978-0-470-09933-9

**Real Estate License Exams For Dummies**
978-0-7645-7623-2

**Six Sigma For Dummies**
978-0-7645-6798-8

**Small Business Kit For Dummies, 2nd Edition*†**
978-0-7645-5984-6

**Telephone Sales For Dummies**
978-0-470-16836-3

## BUSINESS PRODUCTIVITY & MICROSOFT OFFICE

**Access 2007 For Dummies**
978-0-470-03649-5

**Excel 2007 For Dummies**
978-0-470-03737-9

**Office 2007 For Dummies**
978-0-470-00923-9

**Outlook 2007 For Dummies**
978-0-470-03830-7

**PowerPoint 2007 For Dummies**
978-0-470-04059-1

**Project 2007 For Dummies**
978-0-470-03651-8

**QuickBooks 2008 For Dummies**
978-0-470-18470-7

**Quicken 2008 For Dummies**
978-0-470-17473-9

**Salesforce.com For Dummies, 2nd Edition**
978-0-470-04893-1

**Word 2007 For Dummies**
978-0-470-03658-7

## EDUCATION, HISTORY, REFERENCE & TEST PREPARATION

**African American History For Dummies**
978-0-7645-5469-8

**Algebra For Dummies**
978-0-7645-5325-7

**Algebra Workbook For Dummies**
978-0-7645-8467-1

**Art History For Dummies**
978-0-470-09910-0

**ASVAB For Dummies, 2nd Edition**
978-0-470-10671-6

**British Military History For Dummies**
978-0-470-03213-8

**Calculus For Dummies**
978-0-7645-2498-1

**Canadian History For Dummies, 2nd Edition**
978-0-470-83656-9

**Geometry Workbook For Dummies**
978-0-471-79940-5

**The SAT I For Dummies, 6th Edition**
978-0-7645-7193-0

**Series 7 Exam For Dummies**
978-0-470-09932-2

**World History For Dummies**
978-0-7645-5242-7

## FOOD, GARDEN, HOBBIES & HOME

**Bridge For Dummies, 2nd Edition**
978-0-471-92426-5

**Coin Collecting For Dummies, 2nd Edition**
978-0-470-22275-1

**Cooking Basics For Dummies, 3rd Edition**
978-0-7645-7206-7

**Drawing For Dummies**
978-0-7645-5476-6

**Etiquette For Dummies, 2nd Edition**
978-0-470-10672-3

**Gardening Basics For Dummies*†**
978-0-470-03749-2

**Knitting Patterns For Dummies**
978-0-470-04556-5

**Living Gluten-Free For Dummies†**
978-0-471-77383-2

**Painting Do-It-Yourself For Dummies**
978-0-470-17533-0

## HEALTH, SELF HELP, PARENTING & PETS

**Anger Management For Dummies**
978-0-470-03715-7

**Anxiety & Depression Workbook For Dummies**
978-0-7645-9793-0

**Dieting For Dummies, 2nd Edition**
978-0-7645-4149-0

**Dog Training For Dummies, 2nd Edition**
978-0-7645-8418-3

**Horseback Riding For Dummies**
978-0-470-09719-9

**Infertility For Dummies†**
978-0-470-11518-3

**Meditation For Dummies with CD-ROM, 2nd Edition**
978-0-471-77774-8

**Post-Traumatic Stress Disorder For Dummies**
978-0-470-04922-8

**Puppies For Dummies, 2nd Edition**
978-0-470-03717-1

**Thyroid For Dummies, 2nd Edition†**
978-0-471-78755-6

**Type 1 Diabetes For Dummies*†**
978-0-470-17811-9

*** Separate Canadian edition also available**
**† Separate U.K. edition also available**

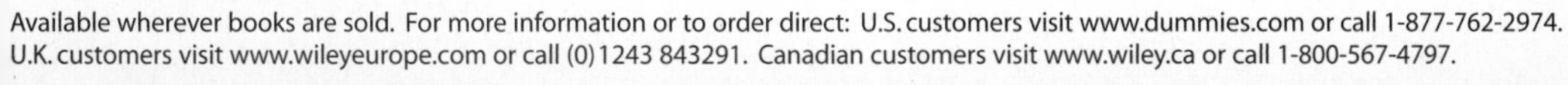